PMP Exam Prep

By

Rita Goodrich

TOC

Introduction

Everyone who has worked in or around project management for any length of time understands that project failures can have consequences that extend far beyond the financial and organizational impact.

Until recently, licensure in the field of project management has lagged behind other industries such as engineering, medicine, transportation, and real estate. However, as companies put more of an emphasis on projects, they are requiring a greater understanding of the process of project management and of the many areas of expertise it touches.

The fact that the vast majority of project managers never even take the first steps toward earning PMI's certification is evidence that the process is not an easy one. This book will cut down on the difficulty factors and demystify the material. In the following chapters, you will find exactly what you need to study for the test, how to learn it, how to apply it, and why it is important. This book is a complete resource to help you prepare fully for the PMP Certification Exam.

One of the recent updates to the PMP examines is the inclusion of Agile issues in its scope. Therefore, this book includes an additional chapter only about the different Agile methods like Scrum and Kanban, what are the differences of this method to the traditional method and why it has become so important in project management.

The Exam

What the Exam Tests?

Before we discuss what the PMP Certification Exam does test, let's clear up a few misconceptions about the exam.

The PMP Certification Exam does not test:

- Your project management experience
- Your common sense
- Your knowledge of industry practices
- Your knowledge of how to use software tools
- What you learned in management school
- Your intelligence

The PMP Certification Exam does test:

- Your knowledge of PMI's processes
- Your understanding of the many terms that are used to describe the processes
- Your ability to apply those processes in a variety of situations
- Your ability to apply key formulas to scheduling, costing, estimating, and other problems
- Your understanding of professional responsibility as it applies to project management

The Exam Material

Your PMP Exam will be made up of exactly 180 questions, covering a broad variety of material, but only 175 of those questions will count. The good news is that these 5 questions do not count toward your grade. The bad news is that you will never know which questions count and which do not.

QUESTION ALLOCATION ON THE PMP EXAM

Domain/Project Management Process Group	Percentage of Questions
Initiating	13%
Planning	24%
Executing	31%
Monitoring	25%
Controlling	7%

The PMP exam results are based on proficiency level i.e. 'Proficient', 'Moderately Proficient', or 'Below Proficient' in each of 5 domains. Hence, there is no fixed PMP passing score.

Getting to the Test (Application)

It is highly recommended that you join the Project Management Institute prior to signing up to take the test. At publication time, the member Exam Fee was $405.00 and non-member 555.00. Application may be made online at www.pmi.org.

In addition to the financial advantage, there are many other benefits that come with joining PMI, including a subscription to PMI's publications, PM Network and PM Today; discounts on books and PMI-sponsored events; and access to a wealth of information in the field of project management.

Earning your PMP Certification is a commitment, and that's why it is valuable. Do you have real-world project management experience that's led to success? Great—you've finished the hardest part. **Before you apply, make sure you meet of the following sets of PMP Certification requirements:**

- A four-year degree
- 36 months leading projects
- 35 hours of project management education/training or CAPM® Certification

— OR —

- A high school diploma or an associate's degree (or global equivalent)
- 60 months leading projects
- 35 hours of project management education/training or CAPM® Certification

Ongoing Education

PMPs are expected to demonstrate not only knowledge and experience but also their ongoing commitment to the field of project management.

To promote such commitment, PMI requires that all PMPs maintain their certification status by completing at least 60 Professional Development Units (PDUs) every 36 months.

The requirements for PDUs are defined in further detail in the PMI Continuing Certification Requirements Program Handbook given to all PMPs, and these requirements are similar in nature to requirements that legal, medical, and other professions have adopted.

In order to maintain the value of this certification, PMI requires its PMPs to maintain a project management focus and a continued commitment to the field of project management.

The Time Limit

230 minutes to complete the Exam! The test can be performed at your home with strict online supervision. During the 230 minutes test period, you can take two 10-minute breaks for computer based tests. No scheduled breaks for paper-based exams.

Question Format

Questions will be a combination of multiple-choice, multiple responses, matching, hotspot and limited fill-in-the-blank.

Many of the PMP questions are quite short in format; however, the PMP Exam is famous (or infamous) for its long, winding questions that are difficult to decipher. To help you prepare, you will see different question styles represented in this book. Going through all the sample questions provided in this volume is an excellent way to prepare for the types of questions you will encounter on the actual exam.

Foundational Concepts

What is a project? How is it different from operations or a program? What is a project manager, and how do different organizational structures change the role and power of a project manager? In order to understand PMI's approach to project management, you need a solid overview to these and other topics.

While much of the rest of this book is focused on PMI's 44 processes, inputs, tools, techniques, outputs, and formulas, this chapter lays the foundation on which those knowledge areas are built.

With the recent inclusion of the agile project management approach in the PMP 2021 Exam, we have also included a chapter dedicated exclusively to this approach in chapter 15. This way, you will have an idea of how project management works in the agile model.

Philosophy

PMI's philosophy of project management does not disconnect projects from the organizations that carry them out. Every project has a context and is heavily influenced by the type of organization in which the project is performed.

Another aspect to be considered in the context of the project is the roles of the different stakeholders.

An effective project manager must identify the different types of project stakeholders (such as customers, the project sponsor, senior management, etc.), understand their needs, and help them all work together to create common and realistic goals that will lead to a successful project.

Importance

Because this chapter is foundational, it is highly important! There will be many questions on the PMP Exam that test your understanding of a project manager, stakeholder, and sponsor.

You must also be able to identify the different types of organizations, and to recognize a project as compared to other types of related endeavors. Spend time making certain you understand these terms and definitions.

Preparation

The volume of material here is significant, and much of the information is marked as important. Test preparation should be focused on memorization of the terms and your ability to apply them. Word for word memorization is not essential, but a solid understanding is.

Essential Terms

Begin your PMP exam study by cementing your understanding of the following terms:

PROCESS

Processes are encountered regularly when studying for the PMP Exam. For purposes of the test, think of a process as a package of inputs, tools, and outputs used together to do something on the project.

For instance, Schedule Development is the process in which the project schedule is created.

Risk Identification is the process in which the list of risks is created, etc. There are 44 unique processes you will need to understand for the exam, all of which are covered in this book.

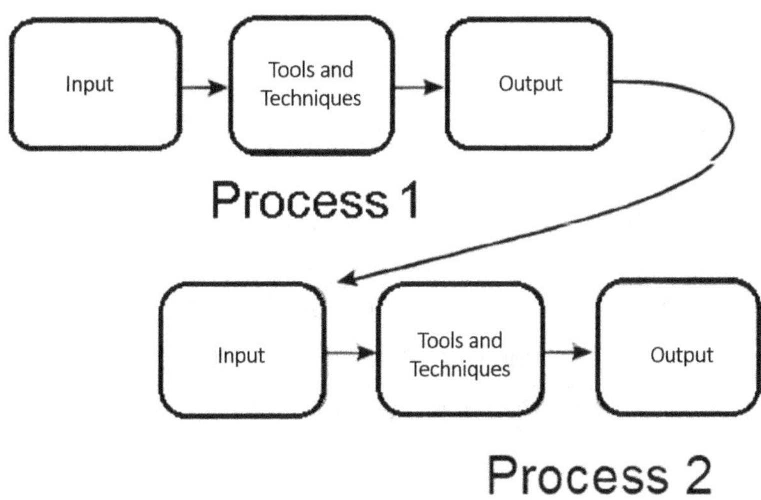

Interactions within a Process

As the preceding diagram illustrates, the outputs from one process are often used as inputs into other processes. Later in this chapter some of the most common inputs, tools, techniques, and outputs are discussed.

- Phases

Many organizations use project methodologies that define project phases. These phases may have names like "requirements gathering," "design," "development," "testing," and "implementation." Each phase of a project produces one or more deliverables.

One of the major problems test takers have when encountering PMI's material is to understand that all of the processes in the PMBOK may take place within each phase of the project. In other words, if your organization's methodology specifies a phase for product design, some or all of the 44 processes (described in chapters 4-13 of this book) may take place in that phase alone, only to be repeated in the subsequent project phases. Keep this fact in mind while reading the remaining chapters.

It is important to understand that PMI does not define what phases you should use on your project. That is because the PMBOK Guide does not describe a project methodology. Instead, processes are defined that will fit into your project methodology.

The example below shows how deliverables are usually associated with each phase. The deliverables are reviewed to determine whether the project should continue. This decision point is known as an exit gate or a kill point, and the decision on whether to proceed with the project is usually made by a person external to the project.

An "Exit Gate," or "Kill Point" is an evaluation of the deliverables of one project phase to determine if the project should continue and the next phase should be initiated.

- **Project**

A project is a temporary (finite) group of related tasks undertaken to create a unique product, service, or result. You may encounter a question on the exam that describes a situation and asks you whether that situation represents a project. If you see such a question, remember that you are looking for the following characteristics:

- A project is time-limited (it has a definite beginning and end).
- A project is unique (it has not been attempted before by this organization).
- A project is comprised of interrelated activities.
- A project is undertaken for a purpose (it will yield a specific product, service, or result).

As the above diagram illustrates, companies set strategic goals for the entire organization.

A company: project portfolio represents all of the project and program investments they make.

Programs represent a group of projects managed together in order to gain efficiencies on cost, time, technology, etc. For instance, by managing three related technology projects as a program, an organization might be able to save time and money by developing several common components only once and leveraging them across all of the projects that use those components.

Project management is the application of resources, time, and expertise to meet the project requirements. Project management usually applies to individual projects.

- **Program**

A program is a larger effort than a project, because it is a group of related projects coordinated together. Programs may also include operations. Organizations often group projects into programs in order to realize some benefit that could not be achieved if those projects were not undertaken in concert.

- **Portfolio**

A company's project portfolio represents the entire investment in projects and programs. Project portfolios should be aligned to the organization's strategic goals. Ideally, the benefit of all project investments should be expressed in how they meet or assist the organization's strategic goals.

- **Progressive Elaboration**

The term "progressive elaboration" simply means that you do not know all of the characteristics about a product when you begin the project.

Instead, they may be revisited often and refined. For instance, you may gather some of the requirements, perform preliminary design, take the results to the stakeholders for feedback, and then return to gather more requirements. The characteristics of the product emerge over time, or "progressively."

Project Management

Project management is using skills, knowledge, and resources to satisfy project requirements.

- **Historical Information**

Historical information is found many places on the exam, usually under the umbrella heading of organizational process assets, and it is almost always used as an input to processes wherever it is found. (Many of these terms will become more clear over the next two chapters). Historical information is found in the records that have been kept on previous projects. These records can be used to help benchmark the current project. They may show what resources were previously used and what lessons were learned. More than anything, historical information is used to help predict trends for the current project and to evaluate the project's feasibility.

Because PMI advocates constant improvement and continuous learning, historical records are extremely important in project management, and they are used heavily during planning activities. They can provide useful metrics, be used to validate assumptions, and help prevent repeated mistakes.

- **Baseline**

The term baseline is used for the project plan, time, scope, and cost. The baseline is simply the original plan plus any approved changes.

Many people who take the exam do not understand that the baseline includes all approved changes, but baselines are used as tools to measure how performance deviates from the plan, so if the plan changes, the new plan becomes the baseline. Suppose you were running a one-mile race, and you considered that distance as your baseline. Your plan was to run at the healthy pace of six minutes per mile.

Now suppose that the race length was changed to a three-mile race. You ran the race and still finished in a respectable twenty minutes. Would you want your progress measured against the original distance or the updated one?

If you did not update your baseline to three miles, your pace of twenty minutes against the original distance of one mile would not be very impressive at all. Your performance measurements would only be meaningful if you had an accurate baseline. Remember - your project's baseline is defined as the original plan plus all approved changes. Even though the baseline changes as the plan changes, it is a good idea to keep records that show how the plan has progressed and changed over time.

- **Lessons Learned**

Lessons learned are documents focused on variances created at the end of each process that detail what lessons were learned that should be shared with future projects. Lessons learned from past projects are an organizational process asset, which is an input into many planning processes. It is important that lessons learned put a special emphasis on variances that occurred on the project between what was planned to happen and what actually happened.

- Regulation

A regulation is an official document that provides guidelines that must be followed. Compliance with a regulation is mandatory (e.g. in the United States, wheelchair ramps are required for ADA compliance).

Regulations are issued by government or another official organization.

  - Standard

A standard is a document approved by a recognized body that provides guidelines. Compliance with a standard is not mandatory but may be helpful.

For example, the size of copy paper is standardized, and it would probably be a very good idea for paper manufacturers to follow the standard, but there is not a law in most countries requiring that copy paper be made the standard size. The PMBOK Guide provides a standard for project management.

- System

There are several instances of "systems" in the PMBOK Guide. A system incorporates all the formal procedures and tools put in place to manage something. The term "system" does not refer simply to computer systems, but to procedures, checks and balances, processes, forms, software, etc.

For instance, the project management information system (discussed in Chapter 4 - Project Integration Management), may include a combination of high-tech and low-tech tools such as computer systems, paper forms, policies and procedures, meetings, etc.

Project Roles

Another area of study regarding the project context is that of the roles and responsibilities found on projects. You should be familiar with the following terms related to project roles:

- **Project Manager**

The project manager is the person ultimately responsible for the outcome of the project. The project manager is:

- Formally empowered to use organizational resources
- In control of the project
- Authorized to spend the project's budget
- Authorized to make decisions for the project

Project managers are typically found in a matrix or projectized organization (more about types of organizations shortly). If they do exist in a functional organization, they will often be only part-time and will have significantly less authority than project managers in other types of organizations.

Because the project manager is in charge of the project, most of the project's problems and responsibilities belong to him. It is typically a bad idea for the project manager to escalate a problem to someone else. The responsibility to manage the project rests with the project manager, and that includes fixing problems.

- **Project Coordinator**

In some organizations, project managers do not exist. Instead, these organizations use the role of a project coordinator. The project coordinator is weaker than a project manager. This person may not be allowed to make budget decisions or overall project decisions, but they may have some authority to reassign resources. Project coordinators are usually found in weak matrix or functional organizations.

- **Project Expeditor**

The weakest of the three project management roles, an expeditor is a staff assistant who has little or no formal authority. This person reports to the executive who ultimately has responsibility for the project. The expeditor's primary responsibility lies in making sure things arrive on time and that tasks are completed on time. An expeditor is usually found in a functional organization, and this role may be only part-time in many organizations.

- **Senior Management**

For the exam, you can think of senior management as anyone more senior than the project manager. Senior management's role on the project is to help prioritize projects and make sure the project manager has the proper authority and access to resources. Senior management issues strategic plans and goals and makes sure that the company's projects are aligned with them. Additionally, senior management may be called upon to resolve conflicts within the organization.

- **Functional Manager**

The functional manager is the departmental manager in most organizational structures, such as the manager of engineering, director of marketing or information technology manager.

The functional manager usually "owns" the resources that are loaned to the project, and has human resources responsibilities for them. Additionally, he may be asked to approve the overall project plan. Functional managers can be a rich source of expertise and information available to the project manager and can make a valuable contribution to the project.

- Stakeholders

Stakeholders are individuals who are involved in the project or whose interest may be positively or negatively affected as a result of the execution or completion of the project. They may exert influence over the project and its results. This definition can be very broad, and it can include a vast number of people!

Often when the term "stakeholders" appears on the exam, it may be referring to the key stakeholders who are identified as the most important or influential ones on the project.

- Sponsor

The sponsor is the person paying for the project. He may be internal or external to the company. In some organizations the sponsor is called the project champion. Also, the sponsor and the customer may be the same person, although the usual distinction is that the sponsor is internal to the performing organization and the customer is external.

The sponsor may provide valuable input on the project, such as due dates and other milestones, important product features, and constraints and assumptions. If a serious conflict arises between the project manager and the customer, the sponsor may be called in to help work with the customer and resolve the dispute.

- **Project Office**

This term refers to a department that can support project managers with methodologies, tools, training, etc., or even ultimately control all of the organization's projects. Usually, the project office serves in a supporting role, defining standards, providing best practices, and auditing projects for conformance.

Project Context

Another major area of study for the PMP exam is the concept of a project context, or organizational environment, in which a project is carried out. A large part of the project context is determined by the organization's structure, which PMI refers to as the type of organization.

- **Types of Organizations**

The type of organization that undertakes a project will have an impact on the way the project is managed and even its ultimate success.

There are three major types of organizations described by PMI: functional organizations, projectized organizations, and a blend of those two called matrix organizations. Furthermore, matrix organizations can be characterized as weak, strong, or balanced.

The chart that follows summarizes essential information regarding these three types of organizations. You should be very familiar with this information before taking the exam, as you may see several questions that describe a project or situation and require you to identify what type of organization is involved.

• Organizational Structures

Type	Description	Who is in Charge?	Benefits	Drawbacks
Functional	Very common organizational structure where team members work for a department, such as engineering or accounting, but may be loaned to a project from time to time. The project manager has low influence or power.	Functional (Departmental) manager	• Deeper company expertise by function. • High degree of professional specialization. • Defined career paths.	• Project manager is weak. • Projects are prioritized lower. • Resources are often not dedicated to a project.
Projectized	The organization is structured according to projects instead of functional departments. The project manager is both the manager of the project and of the people. He is highly empowered and has the highest level of control.	Project manager	• Project manager has complete authority. • Project communication is easiest since everyone is on a single team. • Loyalty is strong, to both the team and the project. • Contention for resources does not exist.	• Team members only belong to a project – not to a functional area. • Team members "work themselves out of a job" - they have nowhere to go when the project is over. Professional growth and development can be difficult.
Matrix	A hybrid organization where individuals have both a functional manager and a project manager for projects. In a strong matrix, the project manager carries more weight. In a weak matrix, the functional manager has more authority. In a balanced matrix, the power is shared evenly between the functional and project managers.	Power shared between project manager and functional manager	• Can be the "best of both worlds." Project managers can get the deep expertise of a functional organization, while still being empowered to manage resources on the project.	• Higher overhead due to duplication of effort on some tasks. • Resources report to a functional manager and they have a "dotted line" to a project manager, sometimes causing conflict and confusion. • High possibility for contention between project managers and functional managers. • Because resources do not report to the project manager, they may be less loyal to him.

Project Manager's Power

Once you are comfortable in your understanding of types of organizations and the roles that different stakeholders play in a project, you can see that the organizational context in which a project is carried out will have a great deal of influence on that project. One way that the type of organization affects the project manager in particular is in how much power he is given. The chart below illustrates the relationship between a project manager's level of empowerment and the type of organization in which he works.

The Project Manager's Power by Organization Type

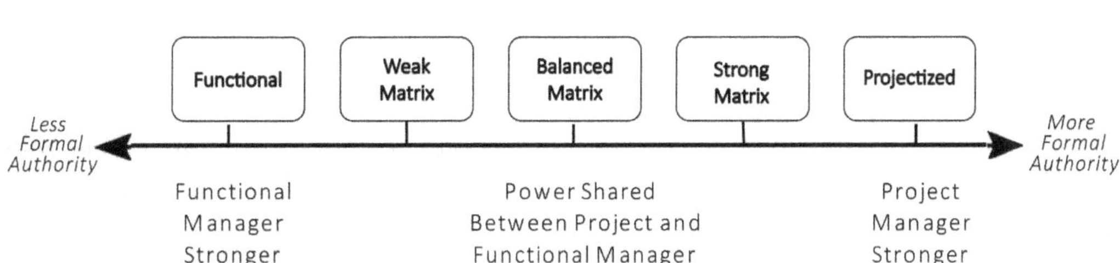

Project Manager's Management Skills

The practice of project management overlaps many other disciplines.

Since most projects are performed within an organization, there are other management skills that make up the foundation of project management. These skills will probably have a significant effect on projects. The project manager should have experience in:

- **Leading**

Motivating people and inspiring them to commit.

- **Communicating**

Exchanging information clearly and correctly. Although communication skills are not emphasized on the exam, they are important to the project manager and critical to the success of the project.

- **Negotiating**

Working to reach a mutual agreement. Negotiations may happen with groups or individuals inside and outside the organization.

- **Problem Solving**

Defining the problem and dealing with the factors that contribute to or cause the problem.

- **Influencing**

Accomplishing something without necessarily having formal power. Influencing the organization requires a keen understanding of the way the organization is structured, both formally and informally.

Project Life Cycle

The project life cycle is simply a representation of the phases that a project typically goes through. These phases are general, but they are representative of the common flow of activities on a project.

The six phases represented at the bottom of the graphic below describe the way in which a project typically progresses. It should be noted, however, that this depiction is very general, and different phases and phase names are used by different industries and projects. The image also shows some other facts about the project life cycle that often appear on the PMP exam in the form of questions. These questions typically focus on the fact that resource and cost levels rise early in the project and drop over time, or how risk and stakeholders' ability to influence the project are highest early in the project and decrease as the project progresses.

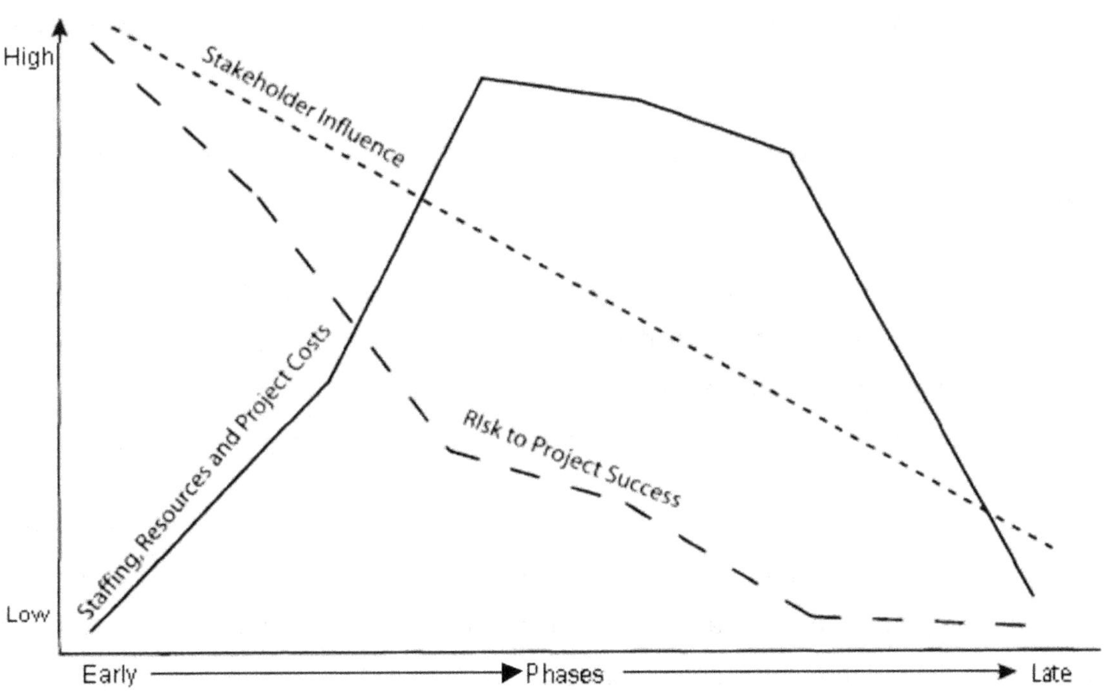

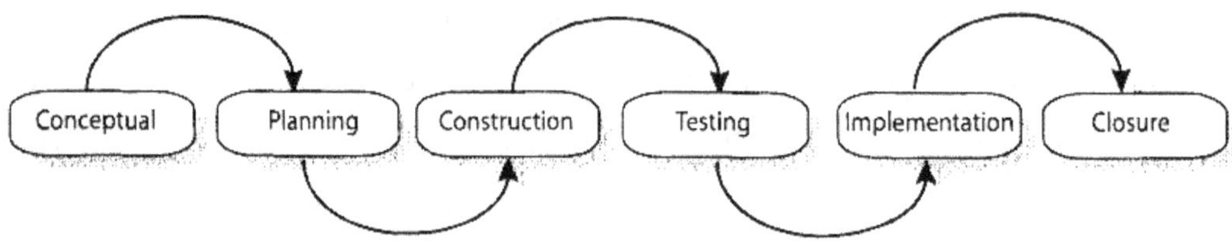

The Triple Constraint

Another fundamental topic in project management is commonly referred to as "the triple constraint." It is based on the realization that while changes do occur during a project, they do not happen in a vacuum. When the scope of a project is changed, time and cost are also affected.

Of course, the same is true when changes are made to cost or time. Those changes will have some impact on the other two areas.

As many different types of changes will be requested in most projects, it is essential in project management to be mindful of the triple constraint and to help keep others aware of it. The project manager should not simply accept all changes as valid; rather, the project manager should evaluate how those changes affect the other aspects of the project.

The triple constraint, or as some know it, the iron triangle, is simply the concept that scope, time, and cost are closely interrelated. Just as you cannot affect one side of a triangle without changing one or both of the other lengths, you cannot simply change one part of the triple constraint without affecting other parts.

However common a practice it may be in some organizations to slash a budget without revisiting the scope or the schedule, the project manager should not simply accept these mandates. The triple constraint is in place whether the organization recognizes and accepts it or not.

The classic approach to the triple constraint is represented in the following diagram:

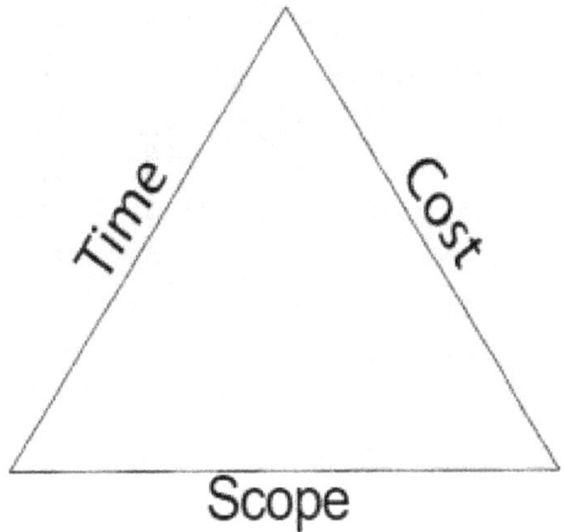

The triple constraint, are called the iron triangle.

But as many topics are interrelated in project management, an expanded view of the "triple" constraint could be represented not as a triangle but as a hexagon as shown below:

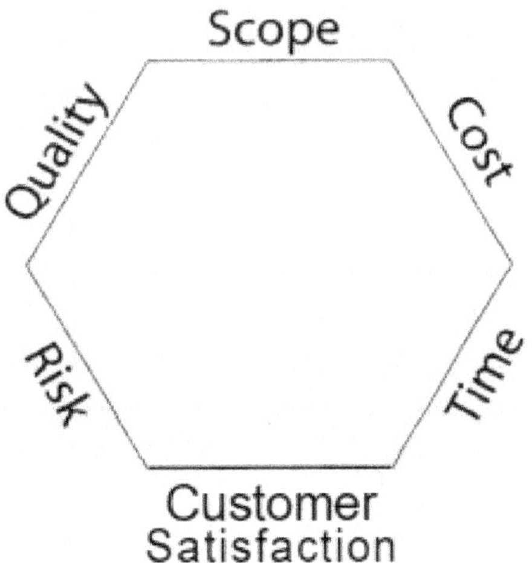

Expanded view of the "triple" constraint

Common Inputs, Tools, Techniques and Outputs

Throughout this book, several inputs, tools and techniques, and outputs of the 44 project management processes are referenced repeatedly. Since there are 592 inputs, tools and techniques, and outputs, the decision was made to discuss several of these in the following paragraphs rather than repeat them time and again throughout the book.

This section should serve as a reference as you encounter these throughout the book, and spending extra time here should help improve your overall understanding of the material.

In fact, this chapter is so important that you should read it now, and then come back to reread it after you have read chapters 4-12.

COMMON INPUTS

- Approved Change Requests

Change requests are common on projects, and they take on many characteristics. You may receive a change request to add functionality to a computer application, to remove part of a building, or to change materials.

The important thing to remember throughout this is that these are only requests until they are approved. If a change is requested, then the change is processed according to the integrated change control system.

This will ensure that the change request is properly understood and considered and that the right individuals or departments are involved before approving or rejecting it.

Approved change requests are used as an input into many processes to make sure that the change gets executed and is properly managed and controlled.

- **Enterprise Environmental Factors**

This input can cover a lot of ground, and it appears as an input into most planning processes. In fact, it is used so frequently that you may be tempted to just skim right over it, but be careful! Enterprise environmental factors are important to your understanding of the exam material, and you should make sure you have a solid grasp of why they are used so commonly.

Consider the things that impact your project that are not part of the project itself. Just a few of these include:

- Your company's organizational structure
- Your organization's values and work ethic
- Laws and regulations where the work is being performed or where the product will be used
- The characteristics of your project's stakeholders (e.g. their expectations and willingness to accept risk)
- The overall state of the marketplace for your project

In fact, enterprise environmental factors can be anything external to your project that affects your project. That is why it is so important to consider these factors when planning your project and to explore how they will influence your project.

- **Organizational Process Assets**

What information, tools, documents, or knowledge does your organization possess that could help you plan for your project? Some of these might be quite obvious, such as the project plan from a previous, similar project performed by your organization, while some may be more difficult to grasp at first, such as company policy.

Consider, however, that both of these assets will help you as you plan. For instance, company policy adds structure and lets you know the limits your project can safely operate within, so you do not have to waste time or resources discovering these on your own.

A few examples of organizational process assets are:

- Templates for common project documents
- Examples from a previous project plan
- Organizational policies, procedures, and guidelines for any area (risk, financial, reporting, change control, etc.)
- Software tools
- Databases of project information
- Historical information
- Lessons learned
- Knowledge bases

Anything that your organization owns or has developed that can help you on a current or future project may be considered an organizational process asset, and part of your job on the project is to contribute to these assets wherever possible on your project.

- **Project Management Plan**

The project management plan is one of the most important documents discussed in this book. It may be thought of as the culmination of all the planning processes. It is crucial that you understand what it is, where it comes from, and how it is used.

For the purpose of the exam, the definition of the project management plan is a single approved document that guides execution, monitoring and control, and closure. The use of the word single in this definition is a bit unusual, since the project management plan is actually made up of several documents; however, once these component documents become approved as the project management plan, they become fused together as one document.

Don't assume that the project management plan is always overly formal or detailed. The project management plan should be appropriate for the project. That means that it may be documented at a summary level, or it may be very detailed.

Following is a list of the components that make up the project management plan. The project management plan is covered in more detail in Chapter 4 - Integration Management under Develop Project Management Plan. Additionally, each of these components is covered in later chapters of this book.

- Project scope management plan
- Schedule management plan
    - The schedule baseline
    - The resource calendar

- Cost management plan
  - The cost baseline
- Quality management plan
  - The quality baseline
- Process improvement plan
- Staffing management plan
- Communications management plan
- Risk management plan
  - The risk register
- Procurement management plan

- **Work performance information**

Work performance information begins to flow as the actual work on the project is executed. Project team members, the customer, and other stakeholders will need to be informed as to the status of the deliverables, how things are performing against cost and schedule goals, how the project team is performing, and how the product stacks up against quality standards.

Although the examples above are the most common types of work performance information, any valuable information about the project that comes about as a result of performing the work would be considered work performance information.

COMMON TOOLS

- **Expert Judgment**

The tool of expert judgment is used time and again throughout the PMBOK Guide. The tool is exactly what it sounds like, and the reason it is so common is that it can be used whenever the project team and the project manager do not have sufficient expertise. You do not need to worry about whether the experts come from inside the organization or outside, whether they are paid consultants or offer free advice.

The most important things to remember for the exam are that this tool is highly favored and is very commonly found on planning processes.

- **Project Management Methodology**

This tool of project management is important. In fact, it's very important because it underscores something that is vital to your understanding The PMBOK Guide does not describe a methodology. The PMBOK Guide describes 44 processes used to manage a project. These processes are used by an organization's project management methodology, but they are not the methodology.

To illustrate the difference between the PMBOK Guide's 44 project management processes and a project management methodology, consider the analogy of two baseball teams. The Atlanta Braves and the New York Mets both have the same set of rules when they play, but they have very different strategies of how they will capitalize on those strengths and use those rules to their advantages. In this analogy, the rules would equate to the processes, and the strategy to methodology.

Different organizations will employ different project management methodologies, while they will all adhere to the 44 processes. And just as a team's strategy may be more nuanced and richer than the simple rules on which it is based, an organization's methodology may be a very rich and detailed implementation of the project management processes.

- **Project Management Information System**

The Project Management Information System (PMIS) is another important tool to know for the exam. It is your system that helps you produce and keep track of the documents and deliverables. For example, a PMIS might help your organization produce the project charter by having you fill in a few fields on a computer screen. It might then create the project charter and set up a project billing code with accounting.

While the PMIS usually consists primarily of software, it will often interface with manual systems.

Another important element of the PMIS is that it will contain the configuration management system, which also contains the change control system. While these will be covered in Chapter 5 - Scope Management, understanding how they fit together now will help.

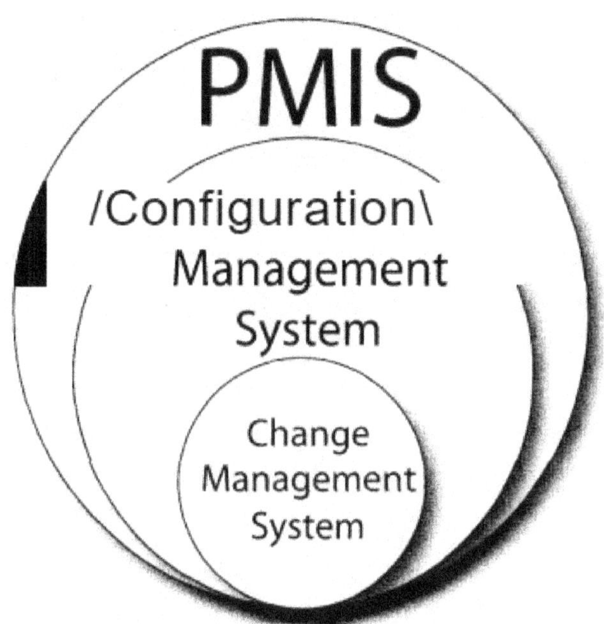

COMMON OUTPUTS

- Recommended Corrective Action

Corrective action is anything done to bring future results in line with the plan. Understanding that definition of corrective action will pay big dividends on the exam. Corrective action is all about the future, and it is the actions you take to make sure that the plan and the future results line up.

For instance, if the testing department has been missing their estimates by 20% ever since the project began, you might take corrective action to fix this. One way might be to go and talk with the department manager and ask her to put someone more experienced on the project.

Another way might be to change the plan for future testing to allow an extra 20%. Either way, you are taking steps to make sure the plan and the results line up. Now that you understand corrective action, be aware that a common output of processes is recommended corrective action.

In other words, you are recommending that steps be taken. These recommendations will be evaluated to determine their impact on the rest of the plan, and will either be approved or rejected.

The process where this occurs, Integrated Change Control, is covered in Chapter 4 - Project Integration Management.

- **Requested Changes**

As work is performed, it is common for changes to be requested. These changes can take on many forms. For instance, there may be change requests to increase the scope of the project or to cut it down in size.

There may be change requests to deliver the product earlier or later, to increase or decrease the budget, or to alter the quality standards.

The point to this is that these change requests are frequent as work is executed, or monitored and controlled. Like the previous example, all requested changes are brought into the process of Integrated Change Control where they will be evaluated for impact on the whole project and ultimately approved or rejected.

- **Updates (All Categories)**

Updates as process outputs occur so often in this edition of the PMBOK as to make it very difficult for the test taker to keep it straight. For purposes of the exam, know that updates to just about every kind of plan come out of planning, executing, and monitoring and controlling processes.

Most of these are common sense, and rather than take up valuable brain space explaining each individual one, the concept is addressed here and referenced throughout the book.

- **Project Management System**

When the term "system" is encountered in this material, you should not assume that it refers to a computer system. Instead, a system is the set of rules, processes, procedures, forms, and technologies, etc., that are used to support something. In this case, the project management system is defined in the project management plan, and it is used to support the project manager's execution and monitoring and control of the project. One example component of the project management system is described below.

- **Work Authorization System**

The work authorization system is actually part of the project management system. The work authorization system is the system used to ensure that work gets performed at the right time and in the right sequence. For example, if you had a tile specialist scheduled to come in next week and lay tile for a building project, you might want that person's manager to contact you before the specialist shows up at the job site. This conversation might be defined as part of the project's work authorization system so that the project manager retains control each time a team member takes on a new work package.

- **Policy**

Although you might think of policies as a headache in your job, they are viewed favorably in project management. In fact, we think of organizational policies as an asset! The reason a policy is an organizational process asset is that it gives guidance to your actions. For instance, you may have a corporate policy that you can only hire contractors for work on internal projects.

If that were the case, then having that policy could save time later in the project by preventing you from doing something your organization would frown upon. Please note that organizational policies may not be broken, even if doing so were to save the company money. You have to follow company policy, especially where the exam is concerned.

- **Approved Change Requests**

Approved change requests may be an output or an input, depending on which process you are performing.

Approved change requests start out as recommended change requests before they are processed in Integrated Change Control. Once they are approved, they are used as inputs to various executing and controlling processes. Approved change requests can also be an output of the Manage Stakeholders process.

- **Recommended Corrective Actions**

Corrective actions are anything done to bring future results in line with the plan. When there is a gap between the plan and the execution, then changes need to be made, whether those changes are to the plan or the way the plan is carried out. Recommended corrective actions are evaluated in the Integrated Change Control process.

- **Approved Corrective Actions**

As referenced above, approved corrective actions start out as recommended corrective actions. They are evaluated in the Integrated Change Control process where the become an output. Approved corrective actions can also be an output of the Manage Stakeholders process. Once they are approved, they are also used as inputs to affect execution so that future results line up with the plan. Keep in mind that corrective action is taken when something in the past has not gone as planned.

- **Recommended Preventive Actions**

Recommended preventive actions are made because a problem is anticipated. Recommendations on how to avoid the problem are made, and these recommendations are evaluated for their overall impact on the project and then approved (or rejected) in Integrated Change Control.

- **Approved Preventive Actions**

Approved preventive actions are taken to keep a problem from occurring in the first place. As was true for the previous input, approved preventive actions start out as recommended preventive actions, and they are approved in Integrated Change Control.

Process Framework

The process framework is the structure on which all of PMI's process material is built. The PMI processes are organized into nine knowledge areas and based on five foundational process groups:

1. Initiating
2. Planning
3. Executing
4. Monitoring and Controlling
5. Closing

Each of the 44 PMI processes performed as part of a project may be categorized into one of these process groups. Additionally, every question on the exam will tie back to one of these five areas or to professional responsibility, discussed in Chapter 13 - Professional Responsibility.

This chapter describes what processes, knowledge areas, and process groups are, and it explains how they are structured, and provides an overview of the project management framework.

Importance

This chapter is essential to your understanding of how this material is organized and structured. Do not be discouraged if you find the material in this chapter somewhat confusing at first.

The more you read and study from this book, the better you will understand these terms and how they are applied.

Preparation

There is significant memorization that accompanies this chapter. Terms must be learned, and more importantly the overall organization of the material needs to be understood. This chapter contains only a little that will actually show up on the exam, but it has to be mastered before chapters 4 - 12 can be fully comprehended.

Essential Terms

The essential information here begins with some key terms that you will need to understand. It is not necessary to memorize all the definitions, but make sure that you do understand them. They are foundational to this book, and highly important for the exam.

PROCESSES

The term "process" is one of the most important and frequently used terms you will encounter when studying for the PMP. Processes are composed of three elements: inputs, tools and techniques, and outputs.

The different inputs, tools and techniques, and outputs are combined to form processes, which are performed for a specific purpose. For instance, Schedule Development is a process, and as its name implies, it is performed to develop the project schedule. Risk Identification is another process, with different inputs, tools, techniques, and outputs, where (you guessed it) you identify the risks which could affect the project. There are 44 unique processes identified and described in the Guide to the PMBOK, and you will need to be familiar with all of them.

- **Inputs**

The inputs are the starting points for the processes. Just as food is a basic building block for the production of energy in living creatures, there are specific and unique inputs into each project management process that are used as building blocks for that process. You might think of inputs as your raw materials.

- **Tools and Techniques**

Tools and techniques are the actions or methods that are used to transform inputs into outputs. Tools can be many things, such as software, which can be used as a tool to help plan the project and analyze the schedule. Techniques are methods, such as flowcharting, which help us to frame, approach, and solve the problem. PMI combines tools and techniques since they are both used to solve problems and create outputs.

- **Outputs**

Every process contains at least one output. The outputs are the ends of our efforts. The output may be a document, a product, a service, or a result. Usually, the outputs from one process are used as inputs to other processes or as part of a broader deliverable, such as the project plan.

KNOWLEDGE AREAS

The knowledge areas in this material have been organized into nine groups. Each of the 44 project management processes defined in the PMBOK Guide fit into one of these nine knowledge areas. They are:

1. Integration Management
2. Scope Management

3. Time Management
4. Cost Management
5. Quality Management
6. Human Resource Management
7. Communications Management
8. Risk Management
9. Procurement Management

Like the PMBOK Guide, this book also has a chapter for each of the knowledge areas as listed above.

Process Groups

The project management processes defined and described in the PMBOK Guide are not only presented according to the nine different knowledge areas; they are also arranged according to process groups. The same 44 processes that are included in the nine knowledge areas are organized into the five process groups.

The five process groups are:

1. Initiating
2. Planning
3. Executing
4. Monitoring and Controlling
5. Closing

Every PMI-defined process that takes place on the project fits into one of those groups.

Organization

As hinted at in the previous paragraphs, each process has two homes. It fits into a process group and a knowledge area. As the chapters in this book are aligned to knowledge areas, you will be able to see how process are associated with different knowledge areas.

UNDERSTANDING THE FLOW

Do not fall into the trap of thinking that the first step is to do the processes in initiation, the second step is to do the processes in planning, and so on. Although projects may flow very roughly that way, you need to understand that the scope of a project is "progressively elaborated," which means that some processes are performed iteratively. Some planning must take place, then some executing, then some controlling processes. Further planning may be performed, further executing, and so on. The five process groups are by no means completely linear.

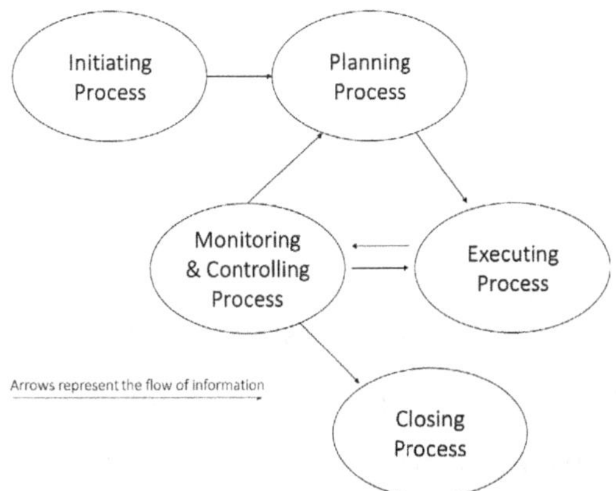

Illustration of the way in which process groups interrelate

Perhaps one of the biggest misconceptions people have of this material is believing that these process groups are the same thing as project phases. Understand that all 44 processes could be performed one or more times in each project phase.

Process Group 1 – Initiating

Inteqration	Scope	Time	Cost	Quality	HR	Communication	Risk	Procurement
✓								

The Initiating Process Group is one of the simpler groups in that it is only made up of only two processes: Develop Project Charter and Develop Preliminary Project Scope Statement. These two processes are described in further detail in Chapter 4 - Project Integration Management. This is the process group that gets the project officially authorized and underway.

The way in which a project is initiated, or begun, can make a tremendous difference in the success of subsequent processes and activities.

Although many processes may not be performed in a strict order, the initiating process should be performed first or at least very early on. In initiation, the project is formally begun, the project manager is named, and the preliminary scope statement is produced.

If a project is not initiated properly, the end results could range from a lessened authority for the project manager to unclear goals or uncertainty as to why the project was being performed. Conversely, a project that is initiated properly would have the business need clearly defined and would include a clear direction for the scope as well as information on why this project was chosen over other possibilities.

Initiation may be performed more than once during a single project. If the project is being performed in phases, each phase could require its own separate initiation, depending on the company's methodology, funding, and other influencing factors.

There is a reason why this might be advantageous. On a longer, or riskier project, requiring initiation to take place on each phase could help to ensure that the project maintains its focus and that the business reasons it was undertaken are still valid.

Process Group 2 – Planning

Inteqration	Scope	Time	Cost	Quality	HR	Communication	Risk	Procurement
✓	✓	✓	✓	✓	✓	✓	✓	✓

Planning is the largest process group because it has the most processes, but do not make the leap that it also involves the most work.

Although this is not a hard and fast rule, most projects will perform the most work and use the most project resources during the executing processes.

Project planning is extremely important, both in real life and on the PMP Exam. The processes from planning touch every one of the knowledge areas! You should be familiar with the 21 processes that make up project planning as shown in the graphic on the left of this page.

The order in which the planning processes are performed is primarily determined by how the outputs of those planning processes are used. The outputs of one process are often used as inputs into a subsequent planning process. This dictates a general order in which they must take place.

For instance, the project scope statement (created during Scope Definition) is used as an input to feed the work breakdown structure (created during the Create WBS process). The work breakdown structure is then used as an input to create the activity list (created during Activity Definition). You may, for example, encounter a question similar to the one below:

What is the correct sequence for the following activities?

A. Create project scope statement. Create work breakdown structure. Create activity list.

B. Create project scope statement. Create activity list. Create work breakdown structure.

C. Create work breakdown structure. Create project scope statement. Create activity list.

D. Create activity list. Create project scope structure. Create work breakdown structure.

In this example, the correct sequence is represented by choice 'A'. This question requires an understanding of concepts introduced throughout this chapter, such as how scope items (the work breakdown structure) are created first, time-related planning processes (the activity durations) are performed second, and cost planning processes (the budget baseline) are performed third.

Process Group 3 – Executing

Integration	Scope	Time	Cost	Quality	HR	Communication	Risk	Procurement
✓				✓		✓		✓

As alluded to earlier, executing processes typically involve the most work. You do not need to memorize a list of executing processes like the one for planning, but you should know that the executing process group is where the work actually gets carried out. In this group of processes, parts are built, planes are assembled, code is created, documents are distributed, and houses are constructed. Other elements are also included here, such as procurement, and team development. These all happen as part of the executing processes.

Some of the processes in the executing process group are intuitive, such as Direct and Manage Project Execution. Others, such as Perform Quality Assurance and Information Distribution often catch test takers by surprise because they had a different preconception of what was involved.

Process Group 4 - Monitoring and Controlling

Inteqration	Scope	Time	Cost	Quality	HR	Communication	Risk	Procurement
✓	✓	✓	✓	✓	✓	✓	✓	✓

Monitoring and controlling processes are some of the more interesting processes. These processes touch each of the nine knowledge areas. Activities that relate to monitoring and controlling simply ensure that the plan is working.

If it is not, adjustments should be made to correct future results. In monitoring and controlling processes, things are measured, inspected, reviewed, compared, monitored, verified, and reported. If you see one of those key words on a question, there is a good chance it is related to a monitoring and controlling process.

Planning processes are easy enough to grasp for most people. Executing processes are simply carrying out the plan, and monitoring and controlling processes are taking the results from the executing processes and comparing them against the plan.

If there is a difference between the plan and the results, corrective action is taken, either to change the plan or to change the way in which it is being executed (or both) in order to ensure that the work results line up with the plan.

Monitoring and controlling processes present another rich area for exam questions. Keep in mind that monitoring and controlling processes look backward over previous work results and the plan, but corrective actions, which often result from these processes are forward-looking.

In other words, monitoring and controlling is about influencing future results and not so much about fixing past mistakes. It is very important that you understand the previous statement for the exam. That concept is reinforced throughout the next several chapters.

Process Group 5 - Closing

Integration	Scope	Time	Cost	Quality	HR	Communication	Risk	Procurement
✓								✓

The closing group is comprised of two very important processes. They are Close Project (covered in Chapter 4 - Integration Management), and Contract Closure covered in Chapter 12 – Procurement Management). These two processes are sometimes difficult for people to master because in their work experience, once the customer signs off and accepts the product, the project is over.

The project does not end with customer acceptance. After the product has been verified against the scope and delivered to the customer's satisfaction, the contract must be closed out (Contract Closure), and the project records must be updated, the team must be released, and the project archives and lessons learned need to be updated (Close Project).

These processes need to be considered as part of the project, since the files, lessons learned, and archives will be used to help plan future projects.

Although there are only two processes in the closing process group, the questions about them make up 9% of the exam (18 questions), so you would do well to build a thorough understanding of what they are, how they work, and how they relate to the other processes.

If you find that you need more help in understanding the content of this chapter, read through it a couple of times before you go on. You can also read through chapter 3 of the Guide to the PMBOK after you have read this chapter.

It will be much easier to comprehend now that you have an understanding of its underlying structure.

Project Management Processes

Process Group / Knowledge Area	Initiating	Planning	Executing	Monitoring & Controlling	Closing
Integration	• Develop Project Charter • **Develop** Prelim Scope Statement	• **Develop Project Mgt.** Plan	• Direct and Manage Project Execution	• Monitor & Control Project Work Integrated Change Control	• Close Project
Scope		• **Scope** Planning Scope Definition Create **WBS**		• Scope Verification • **Scope** Control	
Time		• Activity Definition • Activity Sequencing • Activity Resource Est. Activity Duration Est. Schedule Dev.		• Schedule Control	
Cost		• Cost Estimating • Cost Budgeting		• Cost Control	
Quality		• Quality Planning	• Perform Quality Assurance	• Perform Quality Control	
Human Resource		• Human Resource Planning	.Acquire Project Team • Develop Project Team	• Manage Project Team	
Communi-cations		.Communications Planning	• **Information** Distribution	• Performance Reporting • Manage Stakeholders	
Risk		• Risk **Mgt. Planning** • **Risk** Identification • Qualitative Risk Anal. Quantitative Risk Anal. • Risk Response Plan.		• Risk Monitoring & Control	
Procurement		Plan Purchases & Acquisitions Plan Contracting	• Request Seller Responses • **Select** Sellers	• **Contract** Admin.	-Contract Closure

Integration Management

The processes of Project Integration Management with their primary outputs

Process	Primary output
Integration Management	
Develop Project Charter	Project Charter
Develop Preliminary Scope Statement	Preliminary Project Scope Statement
Develop Project Management Plan	Project Mgt. Plan
Direct and Manage Project Execution	Deliverables
Monitor and Control Project Work	Recom. Corrective Action Requested Changes
Integrated Change Control	Approved Change Req. Deliverables
Close Project	Final Product

When you look at a project, do you see the forest or the trees? In other words, do you look at the big picture, focusing on the deliverables, or the smaller and more numerous tasks that must be performed in order to complete the project?

When it comes to the processes of project management, most of the PMBOK Guide is made up of trees; however, project Integration Management is the whole forest. It focuses on the larger, macro things that must be performed in order for the project to work. Whereas much of the PMBOK Guide is organized into smaller processes that produce a plan or update a document, the processes of integration are larger and more substantial.

Integration Management is the practice of making certain that every part of the project is coordinated. In Integration Management the project is started, the project manager assembles the project plan, executes the plan, and verifies the results of the work, and then the project is closed.

At the same time, the project manager must prioritize different objectives that are competing for time and resources and also keep the team focused on completing the work.

Integration Management focuses on seven processes and how they fit together and interact with each other.

Philosophy

Integration Management takes a high-level view of the project from start to finish. The reason that the word "integration" is used is that changes made in any one area of the project must be integrated into the rest of the project. For instance, the human body's various systems are tightly integrated. What you eat and drink can affect how you sleep, and how much sleep you get can affect your ability to function in other areas.

When viewing your physical health, it is wise to look across your diet, exercise, sleep, stress, etc., since changes or improvements in one area will probably trickle into others. Integration Management is similar. Changes are not made in a vacuum, and while that is true for most of the processes in this book, it is especially true among the processes of Integration Management.

The philosophy behind Integration Management is twofold:

1. During the executing processes of the project, decision-making can be a chaotic and messy event, and the team should be buffered from as much of this clamor as possible. This is in contrast to the planning processes where you want the team to be more involved. You do not want to call a team meeting during execution every time a problem arises. Instead, the project manager should make decisions and keep the team focused on executing the work packages.

2. The processes that make up project management are not discrete. That is, they do not always proceed from start to finish and then move on to the next process. It would be wonderful if the scope were defined and finished and then went to execution without ever needing to be revisited; however, that is not the way these things typically go, and the PMBOK Guide recognizes that.

Importance

The importance of this section is high! You should expect several questions on the exam that relate directly to this chapter.

Preparation

The difficulty factor on this material is considered high, primarily because there is so much information to understand. In previous versions of the exam, Integration Management was not overly difficult, but much of the material that was in other knowledge areas has now been moved into Integration Management. While the material may not present as much technical difficulty as other areas such as time, quality, or cost, it may be new to many project managers and thus present a challenge. This chapter will guide you on where to spend most of your time and how to focus your study effort.

Develop Project Charter

WHAT IT IS

The charter is the document that officially starts the project, and this is the process that creates it.

WHY IT IS IMPORTANT

The charter is one of the most important documents on a project because it is essential for creating the project. If you don't have a charter, you don't have an official project. As you will see later in this section, the lack of a charter can causes problems for the project manager that may not manifest themselves until much later.

WHEN IT IS PERFORMED

This process is one of the first ones performed. It is common for some pre-planning to take place on a project before it becomes official, but it will not be a real project until the charter is issued.

HOW IT WORKS

- Inputs
    - *Contract*

Not all projects are performed under contract, so this input may or may not be relevant. When a project is performed under contract for another organization, it is common for the contract to be signed prior to the project beginning. In this case, as we are ready to start the project and create the charter, the contract provides an essential input.

- *Project Statement of Work*

The statement of work (SOW) is a written description of the project's product, service, or result. This will be supplied by the customer; however, if the customer is your own organization, the project's sponsor should supply this.

If the project is for an external customer, the SOW will typically be attached to the procurement documents and the contract.

The essential elements of the SOW are that it includes what is to be done, the business reason for doing it, and how the project supports the organization's strategy.

- *Enterprise Environmental Factors - See Chapter 2, Common Inputs*
- *Organizational Process Assets - See Chapter 2, Common Inputs*
- Tools
- *Project Selection Methods*

Companies select projects using a variety of methods. The most common methods seek to quantify the monetary benefits and expected costs that will result from a project and compare them to other potential projects to select the ones which are most feasible and desirable. Such methods are called "benefit measurement methods" in the PMBOK Guide.

Other methods apply calculus to solve for maximizations using constrained optimization. Constrained optimization methods are mathematical and use a variety of programming methods. If you see the terms linear programming, or non-linear programming, on the exam, you'll know they refer to a type of constrained optimization method and that the question is referring to techniques of project selection. You do not need to know how to calculate values for constrained optimization or linear programming for the exam, but you do need to know that they are project selection methods.

Following are additional terms in the fields of economics, finance, managerial accounting, and cost accounting that are sometimes used as tools for project selection. These are not listed in the PMBOK, but they may show up on the exam. It is not necessary to memorize these definitions word for word; however, it is important to understand what they are and how they are used.

A. Benefit Cost Ratio (BCR)

The BCR is the ratio of benefits to costs. For example, if you expect a construction project to cost $1,000,000, and you expect to be able to sell that completed building for $1,500,000, then your BCR is 1,500,000 / 1,000,000 = 1.5 to 1.

In other words, you get $1.50 of benefit for every $1.00 of cost. A ratio of greater than 1 indicates that the benefits are greater than the costs.

B. Internal Rate of Return (IRR)

IRR, or "Internal Rate of Return," is a finance term used to express a project's returns as an interest rate. In other words, if this project were an interest rate, what would it be? Calculation of the IRR is no longer required on the exam, but you should understand that just like the interest rate on a savings account, bigger is better when looking at IRR.

C. Net Present Value (NPV)

See Present Value definition below for an explanation of Net Present Value (NPV) and Present Value (PV).

D. Opportunity Cost

Based on the theory that a dollar can only be invested at one place at a time, opportunity cost asks "What is the cost of the other opportunities we missed by investing our money in this project?" For project selection purposes, the smaller the opportunity cost, the better, because it is not desirable to miss out on a great opportunity!

E. Payback Period

The payback period is how long it will take to recoup an investment in a project. If someone owed you $100, you would prefer that they pay it to you immediately rather than paying you $25 per month for 4 months. As you want to recoup your investment as quickly as possible, a shorter payback period is better than a longer one.

F. Present Value (PV) and Net Present Value (NPV)

PV is based on the "time value of money" economic theory that a dollar today is worth more than a dollar tomorrow.

If a project is expected to produce 3 annual payments of $100,000, then the present value (how much those payments are worth right now) is going to be less than $300,000. The reason for this is that you will not get your entire $300,000 until the 3rd year, but if you took $300,000 cash and put it in the bank right now, you would end up with more than $300,000 in 3 years.

PV is a way to take time out of the equation and evaluate how much a project is worth right now. It is important to understand that with PV, bigger is better.

Net Present Value (NPV) is the same as Present Value except that you also factor in your costs. For example, you have constructed a building with a PV of $500,000, but it cost you $350,000. In this case, your NPV would be $500,000 - $350,000 = $150,000.

Note that you are no longer required to calculate Present Value or Net Present Value for the exam. All you need to remember is that a bigger PV or NPV makes a project more attractive, and that NPV calculations have already factored in the cost of the project.

G. Return On Investment (ROI)

Return On Investment is a percentage that shows what return you make by investing in something. Suppose, for example, that a company invests in a project that costs $200,000. The benefits of doing the project save the company $230,000 in the first year alone. In this case, the ROI would be calculated as the (benefit - cost)/cost, or $30,000 / $200.000 = 15%. Note that you no longer need to perform this calculation on the exam, but you do need to understand that for ROI, bigger is better.

- *Project Management Methodology - See Chapter 2, Common Tools*
- *Project Management Information System - See Chapter 2, Common Tools*
- *Expert Judgment - See Chapter 2, Common Tools*

- **Outputs**

  - *Project Charter*

Once an organization has selected a project or a contract is signed to perform a project, the project charter must be created. Following are the key facts you need to remember about the project charter.

- It is created during the Develop Project Charter process.
- It is created based on a business need, a customer request, or market force within the economy or society, and it should explain why the project is being undertaken.
- It is signed by the performing organization's senior management.
- It names the project manager and gives him the authority to direct the project.
- It should include the high-level project requirements.
- It should include a high-level milestone view of the project schedule. Note that you won't be able to develop the detailed schedule until much later.
- It is a high-level document that does not include project derails; the specifics of project activities will be developed during the planning processes, which are carried out after Develop Project Charter is complete.
- It includes a summary-level project budget.
- It should include a milestone-level schedule.

Develop Preliminary Project Scope Statement

WHAT IT IS

It would be wonderful if every project manager were handed a complete definition of the project's scope as soon as the project began, but often one of the most challenging aspects of project management is to develop the project's scope. This process takes a first pass at defining the scope with the understanding that it will be revised in the future.

WHY IT IS IMPORTANT

This process gets the project pointed in the right direction by defining the initial view of the scope. The preliminary project scope statement will give definition to several other processes and will later be revised to become the completed project scope statement.

WHEN IT IS PERFORMED

The preliminary project scope statement is created very early in the project since it sets the initial direction that will govern many of the other activities.

HOW IT WORKS

- Inputs
    - *Project Charter - See Develop Project Charter - Outputs*
    - *Project Statement of Work - See Develop Project Charter - Inputs*
    - *Enterprise Environmental Factors - See Chapter 2, Common Inputs*

- *Organizational Process Assets - See Chapter 2, Common Inputs*

- Tools
  - *Project Management Methodology - See Chapter 2, Common Tools*
  - *Project Management Information System - See Chapter 2, Common Tools*
  - *Expert Judgment - See Chapter 2, Common Tools*

- Outputs
  - *Preliminary Project Scope Statement*

As you can tell by the name of this output, the project's preliminary project scope statement is a document that is meant to be revised. It is not the final word on the project's scope. Instead, it gives direction and information to subsequent processes.

The preliminary project scope statement will explain the project's basic scope, what constraints and assumptions exist, how the project's deliverables will be accepted by the customer, what schedule milestones exist, and even a high-level work breakdown structure (discussed further in the next chapter).

Also, the preliminary scope statement should include high-level preliminary cost estimates. It is developed based on information supplied by the person initiating the project (usually the sponsor or the customer).

It is important to understand that none of this information is intended to be the final word. These estimates, the definitions of scope, the work breakdown structure, and even the acceptance criteria can and probably will be revised later in the project.

This preliminary project scope statement is intended to set an initial direction for the project.

Develop Project Management Plan

WHAT IT IS

When many people think of a project plan, they mistakenly think only of a Gantt chart or a schedule. Many project managers who have carried this misconception into the PMP Exam have been chewed up and spit out by the test! As you will see in this section, the project management plan is a very important document that guides the project's execution and control, and it is much more than a schedule chart.

WHY IT IS IMPORTANT

This one should be easy. The project plan guides your work on the project. It specifies the who, what, when, where, and how. This document is used repeatedly throughout this book, so a solid understanding of it will pay dividends throughout your study.

WHEN IT IS PERFORMED

When the process of Develop Project Management Plan is performed is an interesting point, since the project management plan is not developed all at once. This plan is progressively elaborated, meaning that it is developed, refined, revisited, and updated.

HOW IT WORKS

- Inputs
  - ***Preliminary Project Scope Statement - See Develop Preliminary Project Scope Statement - Outputs***

- *Project Management Processes*

This input refers to the forty-four processes of project management described in the PMBOK Guide and this book. These processes are used as a framework and a guide as the project management plan is developed.

- *Enterprise Environmental Factors - See Chapter 2, Common Inputs*

- *Organizational Process Assets - See Chapter 2, Common Inputs*

- Tools

  - *Project Management Methodology*

The reason this tool is important is that you need to realize that the PMBOK Guide does not describe a methodology. It describes the processes that take place, but it does not tell you how you should carry out these processes. An organization should provide the project manager with a project management methodology. Typically, the more experience an organization has performing projects, the more mature their methodology will be.

A methodology may be comprehensive, covering every detail of how a project is performed, or general and high-level, covering only the major project planning deliverables.

- *Project Management Information System - See Chapter 2, Common Tools*

- *Expert Judgment - See Chapter 2, Common Tools*

- Outputs

  - *Project Management Plan*

The project management plan is the sole output of this process, and it is one of the most important outputs from any process.

To understand the project management plan, let's consider its definition. The PMBOK Guide defines the project plan as "A formal, approved document that defines how the project is managed, executed, and controlled. It may be summary or detailed and may be composed of one or more subsidiary management plans and other planning documents."

The keys to understanding this are broken out below:

1. The project management plan is formal. For the exam, it is important to think of the project management plan as a formal, written piece of communication.

2. The project management plan is a single document. It is not fifteen separate plans. Once those separate documents are approved as the project plan, they become fused into a single document.

3. The project management plan is approved. In other words, there is a point in time at which it officially becomes the project plan. Who approves it is going to differ based on the organizational structure and other factors, but typically it would be:

    - The project manager
    - The project sponsor
    - The functional managers who are providing resources for the project

    It is also important to understand that we do not typically think of the customer or senior management as approving the project plan. The customer will sign a contract, but will often leave the inner workings to the performing organization (ideally, anyway). The organization's senior management usually cannot get down to the level of reviewing every component document and approving the project plan and especially not for each and every project.

4. The project management plan defines how the project is managed, executed, and controlled. This means that the document provides the guidance on how the bulk of the project will be conducted.

5. The project management plan may be summary or detailed. Even though this wording is in the definition, for the exam you will do much better to think of the project management plan as always being detailed!

The project management plan is made up of several components, which you may think of as chapters in the overall plan. More formal and mission-critical projects will have longer and more formal components. On an actual project, not every project management plan will contain every one of the components listed below, but you should be familiar with the components illustrated below before taking the exam.

Components of
the Project Management Plan

Project scope management plan	Schedule management plan	Cost management plan	Quality management plan	Process improvement plan	Staffing management plan	Communication management plan	Risk management plan	Procurement management plan	Milestone list	Resource Calendar	Schedule baseline	Cost baseline	Quality baseline	Risk register

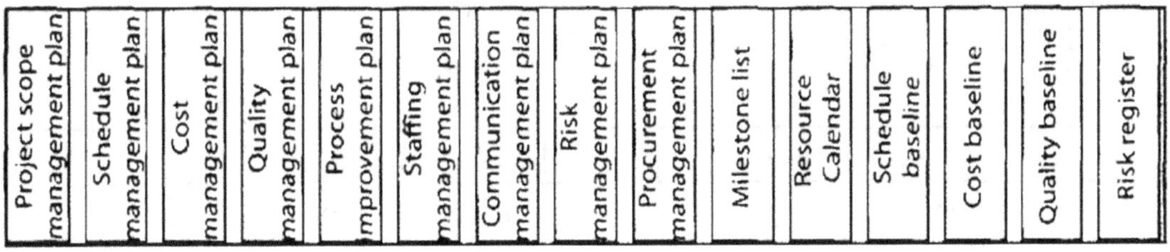

Another important thing to note about the project management plan is that most of its components are developed in other processes.

For instance, the project scope management plan is developed in Scope Planning. There are, however, two very notable exceptions. The schedule management plan and the cost management plan are not created elsewhere. Rather, they are created right here in Develop Project Plan.

The reason why this is true ties all the way back to the project charter. If you recall, when a project is initiated, the charter includes a summary budget and a summary (milestone) schedule. Since you already have these things at the time you begin Develop Project Plan, you can go ahead and develop the schedule management plan and cost management plan instead of waiting.

Later, when you perform Cost Management Planning and Schedule Management Planning, you will revise these components of the project plan with more detail to reflect your deeper understanding of the project.

Component of the Project Plan	Process Where Created
Scope Management Plan	Scope Planning
Schedule Management Plan	Develop Project Plan (revised in Schedule Development)
Cost Management Plan	Develop Project Plan (revised in Cost Estimating and Cost Budgeting)
Quality Management Plan	Quality Planning
Process Improvement Plan	Quality Planning
Staffing Management Plan	Human Resource Planning
Communications Management Plan	Communications Planning
Risk Management Plan	Risk Planning
Procurement Management Plan	Plan Purchases and Acquisitions
Milestone List	Begun in Develop Project Charter and revised in Schedule Development
Resource Calendar	Activity Resource Estimating
Schedule Baseline	Schedule Development
Cost Baseline	Cost Budgeting
Quality Baseline	Quality Planning
Risk Register	Risk Management Planning

Direct and Manage Project Execution

WHAT IT IS

When reading the PMBOK Guide, it is easy to walk away with the impression that the project manager must spend most of his or her time planning. Thankfully, however, that is not the case. Most of a project's time, cost, and resources are expended right here in the Direct and Manage Project Execution process. This is where things get done! In Direct and Manage Project Execution, the team is executing the work packages and creating the project deliverables.

WHY IT IS IMPORTANT

The Direct and Manage Project Execution process is where roads get built, software applications get written, buildings are constructed, and widgets roll off the assembly line.

WHEN IT IS PERFORMED

This process is difficult to put a time frame on, and it is important that you understand why that is true. People mistakenly think about project management as occurring linearly. That is, you plan, you execute, you monitor and control, and then you close, in that order. That is wrong.

On a real project, you may do some planning, some execution, and then monitor and control, only to return to more planning, more execution, and more monitoring and controlling. In reality, you may repeat this cycle numerous times.

Therefore, when looking at the process of Direct and Manage Project Execution, you should not think of it as a single occurrence, but understand that it occurs any time you are following the project management plan to create project deliverables.

HOW IT WORKS

- Inputs

  - *Project Management Plan*

Remember that the project management plan guides the management, execution, and monitoring and controlling of the project. In this process, we are primarily concerned with the execution of the project plan, therefore it is the essential input into this process.

See the preceding process, Project Plan Development, for more details on the project management plan.

  - *Approved Corrective Actions - See Chapter 2, Common Outputs*

  - *Approved Preventive Actions - See Chapter 2, Common Outputs*

  - *Approved Change Requests - See Chapter 2, Common Outputs*

  - *Approved Defect Repair*

At some point, mistakes need to be fixed, and that is what this input is about; however, it is a bit confusing the way it is named. It is not the approved defect repair itself you are using as an input, but the approved request.

Since this process focuses on the deliverables and results, the request is brought in so that the defect can be fixed.

  - *Validated Defect Repair*

This is simply an informational input to let you know that identified defects have been repaired and that those repairs have been inspected.

- *Administrative Closure Procedure*

Before the work actually begins, the procedures by which the deliverables will be accepted and the project will be closed is documented.

These procedures are brought into the Direct and Manage Project Execution process so that the work can be performed with the end goal of closing the project in mind.

- Tools
    - *Project Management Methodology - See Chapter 2, Common Tools*
    - *Project Management Information System - See Chapter 2, Common Tools*
- Outputs
    - *Deliverables*

This is arguably the most important output in the entire book! A deliverable is any product, service, or result that must be completed in order to finish the project. Some projects also must develop capabilities in order to finish a project, and these may be deliverables as well.

For instance, a project may need to develop a new manufacturing technique before they can create a product. In that case, the capability that the team develops could be considered a deliverable.

- *Requested Changes - See Chapter 2, Common Outputs*
- *Implemented Change Requests*

As approved change requests are brought into this process, the implemented changes flow out.

- *Implemented Corrective Actions*

Corrective actions are anything done to bring future results in line with the plan. It may mean either changes to the plan or changes in execution, but since this process focuses on performing the work, the assumption would be that you are making some process adjustment in the way you are performing the work.

- *Implemented Preventive Actions*

Similar to the previous two inputs, this is the fulfillment of a request. Preventive actions are those actions taken to avoid a problem altogether. In this process, you are implementing preventive actions to the way the work is being performed.

- *Implemented Defect Repair*

As requests for defect repair flow into this process, the fixed defects flow out.

- *Work Performance Information*

If the deliverables are the most important output of this process, this one is the second most important. It isn't only the deliverables that flow out of this process, but also the information on the status of these deliverables.

This information is used by several other processes that report on how far along a deliverable is and how it is tracking against the plan.

There could be quite a lot of work performance information of interest on the project. For instance, this process may provide information used by other processes on how the state of the deliverable lines up with the planned schedule, which milestones are being met, or how the actual costs are tracking against the estimated costs. It may also report on quality standards, such as the number of defects per thousand that are occurring.

Any information related to the deliverables being produced here could be considered work performance information.

Monitor and Control Project Work

WHAT IT IS

The process of Monitor and Control Project Work takes a look at all of the work that is being performed on a project and makes sure that the deliverables themselves and the way in which the deliverables are being produced is in line with the plan.

WHY IT IS IMPORTANT

All monitoring and controlling processes fulfill a sort of oversight role on the project. They compare the work results to the plan and make whatever adjustments are necessary to ensure that they match and that any necessary changes in the work or the plan are identified and made.

They also monitor all project information to ensure that risks are being identified and managed properly and to make sure that performance is on track.

WHEN IT IS PERFORMED

Monitor and Control Project Work is closely tied to the previous process (Direct and Manage Project Execution), and it takes place as long as there is work on the project to be done.

HOW IT WORKS

- Inputs
    - *Project Management Plan - See Chapter 2, Common Inputs*
    - *Work Performance Information - See Chapter 2, Common Inputs*
    - *Rejected Change Requests*

Any change requests that were processed and rejected should be brought into this process along with any supporting documentation.

- Tools
    - *Project Management Methodology - See Chapter 2, Common Tools*
    - *Project Management Information System - See Chapter 2, Common Tools*
    - *Earned Value Technique*

Earned value is discussed in detail in this book in Chapter 7 – Cost Management; however, you should understand here that earned value tries to measure how much value you have earned on the project as it progresses. For instance, just because a project has used up 50% of its allotted time and budget does not mean that the work is 50% completed.

The earned value technique listed here is used to measure the work completed and feed that information back into other processes.

- *Expert Judgment - See Chapter 2, Common Tools*

- Outputs

  - *Recommended Corrective Actions - See Chapter 2, Common Outputs*

  - *Recommended Preventive Actions - See Chapter 2, Common Outputs*

  - *Requested Changes - See Chapter 2, Common Outputs*

  - *Forecasts*

One of the tools listed in this process was the earned value technique. By using earned value, the project team can forecast project results. One example of this is the estimate at completion (EAC), which is a forecast of the likely total project costs based on past performance. Another example is the estimate to completion (ETC), which estimates how much more will need to be expended based on project performance thus far. Both of the aforementioned tools are covered in detail in this book in Chapter 7 - Cost Management, and both of them may be used here in the Monitor and Control Project Work process.

  - *Recommended Defect Repair*

Defects that require repair are marked with the status of recommended. These defects will be repaired in the previously mentioned process, Direct and Manage Project Execution.

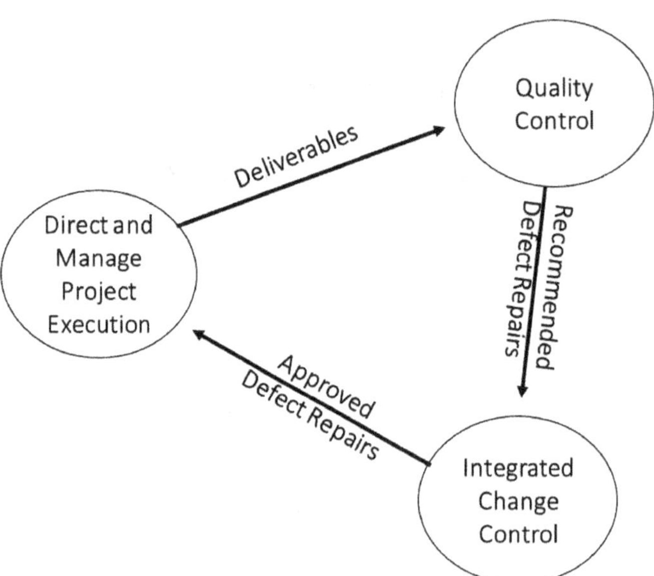

One way in which Direct and Manage Project Execution, Quality Control, and Integrated Change Control are interrelated

Integrated Change Control

WHAT IT IS

Some processes are more important than others for the exam, and this would qualify as one of the most important.

Every change to the project, whether requested or not, needs to be processed through Integrated Change Control. It is in this process where you assess the change's impact on the project.

WHY IT IS IMPORTANT

Integrated Change Control brings together (i.e. integrates) all of the other monitoring and controlling processes. When a change occurs in one area, it is evaluated for its impact across the entire project.

For example, suppose you came in to work one morning and found that a new legal requirement meant that the quality of your project's product needed to be improved. Would you only look at the quality processes on the project? No.

After understanding the impact of this change, you would likely need to evaluate the impact on the scope of the project, the activity duration estimates, the overall schedule, the budget estimates, the project risks, contract and supplier issues, etc. In other words, you would need to integrate this change throughout every area of the project.

Integrated Change Control is unique in that the project's deliverables are both an input and an output to it, which should give you some idea of how changes flow through this process.

One way in which Integrated Change Control differs from the previous process, Monitor and Control Project Work, is that Integrated Change Control is primarily focused on managing change to the project's scope, while Monitor and Control Project Work is primarily focused on managing the way that scope is executed.

For example, consider a new construction project for a hospital. If a change request were submitted that added a new wing to the hospital building, then that change request would be evaluated through Integrated Change Control to understand its impact on the whole project.

If, however, the project team members were performing slower than planned, that would be factored into Monitor and Control Project Work, and corrective action would be taken to ensure that the plan and the execution lined up. Even though both are controlling processes, each has a very different focus.

WHEN IT IS PERFORMED

Like the processes Direct and Manage Project Execution and Monitor and Control Project Work, Integrated Change Control takes place as long as there is work on the project to be performed.

HOW IT WORKS

- Inputs
  - *Project Management Plan - See Chapter 2, Common Inputs*
  - *Requested Changes - See Chapter 2, Common Outputs*
  - *Work Performance Information - See Chapter 2, Common Inputs*
  - *Recommended Preventive Actions - See Chapter 2, Common Inputs*
  - *Recommended Corrective Action - See Chapter 2, Common Outputs*
  - *Recommended Defect Repair - See Chapter 4, Monitor and Control Project Work*
  - *Deliverables - See Chapter 4, Direct and Manage Project Execution*
- Tools
  - *Project Management Methodology - See Chapter 2, Common Tools*
  - *Project Management Information System - See Chapter 2, Common Tools*
  - *Expert Judgment - See Chapter 2, Common*
  - *Change Control Board (Not listed as a tool in the PMBOK Guide)*

The change control board is a formally constituted committee responsible for reviewing changes. The level of authority of a change control board varies among projects and organizations; however, it would have its level of authority spelled out in the project management plan.

- Outputs

    - *Approved Change Requests*

All formally requested changes must be approved or rejected. The approved change requests are channeled back into Direct and Manage Project Execution.

- *Rejected Change Requests*

A requested change may be rejected for any number of reasons. The rejected change, along with any explanations and documentation should be returned to the requestor.

- *Project Management Plan Updates - See Chapter 2, Common Outputs*
- *Project Scope Statement Updates - See Chapter 2, Common Outputs*
- *Approved Corrective Actions*

Once recommended corrective actions have been properly considered and approved in this process, they become a formal output.

- *Approved Preventive Actions*

Once recommended preventive actions have been properly considered and approved in this process, they become a formal output.

- *Approved Defect Repair*

Mistakes must be fixed, and once these recommended defect repair requests have been considered, they change in status from recommended to approved.

- *Validated Defect Repair*

After a defect has been repaired, it is brought back into this process to be reconsidered and validated. That is why deliverables are both an input and an output for this process.

- *Deliverables - See Direct and Manage Project Execution – Outputs*

Close Project

WHAT IT IS

One of the key attributes of a project is that it is temporary. This means that every project eventually comes to an end, and that is exactly where this process comes into play.

Close Project is all about shutting the project down properly. This includes creating the necessary documentation and archives, capturing the lessons learned, ensuring that the contract is properly closed, and updating all organizational processes assets.

WHY IT IS IMPORTANT

Projects that skip this step often are left open, limping along for months without official closure. Taking the time to do this step, and do to it properly, will ensure that the project is closed as neatly and as permanently as possible.

WHEN IT IS PERFORMED

By looking at the name of this process, you can probably deduce that this process is performed at the very end of the project. That is not to say that you wouldn't do some of it before the project ends, but it cannot be completed until the project is finished.

HOW IT WORKS

- Inputs
  - *Project Management Plan - See Chapter 2, Common Inputs*
  - *Contract Documentation*

This input is discussed further in Chapter 12 - Procurement Management. In summary, it is all documentation that needs to be preserved relevant to the contract. Anything that might be of future interest regarding the contract should be documented and archived as part of this process.

  - *Enterprise Environmental Factors - See Chapter 2, Common Inputs*
  - *Organizational Process Assets - See Chapter 2, Common Inputs*
  - *Work Performance Information - See Chapter 2, Common Inputs*
  - *Deliverables - See Direct and Manage Project Execution - Outputs*
- Tools
  - *Project Management Methodology - See Chapter 2, Common Tools*
  - *Project Management Information System - See Chapter 2, Common Tools*

- *Expert Judgment - See Chapter 2, Common Tools*

- Outputs

  - *Administrative Closure Procedure*

Many organizations have very specific procedures that are required in order to formally close a project. This could include filling out forms, getting the customer's signature, gathering necessary project documents, meeting to document variances and the associated lessons learned, and creating the project archives.

- *Contract Closure Procedure*

This is a formal step to confirm that any contract between the performing organization and the customer is complete. This includes completion of the terms and conditions and all necessary payments and delivery of promised items.

Closing contracts is discussed further in Chapter 12 - Procurement Management.

- *Final Product, Service, or Result*

This is the sum of all project deliverables. Acceptance by the customer(s) and project sponsor is implied in this output.

- *Organizational Process Assets Updates*

In the course of a project, information will be gleaned, tools will be purchased or built, knowledge and experience will be gained, and documents (some of which may be reused one day) will be created. All of this should be updated as an organizational process asset and delivered to the appropriate group or individual(s) responsible for maintaining them. Often this will be the project management office.

Integration Management Questions

1. Producing a project plan may BEST be described as:

   A. Creating a network logic diagram that identifies the critical path.

   B. Using a software tool to track schedule, cost, and resources.

   C. Creating a document that guides project plan execution.

   D. Creating a plan that contains the entire product scope.

2. Project management plan updates are an output of:

   A. Integrated Change Control.

   B. Develop Project Plan.

   C. Direct and Manage Project Execution.

   D. Monitor and Control Project Work.

3. You are meeting with a new project manager who has taken over a project that is in the middle of executing. The previous project manager has left the company and the new project manager is upset that change requests are streaming in from numerous sources including his boss, the customer, and various stakeholders. The project manager is not even aware of how to process all of these incoming change requests. Where would you refer him?

   A. Project scope statement.

   B. Project management plan.

C. The previous project manager.

D. Project charter.

4. The preliminary project scope statement is:

   A. Developed before the project charter, and after the project management plan.

   B. Developed after the charter, and before the project management plan.

   C. Developed before the contract, and after the project management plan.

   D. Developed before the contract, and before the project management plan.

5. The work authorization system is used:

   A. So that people know when they will be performing the work.

   B. So that senior management may provide input by authorizing work requests.

   C. To ensure that only people authorized on the project are allowed to do the work.

   D. To ensure work gets performed at the right time in the right order.

6. A defect in the product was brought to the project manager's attention, and now the project team is engaged in repairing it. Which project management process would be the most applicable to this?

   A. Integrated Change Control.

   B. Monitor and Control Project Work.

   C. Direct and Manage Project Execution.

   D. Administrative Closure.

7. If you are creating a single document to guide project execution, monitoring and control, and closure, you are creating:

   A. The execution plan.

   B. The project plan.

   C. The integration plan.

   D. The project framework.

8. The change control system should be created as part of which process group?

   A. Initiation.

   B. Planning.

   C. Executing.

   D. Monitoring and controlling.

9. Which of the following statements is NOT true regarding the project charter?

   A. The project charter justifies why the project is being undertaken.

   B. The project charter assigns the project manager.

   C. The project charter specifies any high-level schedule milestones.

   D. The project charter specifies what type of contract will be used.

10. Which of the following represents the project manager's responsibility in regard to change on a project:

    A. Influence the factors that cause project change.

    B. Ensure all changes are communicated to the change control board.

    C. Deny change wherever possible.

    D. Prioritize change below execution.

11. The project plan is made up of:

    A. The other planning outputs.

    B. The other planning outputs, tools, and techniques.

    C. The aggregate outputs of all software tools.

    D. Scope verification.

12. When changes are approved and made to the project, they should be:

   A. Tracked against the project baseline.

   B. Incorporated into the project baseline.

   C. Included as an addendum to the project plan.

   D. Approved by someone other than the project manager.

13. Work performance information is used for all of the following reasons EXCEPT:

   A. It provides information on resource utilization.

   B. It provides information on which activities have started.

   C. It shows what costs have been incurred.

   D. It is used to help identify defects.

14. You are a project manager, and your team is executing the work packages to produce a medical records archive and retrieval system. Two of the project's customers have just asked for changes that each says should be the number one priority. What would be the BEST thing to do?

   A. Have the project team meet with the customers to decide which would be easiest and prioritize that one first.

   B. Assign someone from the team to prioritize the changes.

   C. Prioritize the changes without involving the team.

D. Deny both changes since you are in project execution.

15. The program manager is asking why your project is scheduled to take sixteen months. He claims that previous projects in the organization were able to be completed in less than half of that time. What would be the BEST thing to do?

    A. Look for historical information on the previous projects to understand them better.

    B. Refer the program manager to the schedule management plan.

    C. Refer the program manager to the project plan.

    D. Explain to the program manager that estimates should always err on the side of being too large.

16. You work for a defense contractor on a project that is not considered to be strategic for the company. Although the project is not the company's top priority, you have managed to secure many of the company's top resources to work on your project. At today's company meeting, you find out that your organization has won a very large, strategic project. What should you do FIRST?

    A. Contact management to find out if you can be transferred to this project because it is strategic to the company.

    B. Contact management to find out how this new project will affect your project.

    C. Hold a team meeting and explain that since the resources have been allocated to your project, they are not eligible to go to the new project.

    D. Fast track your project to accelerate its completion date.

17. Each time that the project sponsor requests a change to the project, the project manager calls a meeting of the project team and several stakeholders. This demonstrates:

   A. A collaborative style of management.

   B. Withdrawal.

   C. The lack of integrated change control system.

   D. The lack of a project management information system.

18. Your organization has a policy that any project changes that increase budget by more than 1.5% should be signed off by the project office. You have a change that was requested by the customer that will increase the budget by 3%;however, the customer has offered to pay for all of this change and does not want to slow it down. Which option represents the BEST choice?

   A. Approve the change yourself and take it to the project office after the work is complete.

   B. Ask the customer to take the change to the project office and explain the situation.

   C. Do not allow the change since it increases the budget by over 1.5%.

   D. Take the change to the project office.

19. The person or group responsible for evaluating change on a project is:

   A. The project manager.

B. The sponsor.

C. The project team.

D. The program manager.

20. The output of the Direct and Manage Project Execution process is:

A. The work packages.

B. The work authorization system.

C. The deliverables.

D. The work breakdown structure.

Answers to Integration Management Questions

1. C. The project plan is a single, approved plan that drives execution, monitoring and control, and closure. Note that the definition in answer 'C' was not perfect, but it was the best choice. 'A' is incorrect since it is only a part of planning. 'B' is incorrect because that will not make up the entire project plan. 'D' is incorrect since scope may or may not be a part of the project plan, but it does not make up all of it.

2. A. This question would be nearly a pure guess unless you understand the purpose of each of the processes listed. You should not go into the exam until you understand each of the 44 processes. Updates to the project plan are an output of the Integrated Change Control process. In general, updates come out of monitoring and controlling processes, which should have narrowed it down to 'A' and 'D'. From there, you should have considered the purpose of each of those two processes. Integrated Change Control changes the project (as in this case).

Monitor and Control Project Execution focuses more on controlling the way in which the project is executed.

3. B. The project plan would contain the methods for processing changes to the project.

4. B. Questions like this will be on the exam, and in order to answer it, you have to understand the rough order in which the deliverables are produced and processes are conducted. In this case, the typical order among the items listed is contract, charter, preliminary scope statement, and project management plan. If you analyze the inputs and outputs used by the integration processes, you will gain a better understanding of this order.

5. D. The work authorization system is covered in Chapter 2 - Foundational Terms and Concepts. The purpose of the work authorization system is to make sure work gets performed in the right sequence and at the right time. 'A' would be referring to the schedule. 'B' is incorrect since senior management should not be involved at that level. That is the job of the functional manager. 'C' is not the purpose of the work authorization system.

6. C. Direct and Manage Project Execution is the only process in the list where defects are repaired. Approved defect repairs are an input, and implemented defect repairs are an output.

7. B. This is the definition of the project plan.

8. B. The change control systems are created during planning processes.

9. D. The project charter does not specify anything about contracts. A contract with your customer would have been an input into the Develop Project Charter process, and any contracts you may use during procurement won't be identified

until later in the project. 'A' is incorrect because the project charter does specify why the project is being undertaken and often even includes a business case. 'B' is incorrect because the project charter is the place where the project manager is named. 'C' is incorrect because the project charter specifies any known schedule milestones and a summary level budget.

10. A. The project manager must be proactive and influence the factors that cause change. This is one of the key tenants of monitoring and controlling processes in general and Integrated Change Control in particular.

11. A. The project plan consists of many things, but the only one from this list that matches is the outputs from the other planning processes, such as risk, cost, time, quality, etc. 'B' is incorrect because the other tools and techniques do not form part of the project plan.

12. B. Did this one fool you? Approved changes that are made to the project get factored back into the baseline. Many people incorrectly choose 'A', but the purpose of the baseline is NOT to measure approved change, but to measure deviation.

13. D. The work performance information is all about how the work is being performed, but it is not used in identifying defects. 'A' is incorrect because it does provide detail on what resources have been used and when. 'B' is incorrect because it provides information on which activities have been started and what their status is. 'C' is incorrect because it provides information on what costs were authorized and what costs have been incurred.

14. C. Prioritizing the changes is the job of the project manager. 'A' is wrong because you do not want to distract the team at this point - they should be doing the work. 'B' is wrong because it is the project manager's responsibility to help prioritize

competing demands. 'D' is incorrect, because changes cannot automatically be denied simply because you are in execution.

15. A. Historical information (an organizational process asset) was covered in Chapter 2 - Foundational Terms and Concepts, and it may provide an excellent justification for why your project is taking sixteen months, or perhaps it will show you how someone else accomplished the same type of work in less time. Either way, it provides a great benchmark for you to factor in to your project. 'B' is incorrect since the schedule management plan only tells how the schedule will be managed. 'C' is incorrect because the project plan will not tell the program manager why the project is taking longer than he expects. 'D' is wrong because estimates should be accurate with a reserve added on top as needed.

16. B. Remember the rule that you should evaluate things first! You need to know if your project is going to be affected before taking action. Many people incorrectly choose 'D', but fast tracking the schedule increases risk, and that would not be necessary or appropriate until you fully understood the situation.

17. C. Team meetings to evaluate a change are generally a very bad idea and show a lack of a good integrated change control system.

18. D. As discussed in Chapter 2 - Foundational Terms and Concepts, organizational policies must be followed. None of the other options presents an acceptable alternative. Choice 'B' would be asking the customer to do the project manager's job.

19. A. The project manager is primarily responsible for evaluating changes to the project, and he is empowered to act on that evaluation.

20. C. The deliverables are an output of Direct and Manage Project Execution.

Scope Management

The processes of Project Scope Management with their primary outputs

Process	Primary Output
Scope Planning	Project Scope Mgt Plan
Scope Definition	Project Scope Statement
Create WBS	WBS
Scope Verification	Accepted Deliverables
Scope Control	Requested Changes Recom. Corrective Action

A solid grasp of project scope management is foundational to your understanding of the material on the PMP. While none of the topics in this book are particularly easy, scope management presents fewer difficulties than most other areas of the test. Most people find scope management to be more intuitive than other areas, since it has no complex formulas to memorize and no particularly difficult theories. Instead, scope management is a presentation of processes to plan, define, and control the scope of the project.

Philosophy

The philosophy behind PMI's presentation of scope management can be condensed down to these two statements: The project manager should always be in control of the scope through rigid management of the requirements, details, and processes, and scope changes should be handled in a structured, procedural, and controlled manner.

It is important to begin with the end in mind when it comes to scope management so that each requirement is documented with the acceptance criteria included.

Good scope management focuses on making sure that the scope is well defined and clearly communicated and that the project is carefully managed to limit unnecessary changes. The work is closely monitored to ensure that when change does happen on the project, it is evaluated, captured, and documented. Project managers should also work proactively to identify and influence the factors that cause change.

The overall goals of scope management are to define the need, to set stakeholder expectations, to deliver to the expectations, to manage changes, and to minimize surprises and gain acceptance of the product.

Importance

The topic of scope is very important on the PMP exam. When PMI refers to the project "scope," they are referring to the work needed to successfully complete the project and only that work. Many companies have a culture in which they try to exceed customer expectations by delivering more than was agreed upon; this practice, often referred to as "gold plating," increases risk and uncertainty and may inject a host of potential problems into the project.

Preparation

While this section requires less actual memorization than some other knowledge areas, many of the test questions can be very tricky, requiring a solid and thorough understanding of the theories and practices of scope management.

Scope Management Processes

As a starting point, you should understand that the knowledge area of scope management consists of the following elements:

- Creating a plan for how scope and changes to the scope will be managed
- Defining and documenting the deliverables that are a part of the project (the scope)
- Creating the work breakdown structure (WBS)
- Checking the work being done against the scope to ensure that is complete and correct
- Ensuring that all of what is "in scope" and only what is "in scope" is completed and that changes are properly managed

Process Group	Scope Management Process
Initiating	(none)
Planning	Scope Planning, Scope Definition, Create WBS
Executing	(none)
Controlling	Scope Verification, Scope Control
Closing	(none)

There are five processes in the scope management knowledge area. These are Scope Planning, Scope Definition, Create WBS, Scope Verification, and Scope Control. Below are the break-outs that show to which process group each item belongs:

In the knowledge area of scope management, it is also essential that you know the main outputs that are produced during each process. The different tasks that are performed in each process are summarized in the chart below.

Process	Key Outputs
Scope Planning	Project Scope Management Plan
Scope Definition	Project Scope Statement
Create WBS	Work Breakdown Structure
Scope Verification	Accepted Deliverables
Scope Control	Requested Changes to the Scope Recommended Corrective Action

Scope Planning

WHAT IT IS:

The process of Scope Planning is all about developing the project scope management plan. This plan, discussed in greater detail below, is your guidebook to the other four scope processes.

Keep in mind that the scope of the product is not actually planned here, as the name of this process might mislead you to believe. Instead, you are planning how you will conduct the scope gathering, definition, monitoring and control, and verification for the entire project. How much time and energy is invested in Scope Planning will vary from project to project, depending on the needs.

WHY IT IS IMPORTANT:

This process lays the groundwork for all of the project scope activities, and the resulting plan sets the stage for how formal or informal the project scope activities will ultimately be. By deciding this in advance and documenting it in the project scope management plan, the team can understand how the scope will be organized, planned, and managed.

In actual practice, the resulting project scope management plan is typically short and borrows heavily from templates and previous examples within the performing organization.

WHEN IT IS PERFORMED:

This process typically takes place quite early in the project, and certainly before any detailed scope or requirements definition takes place. Notice that it is somewhat cyclical, with the project management plan feeding into this process, and the output of this process (the project scope management plan) feeding the project management plan. If this seems confusing to you, bear in mind that projects are not a linear set of activities. In other words, the project management plan and the project scope management plan can be developed together, over multiple iterations.

Before you perform this process, you should perform the following:

1. Develop Project Charter
2. Develop Preliminary Project Scope Statement
3. Develop Project Management Plan

HOW IT WORKS:

- Inputs
    - *Enterprise Environmental Factors*

Enterprise environmental factors were covered in Chapter 2 - Foundational Terms and Concepts. Think of them as a context for your project. Factors such as your organization's culture, the chain of authority, how decisions are made, and whether your organization is a projectized, matrix, or functional organization will all influence how the scope is planned and managed.

The concept here is that you should be conscious of these factors in advance, evaluating how they will influence the activities you will perform as well as your ability to perform them.

- *Organizational Process Assets*

Once thought of as only historical information, organizational process assets make up much more. In addition to historical information gathered from previous projects, the project manager should also factor in any organizational policies and procedures that might influence how the scope is gathered and managed. A project scope management plan from a previous project would be an example of an asset, as would an organizational policy that stated that on-line meetings should be favored over physically co-locating stakeholders for all internal projects.

Project Charter

The charter sets the high-level goals of the project and will be useful in writing the detailed project scope statement. If there is no project charter for this project, then the project manager needs to create a document with similar information, even though it will not carry the same official weight as the project charter.

- *Preliminary Scope Statement*

The preliminary project scope statement simply gives an overview of the scope of the project, and this scope overview will be essential in creating the project scope management plan.

- *Project Management Plan*

There is a bit of circular logic with this input, since the project scope management plan makes up a part of the project management plan, and the project management plan is used as an input into the process that creates the project scope management plan; however, don't despair! It is unlikely that you will have a complete project management plan at the point in time when you perform this process, but the elements you do have should be helpful.

- Tools

  - *Expert Judgment*

As used here, the tool of expert judgment simply refers to involving someone who may have expertise on similar projects or someone who has created a project scope management plan in the past.

- *Templates, Forms, Standards*

Any templates or previous examples of scope documents or forms would be highly useful in developing the project scope management plan.

- Outputs

  - *The Project Scope Management Plan*

The project scope management plan defines what activities the team will perform in order to gather the project requirements, create the work breakdown structure, document the scope, place the scope under control, manage changes to the scope, and verify that all of the work and only the work was performed.

It is important to note that all of the aforementioned activities are not actually performed as part of this process. Instead, we are concerned with creating the plan for how they will be performed.

Once the project scope management plan is created, it becomes part of the overall project management plan.

Scope Definition

WHAT IT IS:

At some point, every project must gain a detailed understanding of the requirements to be executed, verified, and delivered. The scope of the project must be understood and documented in detail, and key stakeholders must be consulted so that their needs are understood and their expectations properly managed.

WHY IT IS IMPORTANT: The scope of the project is what ultimately drives the execution of the project, and Scope Definition is the process where the project's requirements are gathered and documented. The importance of this process is directly related to how important the requirements are. A large, mission-critical project will perform this process very thoroughly, while a smaller project, or one that is highly similar to a project that has been performed previously, will probably be less formal and detailed. Likewise, a project that has extremely high material costs (e.g. an offshore drilling platform) will likely spend more time and effort on Scope Definition than a project with less risk of error.

WHEN IT IS PERFORMED:

This process may be performed when the Scope Planning process has been completed. Because the project scope management plan defines how this process is performed, Scope Planning should be performed before the process of Scope Definition.

Since projects are generally progressively elaborated, this process and the requirements may be revisited many times throughout the life of the project; however, it is generally begun very early in the project lifecycle.

HOW IT WORKS:

Scope Definition takes the preliminary project scope statement, created earlier in the project, and refines it by performing additional analysis, factoring in any approved changes, and adding additional detail. Whereas the preliminary project scope statement was high-level and general, this process produces the actual project scope statement, which has substantially more information about the project scope and requirements.

- Inputs
  - *Organizational Project Assets - See Chapter 2, Common Inputs*
  - *Project Charter*

The project charter is the organization document formally creating the project and outlining its goals. Since this process will be creating a detailed view of the scope, the project charter is needed. If the project charter does not exist, then the project manager should still capture the project's overall goals, a brief description of the scope, and the known constraints and assumptions before performing Scope Definition.

- *Preliminary Project Scope Statement*

The preliminary project scope statement was created earlier in the project. It will be used as the primary input to this process so that the detailed project scope statement can be developed.

- *Project Scope Management Plan*

The project scope management plan is important here because it details how the scope will be gathered and documented for this project.

- *Approved Change Requests*

The project scope can (and usually does) change as soon as the project begins. Approved change requests should be brought into Scope Definition, whether or not this is the first time you are performing it on this project. Once an approved scope change becomes documented in the detailed scope statement and the WBS, it becomes part of the updated scope baseline.

- Tools

  - *Product Analysis*

Product analysis is a detailed analysis of the project's product, service, or result, with the intent of improving the project team's understanding of the product, and helping to capture that understanding in the form of requirements. The tools that may be used in product analysis will vary from industry to industry and organization to organization.

- *Alternatives Identification*

The goal of alternatives identification is to make sure that the team is properly considering all options as they relate to the project's scope. Techniques to generate creative thought are used most often for this.

- *Expert Judgment*

This tool involves having experts work with the project team to develop portions of the detailed scope statement. These are typically experts on the technical matters that need to be documented.

- *Stakeholder Analysis*

Stakeholders have expectations about the project (most often focused on the product itself). These needs must be analyzed, understood, and translated into measurable, objective, and documented requirements. If they are not, then the project stands a high risk of failing product acceptance.

- Outputs

  - *Project Scope Statement*

The project scope statement is the document used to level-set among the project's stakeholders. It typically has considerably more detail than the preliminary project scope statement created earlier in the project.

The project scope statement contains many details pertaining to the project and product scope, including: the goals of the project, the product description, the requirements for the project, the constraints and assumptions, and the identified risks related to the scope.

The objective criteria for accepting the product should also be included in the project scope statement. Finally, a cost estimate should be included with the project scope statement.

This cost estimate may be order of magnitude, conceptual, preliminary, definitive, or control, depending on how much is known at the point in time the estimate is created or revised. (More discussion of these types of estimates can be found in Chapter 6 - Cost Management.)

Although the preliminary project scope statement and the more refined project scope statement have more in common than not, there are some key differences between them. The table below highlights some of the differences.

Differences between Preliminary Project Scope Statement and Project Scope Statement		
	Preliminary Project Scope Statement	**Project Scope Statement**
Process where created	Create Preliminary Project Scope Statement	Scope Definition
Knowledge area where created	Integration	Scope
When it is used	Very early in the project and only for a relatively brief time	Throughout the project
Purpose	Provides a starting point related to the scope. Sets the direction for the project and give rough definition to the scope.	Defines the scope in detail. Contains everything that is in the scope of the project.

- ***Requested Changes***

Because of the amount of analysis performed in this process, it is common for requested changes to the project plan to be created. These should be processed through Integrated Change Control.

- *Project Scope Management Plan Updates*

The project scope management plan details how the scope will be gathered and documented, and how changes to the scope will be managed. As the process of Scope Definition is performed, it is common for changes and updates to be made to this plan to reflect items specific to this project and this organization.

Create WBS

WHAT IT IS:

Once the scope of the project has been defined and the requirements have been documented, it is necessary to create the work breakdown structure (WBS) for this project.

WHY IT IS IMPORTANT:

The work breakdown structure, or WBS, is one of the most important topics for the exam. The reason for its importance is tied to how it is used. After its creation, the WBS becomes a hub of information for the project. Time and cost estimates are mapped back to it, work is measured against it, and deliverables are ultimately compared to it.

WHEN IT IS PERFORMED:

Create WBS is typically performed early in the project, after the scope and requirements have been gathered, but before the bulk of the work is executed.

HOW IT WORKS:

- Inputs
    - *Organizational Process Assets - See Chapter 2, Common Inputs*

- ***Project Scope Statement***

The project scope statement describes the scope of the project in detail. It will be used in this process as a primary starting point from which to create the WBS.

- ***Project Scope Management Plan***

The project scope management plan describes how the scope of the project will be gathered and documented. Since the WBS is a critical component of that documentation, the project scope management plan is brought into this process as a key input. It will give the team guidance as to which approach will be used to create the WBS and how it should be evaluated when it is completed.

- ***Approved Change Requests***

Approved change requests should be brought into this process, whether or not this is the first time you are performing it on this project. Once an approved scope change becomes documented in the detailed scope statement and the WBS, it becomes part of the updated scope baseline.

- Tools

- ***Work Breakdown Structure Templates***

WBS templates may be examples from previous projects or partially- filled-out versions of a WBS from your project office, or even commercially available templates that may be used as a starting point. Regardless of the source or the state of completion, templates can be very helpful for this process.

- ***Decomposition***

Decomposition is one of the tools used in Scope Definition to create the WBS. This involves breaking down the project deliverables into progressively smaller components. In a WBS, the top layer is very general (perhaps as general as the deliverable or product name, or some even go so far as to make the top node the overall program), and each subsequent layer is more and more specific. The key to reading the WBS is to understand that every level is the detailed explanation of the level above it.

Decomposition may be thought of as similar to the arcade game Asteroids™ from years ago. Large pieces are progressively broken down into smaller and smaller pieces.

So, how do you know when you have decomposed your WBS far enough? As you have probably realized, the nodes can be decomposed to ridiculously low levels, wasting time and actually making the project difficult to understand and change. There are many things to consider when deciding how far to decompose work, but two of the best questions to ask are:

1. Are your work packages small enough to be estimated for time and cost?

2. Are the project manager and the project team satisfied that the current level of detail provides enough information to proceed with subsequent project activities?

If you can answer "yes" to those two questions, your work packages are probably decomposed far enough.

- **Outputs**

  - ***Project Scope Statement Updates***

As you decompose the deliverable-based nodes of the work breakdown structure, it often improves your understanding of the scope itself. Updates to the project scope statement are a normal by-product of this process.

- ***Work Breakdown Structure***

The writers of the current edition of PMBOK Guide currently advocate that the work breakdown structure always be based on the project deliverables rather than the tasks needed to create those deliverables and that it be built from the top down.

The WBS is primarily constructed using two techniques:

1. Decomposition is the practice of breaking down deliverables (product features, characteristics, or attributes) into progressively smaller pieces. This process continues until the deliverables are small enough to be considered work packages. A node may be considered a work package when it meets the following criteria:

   - The work package cannot be easily decomposed any further
   - The work package is small enough to be estimated for time (effort)
   - The work package is small enough to be estimated for cost
   - The work package may be assigned to a single person

If the node is being subcontracted outside of the performing organization, that node, regardless of size, may be considered a work package, and the subcontracting organization builds a "sub-WBS" from that node.

2. The other tool is work breakdown structure templates. A WBS template can be an example from a previous project, or a boilerplate example distributed by the project office. The important point is that is provides you with an accelerated starting point and reference when creating your breakdown of the current project's work.

The resulting WBS is a graphical, hierarchical chart, logically organized from top to bottom. Each node on the WBS has a unique number used to locate and identify it.

Elements of a Good Work Breakdown Structure (WBS):

- It must be detailed down to a low level. The lowest level consists of work packages that define every deliverable on the project.
- It is graphical, arranged like a pyramid, where each sub-level rolls up to the level above it.
- It numbers each element, and the numbering system should allow anyone who reads the WBS to find individual elements quickly and easily.
- It should provide sufficient detail to drive the subsequent phases of planning.
- It may often be borrowed from other projects in the organization as a starting point. These starting points are known as templates.
- It is thorough and complete. If an item is not in the WBS, it does not get delivered with the project.
- It is central to the project.
- The project team, and not just the project manager, creates the WBS. Developing the W85 can also be a means of team-building.
- It is an integration tool, allowing you to see where the individual pieces of work fit into the project as a whole.
- It helps define responsibilities for the team.
- It is a communication tool.
    - *WBS Dictionary*

The WBS Dictionary is a document that details the contents of the WBS. Just as a language dictionary defines words, a WBS dictionary provides detailed information about the nodes on a WBS. Because the WBS is graphical, there is a practical limit to how much information can be included in each node. The WBS dictionary solves this problem by capturing additional attributes about each work package in a different document that does not have the graphical constraints that the WBS does.

For each node in the WBS, the WBS dictionary might include the number of the node, the name of the node, the written requirements for the node, to whom it is assigned, and time, cost, and account information.

Project: Online Ordering Application	
Work Package ID: 1.1.3	
Work Package Name: Configure New Hardware	
Work Package Description: All new hardware should be configured, including any hardware settings and preparation such as formatting of storage. The correct operating system should be loaded and the appropriate patches should be applied. Any security settings, including virus scanning software should be applied. The hardware should be added to the company domain and should be compliant with all company policy regarding hardware and security.	
Assigned to: Lee Abbott	**Department:** I.T.
Date Assigned: 2/24/05	**Date Due:** 2/20/06
Estimated Cost: $3,800.00	**Accounting Code:** HMIT-0229

- *Project Scope Management Plan Updates*

The project scope management plan is the plan that specifies how the scope will be created, delivered, verified, and how changes will be made. As the work breakdown structure is created and the team's understanding of the work is improved, it is common for this plan to be updated.

- *Scope Baseline*

A baseline (whether for scope, schedule, cost, or quality), is the original plan plus all approved changes. In this instance, the scope baseline represents the combination of the project scope statement, the WBS and the WBS dictionary. When the scope baseline is created, it is placed under control, meaning that changes to the scope are made according to the scope management plan.

- *Requested Changes*

The scope of a project is usually progressively elaborated, meaning that the team's understanding of the work to be performed progresses in iterations. Because of this, it is normal to have change requests generated as the team's understanding of the work improves. These requested changes may relate to areas of the project other than the scope, such as schedule, cost, or the contract.

Scope Verification

WHAT IT IS:

Scope Verification is easily confused with other processes. Many people fall into the trap of thinking that Scope Verification means verifying that the scope is documented accurately; however, this is not correct. Scope Verification is the process of verifying that the product, service, or result of the project matches the documented scope.

In general, controlling processes compare the plan with the results to see where they differ and take the appropriate action. This fits Scope Verification very well, since the product is compared with the scope to ensure they match.

Scope Verification has quite a few similarities to the process of Quality Control, in that they both inspect the product against the scope; however, there are some key differences between them.

- Scope Verification is often performed after Quality Control, although it is not unusual for them to be performed at the same time.
- Scope Verification is primarily concerned with completeness, while Quality Control is primarily concerned with correctness.
- Scope Verification is concerned with the acceptance of the product by the project manager, the sponsor, the customer, and others, while Quality Control is concerned with adherence to the quality specification.

One more important note about Scope Verification is that if the project is canceled before completion, Scope Verification should be performed to document where the product was in relation to the scope at the point when the project ended.

WHY IT IS IMPORTANT:

If the process of Scope Verification is successful, the product is accepted by the project manager, the customer, the sponsor, and sometimes by the functional managers and key stakeholders. This acceptance is a significant milestone in the life of the project.

WHEN IT IS PERFORMED:

This process would be performed after at least some of the product components have been delivered, although it may be performed several times throughout the life of the project.

As mentioned above, Scope Verification is typically performed at the same time as Quality Control or immediately following Quality Control.

HOW IT WORKS:

- Inputs

  - *Project Scope Statement*

The project scope statement contains a detailed description of the project scope and deliverables, and it is this document that will be used to judge the results. The project scope statement will be used to ensure that the created product matches the scope completely.

  - *WBS Dictionary*

The WBS dictionary contains information on each work package in the work breakdown structure, such as to whom it was assigned, when it is due, where it will be performed, how much it should cost, etc. The WBS dictionary is often used in the process of Scope Verification to go through point by point and confirm that each work package was properly completed.

  - *Project Scope Management Plan*

The project scope management plan is an input into Scope Verification because it specifies how Scope Verification will be performed. It details who should be involved in this process and how it should be conducted.

  - *Deliverables*

Scope Verification is all about comparing the deliverables with the documented scope. This comparison may be performed several times during the life of the project.

- Tools

- *Inspection*

Inspection is the only tool employed in this process. It involves a point-by-point review of the scope and the associated deliverable. For instance, a pre-occupancy walkthrough by a building inspector would be an example of the tool of inspection. User-acceptance testing of a software product could be another example where the tool of inspection was used to make sure the deliverables matched the documented scope.

- Outputs

  - *Accepted Deliverables*

Acceptance of the deliverables is the primary output of the process of Scope Verification. This process is typically performed by the project manager, the sponsor, the customer, and the functional managers.

- *Requested Changes*

Change requests are a normal result of any inspection. When the deliverables are inspected in-depth and compared with the documented scope, change requests will often result.

- *Recommended Corrective Action*

It is important to note that corrective action is not about fixing past mistakes. Instead, it is about asking, "What would we change to prevent this difference from happening again?" Corrective action is anything done to bring future results in line with the plan. It compares the plan with the results or deliverables and evaluates differences.

Corrective action could be about changing the plan or changing the way in which the work was performed. The recommended corrective actions are a resulting output of Scope Verification.

Scope Control

WHAT IT IS:

Scope Control is a process that lives up to its name. This process is about maintaining control of the project by preventing scope change requests from overwhelming the project, and also about making certain that scope change requests are properly handled.

One of the more challenging concepts behind Scope Control can be that of resolving disputes. Although many disputes over project scope or product requirements are not simple, keep in mind that the customer's interests should always be weighed heavily. That does not mean that the customer is always right! Instead, it means that the customer is one of the most important stakeholders. All other things being equal, disputes should be resolved in favor of the customer.

WHY IT IS IMPORTANT:

Anyone who has managed a project where scope change was a problem knows the importance of Scope Control. This process makes certain all change requests are processed, and also that any of the underlying causes of scope change requests are understood and managed. It is important not only to manage scope change requests, but to prevent unnecessary ones.

WHEN IT IS PERFORMED:

Scope Control is an ongoing process that begins as soon as the scope baseline is created. Until that point, the scope is not considered stable or complete enough to control; however, once the scope baseline is created, each scope change request must be carefully controlled and managed. Additionally, any time the work results are known to differ from the documented scope, this process should be performed, whether or not the scope change was requested in advance.

HOW IT WORKS:

- Inputs
  - *Project Scope Statement - See Chapter 5, Scope Definition, Outputs*
  - *Work Breakdown Structure - See Create WBS - Outputs*
  - *WBS Dictionary - See Create WBS - Outputs*

As a general note regarding the three preceding inputs, you should understand that the project scope statement, work breakdown structure, and WBS dictionary all make up the scope baseline, and the process of Scope Control is all about controlling, managing, and influencing changes to this baseline.

  - *Project Scope Management Plan*

The project scope management plan, among other things, defines the scope change control system which specifies how changes and change requests to the scope will be managed. The project scope management plan gives us a road map for managing this process.

- *Performance Reports*

The project's performance reports, an output of the Performance Reporting process discussed in Chapter 10 - Communications Management, give information on how the project is progressing, including components of the project that have been delivered.

- *Approved Change Requests*

Approved changes to be made to the project scope statement, the WBS, or the WBS dictionary should be brought into this process as part of the overall effort to manage and control changes. Once the change is approved, it becomes part of the scope baseline.

- *Work Performance Information*

This input is very similar in nature to the performance reports. It provides information on all aspects of the work completed as they relate to the project plan.

- Tools

- *Change Control System*

The scope change control system is the set of forms, tools, people, and procedures used to control changes to the scope baseline. The system, as it is defined here, is focused on the scope; however, it is also part of a larger change control system that controls changes across all aspects of the project.

- *Variance Analysis*

Variance analysis can be used to measure differences between what was defined in the scope baseline and what was created. Variance analysis can be particularly useful in the process of Scope Control as a way to investigate and understand the root causes behind these differences.

- *Replanning*

When changes are made to the project's scope baseline, additional planning must be performed in order to evaluate these changes in light of the cost, schedule, risk, quality, customer satisfaction, and other potential areas of the project management plan.

- *Configuration Management System*

The configuration management system is the overall system used to receive change requests, review them, track them, and ultimately approve or reject them. The configuration management system may be thought of as the umbrella under which the other change control systems fit (i.e. Integrated Change Control system, scope change control system, schedule change control system, cost change control system, contract change control system).

One key to understanding the configuration management system is that it is focused on the physical characteristics of the product and not the overall project scope.

- Outputs

- *Project Scope Statement Updates*

To a large degree, the process of Scope Control is about managing change to the scope baseline, and as changes are made, updates are added to the project scope statement so that it maintains a current and accurate picture of the scope.

- *Work Breakdown Structure Updates*

As scope changes are made to the scope baseline, these must be factored into updates to the WBS. If the change represents a new piece of scope, the work must be decomposed in the WBS down to work package level.

- *WBS Dictionary Updates*

Any changes to the scope baseline that affect the work breakdown structure should also be updated in the WBS dictionary so that each work package description is current and accurate.

- *Scope Baseline Update*

The scope baseline is comprised of the project scope statement, the work breakdown structure, and the WBS dictionary. Once the baseline is created, it may only be changed in a controlled manner. Any time a project scope change is approved, a corresponding update to the scope baseline must be made.

- *Requested Changes*

As the project scope is controlled, change requests to the scope or to other areas of the project normally result. These change requests are funneled into the Integrated Change Control system, discussed previously in Chapter 4.

- *Recommended Corrective Action*

As discussed earlier, corrective action is not about fixing past mistakes. Corrective action is anything done to bring future results in line with the plan. It compares the plan with the results or deliverables and evaluates differences.

Again, corrective action is about asking, "What would we change to prevent this difference from happening again?" Corrective action could be about changing the plan or changing the way in which the work was performed. The recommended corrective actions are a resulting output of this process.

- *Organizational Process Assets Updates*

Any time corrective action is implemented, changes may need to be made to the organizational process assets. The reason behind this is that the change or corrective action may have demonstrated that the organizational process assets you used (e.g. a previous project scope management plan or an organizational policy) were not wholly adequate for this project and need to be updated for future projects.

- *Project Management Plan Updates*

The project management plan must be kept up to date throughout the life of the project. Any change in scope should be reflected in the project management plan, along with any other resulting changes in cost, schedule, risk, quality, contract, etc.

Scope Management Questions

1. Your project team is executing the work packages of your project when a serious disagreement regarding the interpretation of the scope is brought to your attention by two of your most trusted team members. How should this dispute be resolved?

   A. The project team should decide on the resolution.

   B. The dispute should be resolved in favor of the customer.

   C. The dispute should be resolved in favor of senior management.

   D. The project manager should consult the project charter for guidance.

2. Which of the following statements is FALSE regarding a work breakdown structure?

   A. Activities should be arranged in the sequence they will be performed.

   B. Every item should have a unique identifier.

   C. The work breakdown structure represents 100% of the work that will be done on the project.

   D. Each level of a work breakdown structure provides progressively smaller representations.

3. Mark has taken over a project that is beginning the construction phase of the product; however, he discovers that no work breakdown structure has been created. What choice represents the BEST course of action?

   A. He should refuse to manage the project.

   B. He should stop construction until the work breakdown structure has been created.

   C. He should consult the WBS dictionary to determine whether sufficient detail exists to properly manage construction.

   D. He should document this to senior management and provide added oversight on the construction phase.

4. The project has completed execution, and now it is time for the product of the project to be accepted. Who formally accepts the product?

   A. The project team and the customer.

   B. The quality assurance team, senior management, and the project manager.

   C. The sponsor, key stakeholders, and the customer.

   D. The project manager, senior management, and the change control board.

5. Creating the project scope statement is part of which process?

   A. Project Scope Management.

   B. Scope Planning.

   C. Scope Definition.

   D. Scope Verification.

6. The project scope statement should contain:

   A. The work packages for the project.

   B. A high-level description of the scope.

   C. The level of effort associated with each scope element.

   D. A detailed description of the scope.

7. The most important part of Scope Verification is:

   A. Gaining formal acceptance of the project deliverables from the customer.

   B. Checking the scope of the project against stakeholder expectations.

   C. Verifying that the project came in on time and on budget.

   D. Verifying that the product met the quality specifications.

8. The organizational process assets would include all of the following except:

   A. Historical information.

   B. Organizational policies.

   C. Lessons learned from previous projects.

D. The project management information system.

9. Which of the following is NOT part of the scope management plan?

   A. The senior management statement of fitness for use.

   B. The scope change control system.

   C. A description of how the WBS will be created.

   D. A description of how acceptance will be handled.

10. You have taken over as project manager for a data warehouse project that is completing the design phase; however, change requests are still pouring in from many sources, including your boss. Which of the following would have been MOST helpful in this situation:

    A. A project sponsor who is involved in the project.

    B. A well-defined scope management plan.

    C. A change control board.

    D. A change evaluation system.

11. What is the function of the project sponsor?

    A. To help manage senior management expectations.

    B. To be the primary interface with the customer.

    C. To fund the project and formally accept the product.

    D. To help exert control over the functional managers.

12. The project manager and the customer on a project are meeting together to review the product of the project against the documented scope. Which tool would be MOST appropriate to use during this meeting?

    A. Verification analysis.

    B. Inspection.

    C. Gap analysis.

    D. Feature review.

13. You have just assumed responsibility for a project that is in progress. While researching the project archives, you discover that the WBS dictionary was never created. Which of the following problems would LEAST likely be attributable to this?

    A. Confusion about the meaning of specific work packages.

    B. Confusion about who is responsible for a specific work package.

    C. Confusion about which account to bill against for a specific work package.

    D. Confusion about how to change a specific work package.

14. A team member makes a change to a software project without letting anyone else know. She assures you that it did not affect the schedule, and it significantly enhances the product. What should the project manager do FIRST?

    A. Find out if the customer authorized this change.

    B. Submit the change to the change control board.

    C. Review the change to understand how it affects scope, cost, time, quality, risk, and customer satisfaction.

    D. Make sure the change is reflected in the scope management plan.

15. The product you have delivered has been reviewed carefully against the scope and is now being brought to the customer for formal acceptance. Which process is the project in?

    A. Scope Verification.

    B. Scope Auditing.

    C. Scope Closure.

    D. Scope Control.

16. The scope baseline contains all of the following EXCEPT:

    A. The project scope statement.

    B. The project scope management plan.

    C. The work breakdown structure.

    D. The WBS dictionary.

17. You are the project manager for a large construction project, and you identify two key areas where changing the scope of the product would deliver significantly higher value for the customer. Which of the following options is MOST correct?

    A. Make the changes if they do not extend the cost and timeline.

    B. Make the changes if they do not exceed the project charter.

    C. Discuss the changes with the customer.

    D. Complete the current project and create a new project for the changes.

18. Which of the following activities is done FIRST?

    A. Creation of the preliminary project scope statement.

    B. Creation of the work breakdown structure.

    C. Creation of the project scope management plan.

    D. Creation of the scope baseline.

19. Which of the following statements is TRUE concerning functionality that is over and above the documented scope?

    A. It should be channeled back through the change control board to ensure that it gets documented into the project scope.

    B. Additional functionality should be leveraged to exceed customer expectations.

    C. The final product should include all the functionality and only the functionality documented in the scope baseline.

    D. Additional functionality should be reviewed by the project manager for conformity to the product description.

20. A project manager has been managing a project for six months and is nearing completion of the project; however, change requests are still pouring in. The project is ahead of schedule but over budget. Which of the following statements is TRUE?

    A. The project manager should influence the factors that cause change.

    B. Changes should only be evaluated after the original scope baseline has been delivered and accepted.

    C. Changes introduced at this point in the project represent an unacceptable level of risk.

    D. Changes should be evaluated primarily on the basis of how much value they deliver to the customer.

Answers to Scope Management Questions

1. B. In general, disagreements should be resolved in favor of the customer. In this case, the customer is the best choice of the four presented. 'A' is not a good choice because it is your job to keep the team focused on doing the work and out of meetings where they are arguing about the scope. Besides, the team brought you this problem, so their ability to resolve it is already in question. 'C' is incorrect because all things being equal, project disputes should be resolved in favor of the customer and not in favor of senior management. Since you don't have enough information to steer you toward senior management, resolving it in favor of the customer was the right choice here. 'D' is incorrect because the project charter is a very general and high-level document. As it is issued before either the scope statement or the work breakdown structure is created, it would be of little use in resolving an issue of scope dispute that occurred during execution.

2. A. You don't tackle activity sequencing as part of the work breakdown structure. That part comes later. The WBS has no particular sequence to it, not to mention that it is not decomposed to activity level. 'B' is incorrect since every WBS element does have a unique identifier. 'C' is incorrect since the WBS is the definitive source for all of the work to be done. Remember - if it isn't in the WBS, it isn't part of the project. Choice 'D' is incorrect because the WBS is arranged as a pyramid with the top being the most general, and the bottom being the most specific. The lowest level of the WBS would also be the smallest representation of work.

3. B. In this situation, you cannot simply skip the WBS, as you may be tempted to do. Mark should take time to create the WBS which is usually not a lengthy process. 'A' may sound good, but in reality, a PMP needs to be ready to work to solve most problems. You might refuse to manage a project if there is an ethical dilemma or a conflict of interest, but not in other circumstances. 'C' is incorrect since the WBS

dictionary cannot be created properly unless the WBS was created first. If there is a WBS dictionary and no WBS, that would be a big red flag. 'D' is incorrect since merely documenting that there is a serious problem is not a solution. Additionally, providing more oversight would not solve the problem here. The real problem is that the WBS has not been created, and that will trickle down to more serious problems in the future of the project.

4. C. The project manager verifies the product with the key stakeholders, the sponsor, and the customer.

5. C. The project scope statement is created as part of two processes. The first is "Create Preliminary Scope Statement," which is part of integration management, and the second is Scope Definition, where the preliminary scope statement is elaborated into the project scope statement. 'C' is the only one of those processes of the four choices provided.

6. D. The project scope statement needs to include a detailed description of the scope. Choice 'A' is incorrect as the WBS is created later as part of the Create WBS process. 'B' is tricky, but it is incorrect. The preliminary scope statement contains a high level description of the scope, but the project scope statement is detailed. 'C' is incorrect, because the level of effort is estimated after the scope has been defined.

7. A. It is important to understand the processes and their inputs and outputs! Whereas all of these choices may be important, the only one that is listed as a part of Scope Verification is to get customer acceptance of the product. The other activities may be done during the project, but they aren't part of the Scope Verification process. 'D' is close, but that is formally part of the Quality Control process.

8. D. Organizational process assets include historical information, organizational policies (since those policies may help or constrain your activities), and lessons learned from

previous projects. The PMIS (project management information system) is a group of tools used together as a system, but it is not considered part of organizational process assets.

9. A is the only choice that is not part of the scope management plan. It is a made up term. 'B', 'C', and 'D' all belong.

10. B. This one is tricky. If you missed it, don't feel bad, but it is important to know that questions like this are on the exam. The reason 'B' is correct is that the scope management plan contains a plan for how changes will be handled. If too many changes are pouring in, it is likely that the scope management plan was not well defined. 'A' is incorrect because it is not the sponsor's role to control change. He or she is paying you to handle that. 'C' is incorrect because if the change control board exists on your project, it only evaluates changes. The board is almost always reactive, not proactive. 'D' is incorrect since the change evaluation system is a made up term not found in PMI's processes.

11. C. It is the sponsor's job to pay for the project and to accept the product. Choice 'A' is really the project manager's job. 'B' is the project manager's job as well. It is not a clearly defined job for the sponsor. 'D' is not a function of the sponsor. If more control were needed over the functional managers, that would be the role of senior management.

12. B. The project manager and customer are involved in the Scope Verification process, and the tool used here is inspection. The product is inspected to see if it matches the documented scope. 'A', 'C', and 'D' are not documented as part of the processes.

13. D is the best choice here. The WBS dictionary contains attributes about each work package such as an explanation of the work package (which invalidates choice 'A'), who is assigned responsibility for the work package (which invalidates choice 'B'), and a cost account code (which invalidates choice 'C'). If a work package were changed, that would most likely alter the scope baseline, and information on how to go about this would be found in the scope management plan and not the WBS dictionary.

14. C. Notice the use of the word 'FIRST'. 'A' is wrong because the customer should never bypass the project manager to authorize changes directly. It is the project manager's job to authorize changes on the project. 'B' is incorrect since all changes might not go to the change control board. Even if a change control board exists on the project, the project manager doesn't automatically just send everything their way. The project manager should deal with it first. 'D' is incorrect because the scope management plan is not even the place this would be reflected. The scope baseline would need to be updated, but only after the change had been properly evaluated.

15. A. The customer accepts the scope of the product in Scope Verification.

16. B. The scope baseline represents the documented scope once it has been defined and put under control. The three primary ingredients of the scope baseline are the scope statement, the WBS, and the WBS dictionary. The scope management plan is not part of the scope baseline.

17. C. Choices 'A' and 'B' are no-nos on the PMP exam. You don't just make changes because they "add value." Does the customer want the change? Does the change increase the project risk or put the quality in jeopardy? Between 'C' and 'D', the best answer is 'C'. The reason is that just because the project manager thinks this is a good piece of functionality doesn't mean that he should automatically add it. The customer should have input into this decision as well. Choice 'D' might be correct in limited circumstances if you knew that you were at or near the end of the project, but changes on a project rarely require the automatic initiation of a new project.

18. A. The preliminary scope statement is typically created quite early on the project. In this case, it would be created well before the work breakdown structure and the scope baseline (which is made up of the scope statement, the WBS, the WBS dictionary). The project scope management plan is typically created shortly after the preliminary scope statement.

19. C. Answers 'A', 'B', and 'D' are all incorrect, since they encourage adding or keeping the additional functionality. It is important not to add extras to the project for many reasons. The final product should be true to the scope. If you missed this, reread the section on scope management and gold plating.

20. A. The project manager needs to be proactive and influence the root causes of change. 'B' would be ridiculous in the real world. Imagine getting a change request near the end of the project that is good for the project and refusing to do it until you completed the original scope. Some change is good! 'C' is incorrect because although some change may introduce an unacceptable level of risk, all change certainly does not. Some changes could dramatically reduce the project risk and help the project. 'D' is incorrect, because value is not the primary criterion for evaluating change. A change may deliver high value, but also introduce too much risk or cost, or delay the project unacceptably.

Time Management

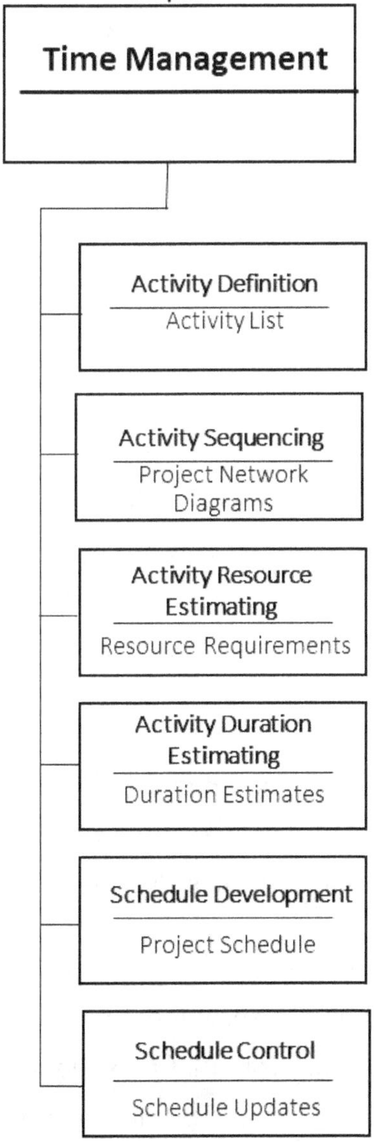

The processes of Project Time Management with their primary outputs

Time Management

- **Activity Definition** — Activity List
- **Activity Sequencing** — Project Network Diagrams
- **Activity Resource Estimating** — Resource Requirements
- **Activity Duration Estimating** — Duration Estimates
- **Schedule Development** — Project Schedule
- **Schedule Control** — Schedule Updates

From the indicators at the top of the page, you can tell that you may find this chapter to be slightly more difficult than the previous chapter on scope management.

In order to help you prepare for this topic, this book has clearly broken down the practices and outlined the techniques and formulas you need to know in order to ace the questions on the exam.

By the time you are through with this chapter, you may well have a higher level of confidence here than on any other section of the test.

Spending extra time in this chapter will yield direct dividends on the exam! Make sure that you learn both the processes and the techniques so that you may approach these questions with absolute confidence.

Philosophy

Project Time Management is concerned primarily with resources, activities and schedule management. PMI's philosophy here, as elsewhere, is that the project manager should be in control of the schedule, and not vice versa.

The schedule is built from the ground up, derived from the scope baseline and other information, and rigorously managed throughout the life of the project.

There are six processes related to time management that you should understand. They are:

1. Activity Definition (list the project activities)
2. Activity Sequencing (order the activities and create the project network diagram)
3. Activity Resource Estimating (estimating the resources needed to complete each activity)
4. Activity Duration Estimating (determine time estimates for each activity)
5. Schedule Development (create the schedule)
6. Schedule Control (monitor schedule performance)

You will see from reading this chapter that the driving philosophy behind time management is mathematical; it is primarily cold, hard analysis. The project manager does not merely accept whatever schedule goals are handed down or suggested. Instead, he builds the schedule based on the work to be done and then seeks to make it conform to other calendar requirements, constraints, and strategic goals.

Additionally, while most of the topics within the PMBOK Guide are related, the topics of scope, time, and cost (and to a slightly lesser degree, quality and risk) are particularly tightly linked. Changes made to one area will almost certainly have impacts elsewhere.

As in other areas, PMI prefers that project managers begin from the bottom up. The WBS, which was covered in the previous chapter, is a key input to time management processes. Using this comprehensive list of deliverables, you now define the work that must be done in order to produce these deliverables. Like the items in the WBS, the individual activities are then sequenced, and the resource and duration estimates are applied to these activities. This approach has many similarities to the practices in scope management. Most similar is that your analysis and outputs are comprehensive and complete.

Importance

Time management is of high importance, both in the application of PMI's processes and on the exam. You should expect to see several questions on the exam where you will need to apply formulas, calculate the critical path, and determine the effect of a change to the schedule.

The difficulties that many people encounter here fall into three categories:

1. Modern project managers typically rely on software to perform schedule and time calculations. While this is not a bad thing, being a PMP requires that you understand the theories and practices for time management that underlie the software.

2. Some people are intimidated by the mathematical and logical aspects of this section. Although the math is relatively simple, it does require memorization of a few key formulas.

3. Some project managers do not understand the diagramming techniques and processes. A reliance on intuition will not get you very far on these questions. You either know how to calculate them or you do not. It is far better to spend time learning the techniques than to try to fumble your way through them on the exam.

Preparation

A heavier reliance on memorization is necessary here. The memorization for time management falls into two categories:

1. There are several key terms that may be new to you, or they may have slightly different meanings than you are used to. Many of the questions on the exam will test your knowledge of specific terms and nuances, so it is important to be able to recognize and clearly understand them even if you do not memorize the definitions word for word.

2. Second, there are formulas and techniques for diagramming that you will need to be able to apply on the exam. Memorization alone is not sufficient here. You must be able to calculate standard deviations (using a highly simplified formula), float, and critical path, among others.

Although the logic and math portions can be daunting to some people as they approach the exam, it should be pointed out that with each of the hard logic questions, the right answer is often much easier to determine than in the situational questions. Once you master the techniques presented in this chapter, you can work out the right answer on your own and simply match it to the list of choices. For many, these questions become favorites on the exam. There are six processes in the time management knowledge area. These are Activity Definition, Activity Sequencing, Activity Resource Estimating, Activity Duration Estimating, Schedule Development, and Schedule Control. Below are the break-outs that show to which process group each item belongs:

Process Group	Time Management Process
Initiating	(none)
Planning	Activity Definition, Activity Sequencing, Activity Resource, Estimating, Activity Duration Estimating, Schedule Development
Executing	(none)
Controlling	Schedule Control
Closing	(none)

In the knowledge area of time management, it is also essential that you know the main outputs that are produced during each process. The different key outputs that are created in each process are summarized in the chart below.

Process	Key Output (s)
Activity Definition	Activity list, Milestone list
Activity Sequencing	Project schedule network diagrams
Activity Resource Estimating	Activity resource requirements, Resource breakdown structure
Activity Duration Estimating	Activity Duration Estimates
Schedule Development	Project schedule, Schedule baseline
Schedule Control	Performance measurements, Recommended corrective actions

Activity Definition

WHAT IT IS:

Once the scope baseline has been created it is used to decompose the work into activity detail. The main result of this planning process is the activity list. This list represents all of the schedule activities that need to take place for the project to be completed. This is primarily accomplished by taking the WBS and decomposing the work packages even further until they represent schedule activities. The difference between work packages in a WBS and an activity list is that the activity list is more granular and is decomposed into individual schedule activities. Work packages will often contain bundles of related activities that may involve multiple groups of people. It is these activities that comprise the activity list. The activity list is used as the basis for the next two planning processes, Activity

Sequencing and Activity Duration Estimating.

If the project is being performed under procurement, this planning process will most likely be performed by the subcontracting organization, with the results being provided to the organization that is responsible for the management of the overall project.

WHY IT IS IMPORTANT:

Earlier in project scope management, we focused on what work needed to be performed on the project. Now, in time management, we need to focus on how and when it is accomplished. The activity list will be an essential input into building the schedule, so it is important that it be both complete and correct.

WHEN IT IS PERFORMED:

The process of Activity Definition is often performed as soon as the scope has been baselined. In other words, it is common to create the activity list after the project scope statement, the work breakdown structure, and the WBS dictionary have been created and are in a stable form; however, it is also acceptable to create the activity list at the same time as the work breakdown structure and the WBS dictionary.

HOW IT WORKS:

- Inputs
  - *Enterprise Environmental Factors - See Chapter 2, Common Inputs*
  - *Organizational Process Assets - See Chapter 2, Common Inputs*
  - *Project Scope Statement*

The project scope statement is part of the scope baseline. It contains a thorough description of the project's deliverables which will be used here in Activity Definition. Another important feature of the project scope statement is that it contains the constraints and assumptions. As the scope is being translated into activities in this process, anything that constrains your decision making, or any assumptions you are making, should be carefully considered.

  - *Work Breakdown Structure*

The WBS is the most important input into this process, since its work packages will be decomposed further down to activity level. Each work package at the bottom of the WBS will be decomposed into one or more activities.

- Tools
  - *Decomposition*

If you understood decomposition as it was used in the Create WBS process, you should have no problems understanding it here. Each work package at the bottom of the WBS is simply decomposed into smaller pieces, known as schedule activities. The project manager should solicit heavy involvement from the project team or the functional managers in this process.

- *Templates*

Similar to the Create WBS process, templates for Activity Definition provide a starting point for the project manager. These templates may exist as examples from previous similar projects, or they may be commercially available activity list examples or even basic skeletal templates provided by the organization's project office.

Regardless of the format or the source of these templates, they provide the project team with a starting point rather than forcing them to build the activity list from scratch.

- *Rolling Wave Planning*

The concept of rolling wave planning is a form of progressive elaboration that models project planning the way we see things in the real world.

Suppose, for example, that you are standing at the edge of a field. You may be able to see some things in great detail, even counting individual blades of grass at your feet. The further off you look, however, the less detail you will be able to perceive. Distant objects, such as a mountain range miles away, would appear general and hazy.

Rolling wave planning mirrors this construct by assuming that things in the near future should be relatively clear, while project activities in the future may not be as detailed or as easily understood. Armed with that perspective, a project manager may choose to carefully decompose certain work packages, with anticipated execution in the near term, in great detail, while delaying analysis on work packages that will not be accomplished until later in the project. This type of planning must be revisited throughout the project, much as waves continually pound the shore.

Rolling wave planning is used more frequently on projects like the creation of an I.T. system, and less often on construction projects within the construction industry where unknowns may cost millions of dollars.

- *Expert Judgment*

Expert judgment in decomposing activities may come from numerous sources, including team members, consultants, and functional managers.

One of the best sources for expertise in decomposing a work package into an activity may be the person who will ultimately be responsible for executing the work package or the schedule activity, although that may not be known at this point.

- *Planning Component*

There are times when the project team cannot decompose the work breakdown structure down to a low level because they do not have enough information. When this situation occurs, a node on the WBS will not be decomposed down to work package level.

So, if no work package exists, how will the project team decompose activities? The answer is that the team does not create an activity list from these nodes. Instead, they create a planning component, which is simply a sort of summary activity that will be revisited once its parent WBS node has been defined in sufficient detail.

A control account is one type of planning component. This control account is represented as a node on the work breakdown structure.

Control accounts are typically placed on nodes on the work breakdown structure where no actual work packages have been defined yet.

Higher level planning is performed against the control accounts and is a way for the team to continue planning when one or more components of the project is not clear at a point in time.

- Outputs

- ***Activity List***

All of the activities that need to be performed in order to complete the project are compiled into the activity list. Each activity in the list should map back to one and only one work package (a work package, however, typically has more than one activity belonging to it). The activities should include enough information to transition them to the project team so that the work may be performed. The activity list's usefulness is tied to its completeness and accuracy. It is important to identify and document each activity that must be performed in order to complete the project.

It is important to note that a line exists between the work breakdown structure and the activity list. Although the activity list is an extension of the work breakdown structure, it is not a part of the work breakdown structure. The activities are used to create the project schedule. The activity list by itself is generally limited only to an identifier, such as an activity name or a unique numeric identifier, and a description.

- ***Activity Attributes***

As planning progresses, there will be a need to store additional information about activities. For example, the person responsible for this activity, the parts that need to be procured before this activity may be started, and the location the work will be performed could be highly important. Activity attributes may be stored with the activity list or in a separate document and are typically added after the initial activity list has been created.

Note that any time you see the activity list, you will also see the activity attributes. The activity attributes may be thought of as an expansion of the activity list.

- ***Milestone List***

The key project milestones are produced as a part of this plan. These milestones may be related to imposed dates such as a contractual obligation, or projected dates based on historical information. This list of milestones will become essential in building the schedule.

- *Requested Changes*

As with many other key planning processes, as the process of Activity Definition is performed, change requests are a normal output. These change requests will relate to previous planning outputs that are used in this process, such as the scope baseline.

Activity Sequencing

WHAT IT IS:

The Activity Sequencing process is primarily concerned with taking the activity list defined earlier and arranging the activities in the order they must be performed.

This process is all about understanding and diagramming the relationships that schedule activities have with each other.

Schedule activities logically sequenced

WHY IT IS IMPORTANT:

Project schedule network diagrams are an important part of the PMP exam, and you should expect to see several questions on the PMP that pertain directly to this topic. A network logic diagram is a picture in which each activity is drawn in the order it must be performed, and the amount of time each activity takes is represented with numbers.

Activity Sequencing is the planning process in which network diagrams are produced. Network diagramming is the best method for representing activities, their dependencies, and sequences.

Two examples of network logic diagramming techniques, the precedence diagramming method and the arrow diagramming method, are discussed later in this chapter.

WHEN IT IS PERFORMED:

Because of the flow of inputs and outputs between other processes, the process of Activity Sequencing must be performed after Activity Definition and before Schedule Development.

Activity Definition must be performed before Activity Sequencing or Schedule Development.

HOW IT WORKS:

- Inputs
    - *Project Scope Statement*

The description of the project's scope will often give you information about the project that will influence the order in which certain activities are performed. For instance, if a computer system contained hardware and software, it might be necessary to procure and configure the hardware before constructing the software. This may seem like common sense, but the project scope statement is used to ensure that nothing is missed.

- *Activity List*

The activity list is the most important input into Activity Sequencing. It is in this process that the schedule activities from the activity list will be arranged, or sequenced, into a diagram that represents the order in which they must be performed.

- *Activity Attributes*

The activity attributes, produced in Activity Definition, are brought into this process since they contain additional information about each activity that may influence how it is sequenced.

- ***Milestone List*** Milestones are events that must be considered in the life of the project. There is an interesting relationship between milestones and the process of Activity Sequencing. Milestones are often imposed from outside the Project (e.g. the project sponsor indicates an overall deliverable date), and activities typically come from within the project (e.g. the decomposition of work packages). Because of this, activities will often need to be arranged in a specific way in order to meet key milestones.

For instance, if the customer specified a milestone of a pre-construction walkthrough at a certain point in time, then the activity of cleaning up the worksite may need to be sequenced in earlier than would otherwise be necessary.

- *Approved Change Requests*

Any approved change requests may affect Activity Sequencing and must be factored in for planning.

- Tools

  - *Precedence Diagramming Method (PDM)*

The precedence diagramming method creates a graphical representation of the schedule activities in the order in which they must be performed on the project.

Activities are represented by the nodes (rectangles), with arrows representing the dependencies that exist between the activities.

The project network diagram illustrated below uses the activity on node convention to represent the activities. In this case, the nodes are shown as rectangles, and the activities are represented inside the node, usually by letters of the alphabet. Units of duration are shown above the nodes.

  - *Arrow Diagramming Method (ADM)*

The arrow diagramming method is another way of graphically representing the activities and their dependencies on a project. The arrow diagramming method, however, differs from the precedence diagramming method described earlier. When using the arrow diagramming method, the activities are not on the nodes. Instead, the activities are represented on the arrows, with the nodes being connecting points. The nodes for activity on arrow are typically represented as circles, which helps to visually differentiate them from the activity on node diagrams.

The diagrams that result from the arrow diagramming method are known as "activity on arrow" diagrams. The illustration below represents a project network diagram for activity on arrow.

Note that a key difference between the Arrow Diagramming Method and the Precedence Diagramming Method is the ability of Activity on Arrow diagrams to show an activity with a duration of 0, such as the one depicted above in activity B-D.

These activities, shown with a dashed line, are called “dummy activities." Dummy activities force the creation of new paths through the network. In the example above, activity D-Finish cannot begin until activity A-B (as well as activity F-D) is complete.

Activity on arrow diagrams can contain these dummy activities with a duration of 0, while activity on node diagrams cannot.

- *Schedule Network Templates*

Templates may be anything from an example on a previous project to a standard format issued by the project management office. Templates may contain a section from a previous project. This segment is known as a subnetwork or a fragment network.

Subnetworks or fragment networks may be particularly helpful when projects are highly similar. For instance, the activities and the sequences involved in building and testing an airplane may be highly similar from one project to another within the same organization. Using subnetworks or fragment networks prevents having to "reinvent the wheel" each time the process of Activity Sequencing is performed.

- *Dependency Determination*

Dependencies are those things that influence which activities must be performed first. For example, a road must be graded before it can be paved. If grading and paving the road were two activities on the project, then we would say that the start of the activity of paving the road is dependent upon the finish of the activity of grading the road, thus these two activities have a finish-to-start relationship.

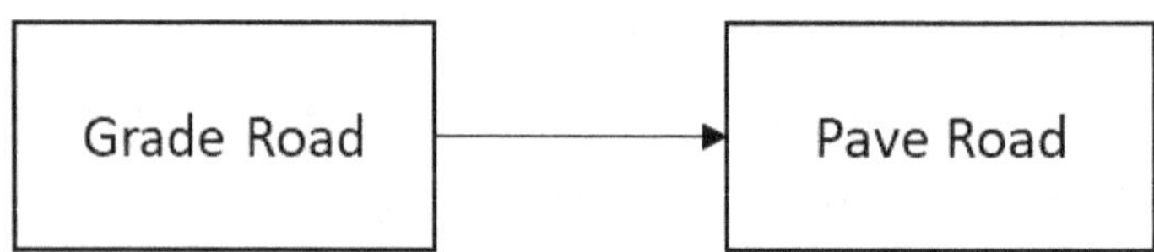

There are three kinds of dependencies that may exist among activities.

A. Mandatory Dependencies

A mandatory dependency is one that cannot be broken. Given the example where paving the road is dependent upon grading it, the dependency is unavoidable, or mandatory. Mandatory dependencies are traditionally known as hard logic, since a mandatory dependency is considered unmovable and always true.

B. Discretionary Dependencies

Discretionary dependencies, unlike the preceding example, are not always true. These would often be the result of best practices, and may vary organization to organization and even project to project. For instance, the project of managing the remodeling of a house may have a discretionary dependency between painting the walls and carpeting the floors, where painting must be completed before carpeting could be installed. There is no absolute rule that says the carpet could not be installed before the painting begins.

Discretionary dependencies are also known as soft logic or preferred logic and are typically based on historical information, expert judgment, and best practices.

C. External Dependencies

External dependencies are those dependencies that must be considered but are outside of the project's control and scope. For instance, if an automobile is being developed to use alternative technology, there may be an external dependency on a supplier providing a battery that meets certain specifications before the project can meet its schedule. Because these are dependencies, they must be identified and documented as part of this process.

- *Apply Leads and Lags*

A lead is simply one activity getting a jump start on another. Consider, for instance, a software project that has a dependency between finishing the development of a section of code and beginning the quality inspection. Since the development has to finish before the testing begins, we would say that a mandatory finish-to-start dependency exists between the two activities.

An example of a lead would be if the individual performing the quality inspection gets an unfinished beta copy of the software in order to get a head start. Note that leads do not do away with the finish to start relationship that exists. Instead, it simply "cheats" that relationship.

Leads, and the rationale behind them, must be clearly explained and documented.

Think of a lag as a waiting period that exists between two activities. A lag is a situation where a waiting period must occur between one activity and another dependent activity. An example of a lag would be if one activity was to order a computer server and another activity was to configure the server, then a lag might exist between the time the server is ordered and the time it arrives and can be configured. During this lag, there is no work being performed by the organization against this activity; no resources are being expended. They are simply waiting.

- **Outputs**
    - ***Project Schedule Network Diagrams***

The name of this output may be slightly misleading. Although it is used later in creating the schedule, the schedule network diagram is not the schedule. In other words, no start or finish dates are assigned to the activities yet. They are simply arranged in the order they need to be performed on the project. A project schedule network diagram may include a full representation of every activity in the project, or it may include summary nodes. In the event that a summary node is used, enough documentation should be included so that the basic flow of activities may be understood.

- *Activity List Updates*

As activities are arranged and their relationships understood, changes to the activity list are a normal by-product.

- *Activity Attributes Updates*

As updates to the activity list are made, corresponding updates to the activity attributes must be made as well.

- *Requested Changes*

As the activities, their relationships, and their dependencies are understood in greater detail, change requests to the project are a normal by-product.

Activity Resource Estimating

WHAT IT IS:

How long an activity takes is usually a function of determining the size of the activity, the number of resources that will be applied to it, and the resource availability. This process is all about analyzing the project activity to determine the resource requirements.

WHY IT IS IMPORTANT:

Understanding the number of resources required to complete an activity and determining how long they will be used for that activity is an important step in project planning and an essential ingredient to the schedule, which will be developed later.

WHEN IT IS PERFORMED:

Because the process of Activity Resource Estimating uses the activity list and activity attributes, it must be performed after Activity Definition. Additionally, since the output of this process is used to build the project schedule, this process must be performed before Schedule Development. This process often goes hand in hand with Cost Estimating, since cost and time are closely linked.

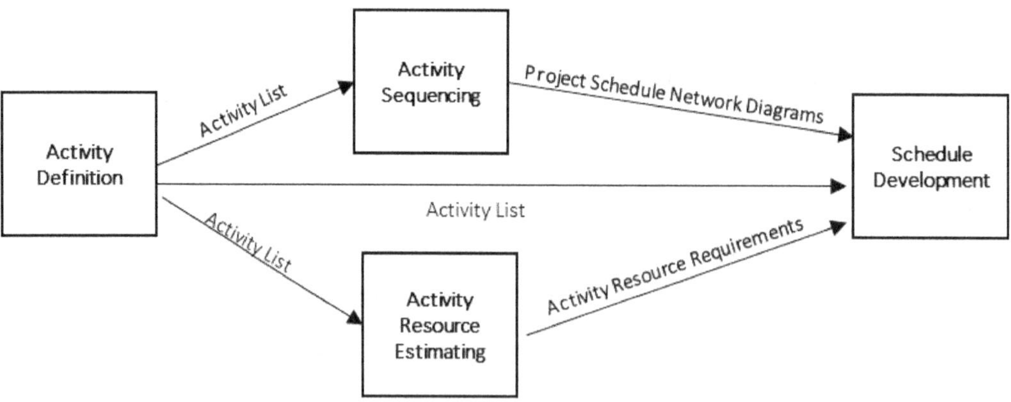

HOW IT WORKS:

- Inputs

  - *Enterprise Environmental Factors*

Every project environment is different, and these different factors will influence how long an activity will take.

For instance, an activity could require access to an aircraft manufacturer's flight simulation systems, or a project that had some manufacturing as an activity might require access to the organization's facilities in order to perform this activity.

- ***Organizational Process Assets - See Chapter 2, Common Inputs***

- ***Activity List***

The activity list is the most important input into this process. Each activity in the list will be evaluated and the resources will be estimated for it.

- ***Activity Attributes***

In your mind, the activity attributes should be virtually indistinguishable from the activity list. The two always go together. In this case, the activity attributes give expanded information on each activity in the activity list.

- ***Resource Availability***

Resources may include both physical resource and human resources, and their availability needs to be factored into the process of Activity Resource Estimating. For instance, a piece of heavy machinery may currently be available only during the months of April and May, which must be considered.

- ***Project Management Plan***

While several pieces of the project management plan may be helpful here, it is the schedule management plan, a component of the project management plan, created in Develop Project Management Plan, which is particularly helpful.

The schedule management plan specifies how the schedule will be developed and managed and how changes to the schedule will be managed.

- Tools

  - *Expert Judgment*

There is no substitute for asking for expert opinion on how to estimate resource needs for an activity. Asking someone who has performed this type of activity previously, a functional manager, or even the resource who will be performing the work can bring insight into the resource needs for the activities.

  - *Alternatives Analysis*

The old saying goes that "there is more than one way to skin a cat," implying that just because one way has been identified, alternatives may still be helpful. This could include outsourcing an activity, purchasing a software component off the shelf rather than building one, or using a totally different approach to complete the activity.

  - *Published Estimating Data*

Some industries have extensive data available through published, recognized sources that can help in estimating. For instance, if you take a car to a body shop for repair, they often have books provided by the insurance industry with almost every conceivable repair listed, along with how long the repair should take for an experienced person to complete.

This type of data can help give insight into the Activity Resource Estimating process.

  - *Project Management Software*

Software is a means, not the end. It can help the project manager store and organize information, experiment with alternatives, and rapidly perform the routine calculations.

- *Bottom-Up Estimating*

It may be that you encounter an activity that cannot be estimated, either because it has not been broken down enough, or because it is simply too complex. In this case, it is appropriate to break down the activity further into progressively smaller pieces of work until these pieces may be estimated for their resource requirements. Once these estimations have been performed, the pieces may be summed up from the bottom back to activity level.

- Outputs

  - *Activity Resource Requirements*

The resources required for each schedule activity are the primary output of Activity Resource Estimating. These resources include the kind of resource and the number of these resources. The activity resource requirements need to specify, for instance, if two senior programmers are required for four months or if three junior programmers are required for five months. Each activity resource requirement should be documented with sufficient detail to explain the decision-making process used to arrive at these estimates.

- *Activity Attribute Updates*

As the resources are analyzed for each activity, updates to the activity attributes are a normal by-product.

- *Resource Breakdown Structure*

The resource breakdown structure, or RBS, is similar in many ways to the WBS. It is graphical and hierarchical, logically arranged from top to bottom, and it arranges the resources by category and type.

- *Resource Calendar (Updates)*

The resource calendar, which is first produced in Develop Project Management Plan, is focused on resource utilization. It shows when the resources will be used and when they will be available for use elsewhere. For instance, if a mobile construction crane were going to be used in the first and fourth weeks of June, that would be reflected as an update on the resource calendar so that functional managers or other projects in the organization might be able to utilize it during the second and third weeks of June.

The resource calendar applies to both human and physical resources, and it applies to all resources to be used throughout the life of the project. It is first developed as part of the project management plan, but it is revised here when more details are known about the resources.

- *Requested Changes*

As progressively detailed planning is performed, change requests to the project's scope or schedule are normal outputs of Activity Resource Estimating.

Activity Duration Estimating

WHAT IT IS:

This process is exactly what it sounds like. Each activity in the activity list is analyzed to estimate how long it will take. There is an important difference between duration and level of effort, and this process focuses on determining duration.

The duration of an activity is a function of many factors, including who will be doing the work, when they are available, how many resources will be assigned to this activity, and the amount of work contained in the activity.

WHY IT IS IMPORTANT:

These activity duration estimates will become a primary input into creating the schedule when the overall project timeline has been created.

WHEN IT IS PERFORMED:

Activity Duration Estimating is performed after the activity resource requirements have been gathered and before the schedule is developed.

A diagram showing the order of Time Management's planning processes

HOW IT WORKS:

- Inputs
  - *Enterprise Environmental Factors*

Factors such as what records an organization requires, safety standards, and regulations can all affect how long an activity takes. For instance, it may take far longer to perform an activity in a nuclear power plant than it does in a conventional power plant due to enterprise environmental factors, and these must be considered in order to accurately estimate the duration of activities.

- *Organizational Process Assets*

Organizational process assets could take the form of a rich database of historical information that shows the estimated and actual durations for activities for a previous project, while another organization may have specific calendar requirements when shifts or resources are available.

Anything that gives structure or guidance to your Activity Duration Estimating would be considered an organizational process asset.

- *Project Scope Statement*

The project scope statement, created earlier in Scope Definition, has the constraints and assumptions for the project that can affect this process. For instance, a constraint that a particular component of a house had to be built at a different facility and transported to the construction site would influence the activity's duration.

Likewise, an assumption that a particular component of a software product could be completed by a subcontracting firm faster than it could be built internally would be helpful in performing for Activity Duration Estimating.

- *Activity List*

The activity list is a primary input for this process. Every activity in the list should be estimated to determine its duration.

- *Activity Attributes*

The activity attributes always accompany the activity list. These attributes provide additional information about each activity in the list.

- *Activity Resource Requirements*

Since the duration is a function of the amount of work associated with this activity and the resources assigned to perform that work, the activity resource requirements need to be brought into this process.

- *Resource Calendar*

This is the resource calendar that was updated in the Activity Resource Estimating process. This calendar shows physical and human resource usage across the entire project.

- *Project Management Plan*

The project management plan is the single unified plan that details how the project will be executed, monitored, and controlled. For the process of Activity Duration Estimating, there are two components of particular interest. They are the risk register and the activity cost estimates.

The risk register provides details on the identified risks. At this point it is appropriate to review it for risks that might affect the duration of an activity. For instance, if weather is identified as a risk on a construction project, it may be appropriate to update the activity durations to reflect the risk.

If the activity cost estimates have been created at this point, then they may provide some guidance to the process of Activity Duration Estimating.

- Tools

  - *Expert Judgment*

Anyone who has managed a project will attest that the duration of an activity can be notoriously hard to estimate in advance. The expert providing the judgment should follow some basis, such as historical information, whether documented or experiential.

  - *Analogous Estimating*

Analogous estimating, also known as top-down estimating, is where an activity from a project previously performed within the organization is used to help estimate another activity's duration.

Typically, the previous actual time spent on the similar activity is used as the estimate for another similar activity. The technique is combined with expert judgment to determine if the two activities are truly alike.

  - *Parametric Estimating*

If one team can install 100 feet of fence in one day, then it would take 10 teams to install 1,000 feet of fence in one day. This kind of linear extrapolation is an example of parametric estimating.

Parametric estimating can work well for activities that are either linear or easily scaled. It is not as effective for activities that have not been performed before or those for which little or no historical information has been gathered.

- *Three-Point Estimates*

Three-point estimates, also called PERT estimates, use three data points for the duration instead of simply one. These are pessimistic, most likely (also known as realistic) and optimistic estimates.

As an example of how this is used, suppose a developer estimates that it will most likely take 9 days to write a module of code; however, he also supplies an optimistic estimate of 7 days and a pessimistic estimate of 17 days.

The project manager then applies a formula, usually in the form of a weighted average to these estimates to distill them down to a single estimate. The traditional formula for this is:

$$\frac{\text{Pessimistic} + 4\text{X Realistic} + \text{Optimistic}}{6}$$

Where P is pessimistic, O is optimistic, and R is the most likely, or realistic estimate.

In the example above, the numbers would be substituted as (17 + 4*9 + 7) / 6 = 10 days. This number is used as the activity duration estimate for this schedule activity.

Another important formula to memorize for the exam is used to calculate the standard deviation for an estimate.

$$\sigma = \frac{\text{Pessimistic} - \text{Optimistic}}{6}$$

The value for standard deviation tells us how diverse our estimates are. If an activity has a pessimistic and an optimistic estimate that are very far apart, the standard deviation (σ) will be very high, indicating a high degree of uncertainty and consequently a high degree of risk for this estimate. Note that this is not a real standard deviation; however, it is a widely used substitute formula in estimating activity durations.

Given these formulas, let us consider the following estimates for activities A, B, and C

Activity	Optimistic	Pessimistic	Realistic
A	22	35	25
B	60	77	70
C	12	40	20

Now, using the formulas above, we will calculate the three-point estimate for each of these activities.

Activity	Optimistic	Pessimistic	Realistic	3 Point Estimate
A	22	35	25	26.17
B	60	77	70	69.5
C	12	40	20	22

- *Reserve Analysis*

Reserve time, also called contingency, is extra time added to a schedule activity duration estimate.

Reserve time estimates are revisited throughout the life of the project, being revised up or down as more information on schedule risk becomes available.

- Outputs
  - *Activity Duration Estimates*

All of the preceding inputs and tools for this process are used together to produce one main output: the activity duration estimates.

The activity duration estimates contain an estimated duration for each activity in the activity list. Ideally these estimates represent a range such as the optimistic, pessimistic, realistic one represented in the three-point estimate technique discussed earlier in this process.

- *Activity Attribute Updates*

As each activity is being estimated at a low level, updates to the activity attributes are a normal byproduct of this process.

Schedule Development

WHAT IT IS:

The process of Schedule Development is one of the largest of the 44 PMBOK Guide processes, containing a whopping 27 combined inputs, tools, and outputs.

As anyone who has managed a complex project will attest, developing the schedule can be one of the most daunting parts of the project.

WHY IT IS IMPORTANT:

The schedule is one of the most visible and important parts of the project plan. In fact, many inexperienced project managers often mistakenly refer to the schedule as the project plan or use the two terms interchangeably.

As you could guess by the name, this process is the one where the project's schedule is developed.

WHEN IT IS PERFORMED:

The process of Schedule Development is typically performed after the processes of Activity Resource Planning, Activity Duration Estimating, and Activity Sequencing have been performed and before Cost Budgeting is performed.

HOW IT WORKS:

- Inputs
  - *Organizational Process Assets*

As is true with other processes, assets come in many forms. For this process, organizational process assets include anything the performing organization owns that would help with this process. Items such as an overall resource calendar or examples from previous projects would be particularly helpful in performing Schedule Development.

  - *Project Scope Statement*

The project scope statement, created earlier in the Scope Definition process, is the document that defines the scope of the project. It is particularly helpful here in Schedule Development because it also contains constraints and assumptions for the project.

Since most constraints and assumptions ultimately relate to cost and time, they need to be factored into this process.

  - *Activity List*

The activity list is the list of all activities that need to be scheduled and performed on the project. It is a bit redundant here as an input, since project schedule network diagrams (see later in the list of inputs to this process) implicitly contain the activity list.

  - *Activity Attributes*

The activity attributes always accompany the activity list. They provide expanded information on each activity in the list, and these details may be important when scheduling the activity.

- ***Project Schedule Network Diagrams***

The project schedule network diagrams were created earlier in Activity Sequencing, and they now become a primary input into Schedule Development. The project schedule network diagrams show the order in which activities must be completed, while the schedule assigns dates to each of these activities.

- ***Activity Resource Requirements***

The activity resource requirements were developed previously in Activity Resource Estimating. They show which physical and human resources will be required for each activity.

- ***Resource Calendars***

The resource calendars show resource usage across the organization and will assist in Schedule Development since the resource calendar may put additional constraints on when resources are available to be scheduled.

- ***Activity Duration Estimates***

This is another important input into the process of Schedule Development. The activity duration estimates, created earlier in Activity Duration Estimating, specify how long an activity will take, which has a direct bearing on the schedule.

- ***Project Management Plan***

The project management plan is the single, unified document that guides execution, monitoring, control, and closure. Of particular interest in the process of Schedule Development are the schedule management plan and the risk register. The schedule management plan describes how the schedule will be managed and how changes to the schedule will be managed. The risk register lists specific risks and responses. Because this process is all about developing the schedule, the risks related to the schedule would be of highest interest here.

- Tools
  - *Schedule Network Analysis*

This technique actually refers to a group of techniques used to create the schedule. Any of the other specific tools or techniques that are part of the process of Schedule Development may be used as part of this general tool.

- *Critical Path Method (CPM)*

Before trying to understand the critical path method, it is important to understand what the critical path is. A project's critical path is the combination of activities that, if any are delayed, will delay the project's finish.

The critical path method is an analysis technique with two main purposes:

1. To calculate the project's finish date.
2. To identify how much individual activities in the schedule can slip (or "float") without delaying the project.

A more detailed explanation of the critical path method appears later in this chapter under the heading, "Special Focus: Critical Path Method." The section shows specific techniques relevant to the exam and provides exercises that you will need to be able to perform to pass the exam.

- *Schedule Compression*

On many projects, there are ways to complete the project schedule earlier without cutting the project's scope. That is the purpose of schedule compression.

Two types of questions that you will probably encounter on the exam involving schedule compression are crashing and fast tracking. Crashing involves adding resources to a project activity so that it will be completed more quickly. Crashing almost always increases costs.

A. Original Estimate

Activity	Resources	Estimated days
200 yards of pipeline constructions	1	12

B. Crashed Estimate

Activity	Resources	Estimated days
200 yards of pipeline constructions	4	4

An example of crashing the schedule by adding more resources to an activity

Note that in the example above, as is often the case, increasing the number of resources does decrease the time but not by a linear amount. This is because activities will often encounter the law of diminishing returns when adding resources to an activity. The old saying "Too many cooks spoil the broth" applies to projects as well as cooking.

Fast tracking means that you re-order the sequence of activities so that some of the activities are performed in parallel, or at the same time. Fast tracking does not necessarily increase costs, but it almost always increases risk to the project since discretionary dependencies are being ignored and additional activities are happening simultaneously. (Discretionary dependencies were discussed under the process of Activity Sequencing.)

A. Original Estimate

B. Fast Tracked Estimate

Note that fast tracking often results in some individual activities taking longer, and it increases the risk. In the example above, the workers may have a harder time moving around each other, thus increasing the time to paint and install carpet. Also, there is an increased risk associated with these activities. For example, the painters could damage the carpet, or the carpet installers could be hampered by the fresh paint.

- *What-If Schedule Analysis*

"What-if analysis" typically uses Monte Carlo analysis to predict likely schedule outcomes for a project and identify the areas of the schedule that are the highest risk.

This analysis is performed by computer and evaluates probability by considering a huge number of simulated scheduling possibilities, or a few selected likely scenarios. A computer employing Monte Carlo Analysis can perform what-if analysis and identify the highest risk activities that may not otherwise be apparent, showing the impact of these changes on the schedule.

- *Resource Leveling*

When many people think about the technique of resource leveling, they may mistakenly consider only what their project management software does to level resources. Resource leveling is when your resource needs meet up with the organization's ability to supply resources.

In order to resource-level the project, you first use the critical path method to calculate and analyze all of the network paths for the project. Then you apply resources to that analysis to see what effect it has on schedule outcome.

Consider the following scenario. After performing the processes of Activity Duration Estimating and Activity Resource Estimating, you end up with the following project schedule network diagram.

Activity ID	Activity	Preceding Activity	Resources needed	Quantity of Resource Required	Estimated Duration (in days)
A	Database Design	Start	Database Administrator	2	8
B	Data Entry	A	Data Entry Clerk	10	5
C	Write stored procedures	A	Programmer	8	5
D	Test stored procedures	B, C	Quality Control Engineer	1	2

The project network diagram above was created using the critical path method. Now, however, consider what would happen if the organization could not provide eight resources as reflected in this scenario. Instead, they could only supply two. The scenario must be resource leveled, resulting in a longer overall network diagram as follows:

After the resource calendar has been applied and the schedule has been leveled, a schedule is created based on the resources applied, where each activity is assigned a projected start and finish date.

- *Critical Chain Method*

Although it is not yet widely used, the critical chain method is considered to be one of the most significant potential advancements to project management theory in the past thirty years. Based on the theories of Eliyahu Goldratt, critical chain provides a new way to view and manage uncertainty when building the project schedule.

The traditional way, using the critical path method, puts the primary managerial focus on making sure no activity exceeds its float. If you are successful at that, the project will finish on time. With the critical chain method, you first determine latest possible start and finish date for each activity, and then add schedule "buffers" between activities.

The goal is to manage the project so that no matter what uncertainties or problems occur, you do not exceed your buffers. If you are successful, the project will finish on time or early. To use the critical chain method, you first perform normal critical path analysis and then analyze resource constraints and probabilities in order to build the schedule and the buffers.

- ***Project Management Software***

The tool of project management software simply helps to facilitate the other tools and techniques listed in this section (in addition to cost and other knowledge areas). Project management software can relieve the project manager of many of the routine calculations.

- ***Applying Calendars***

Calendars may be thought of as a sort of constraint. These may be calendars for the organization, such as corporate holidays or periods when work cannot be performed, or they could be calendars related to specific resources or groups.

For instance, a large manufacturing organization may have the ability to run shifts around the clock, while a project spanning international boundaries will probably have non-working holidays in one country that do not apply in other countries. Another example might be an engineering department's unavailability during a particular week due to training.

- ***Adjusting Leads and Lags***

Leads and lags can significantly affect the critical path as well as other components of the schedule. Adjusting leads and lags amounts to fine tuning the schedule so that they are as accurate and realistic as possible.

- *Schedule Model*

The schedule model precedes the project schedule. The project manager uses the activity attributes and estimates along with a schedule tool or method to create the project's schedule. The schedule model allows the project manager to experiment with different allocations and scenarios in order to produce the project schedule.

- Outputs

  - *Project Schedule*

The project schedule shows when each activity is scheduled to begin and end, as well as showing a schedule start and finish for the overall project. The schedule is typically represented graphically, and there are different forms it may take. The most common forms are covered as follows:

A. Project Network Diagram

The project network diagram is a useful detail-driven tool that provides a powerful view of the dependencies and sequences of each activity. It is the best representation for calculating the critical path and showing dependencies on the project.

B. Bar Charts (also called Gantt Charts)

Bar charts, or Gantt charts, show activities represented as horizontal bars and typically have a calendar along the horizontal axis. The length of the bar corresponds to the length of time the activity should require.

A bar chart, or Gantt chart, can be easily modified to show percentage complete (usually by shading all or part of the horizontal bar).

It is considered to be a good tool to use to communicate with management, because unlike the project network diagram, it is easy to understand at a glance.

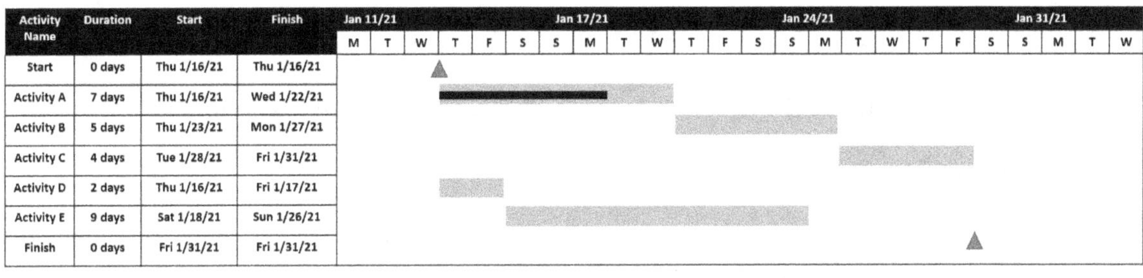

Activity Name	Duration	Start	Finish
Start	0 days	Thu 1/16/21	Thu 1/16/21
Activity A	7 days	Thu 1/16/21	Wed 1/22/21
Activity B	5 days	Thu 1/23/21	Mon 1/27/21
Activity C	4 days	Tue 1/28/21	Fri 1/31/21
Activity D	2 days	Thu 1/16/21	Fri 1/17/21
Activity E	9 days	Sat 1/18/21	Sun 1/26/21
Finish	0 days	Fri 1/31/21	Fri 1/31/21

C. Milestone Chart

A milestone chart, as the name implies, only represents key events (milestones) for the project.

Milestones may be significant events or deliverables by you or external parties on the project.

Milestone charts, because of the general level of information they provide, should be reserved for brief, high level project presentations where a lot of schedule detail would be undesirable or even distracting.

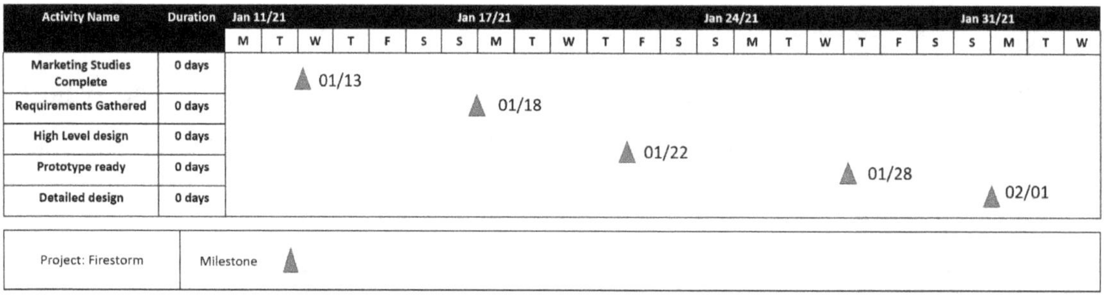

Activity Name	Duration	Jan 11/21						Jan 17/21							Jan 24/21							Jan 31/21		
		M	T	W	T	F	S	S	M	T	W	T	F	S	S	M	T	W	T	F	S	S	M	T	W
Marketing Studies Complete	0 days			▲ 01/13																					
Requirements Gathered	0 days								▲ 01/18																
High Level design	0 days												▲ 01/22												
Prototype ready	0 days																		▲ 01/28						
Detailed design	0 days																						▲ 02/01		

Project: Firestorm	Milestone ▲

- *Schedule Model Data*

The schedule model data means the information the project team used to model and create the project. It would include schedule templates that were used, the activities and their attributes, estimated durations, and any constraints and assumptions.

This output is simply the data which supports how this schedule was developed.

- *Schedule Baseline*

Any baseline (whether scope, schedule, or cost) is the original plan plus all approved changes. In this example, the schedule baseline is created at the point at which the schedule is approved by the customer, the sponsor, and the project manager (as well as possibly others such as the functional managers and the team). When schedule change requests are approved, they become part of the schedule baseline.

- *Resource Requirement Updates*

Updates to the resource requirements are a very common output in Schedule Development, especially in the case of the tool of resource leveling, which adjusts the resources that are required in light of the organization's ability to supply them.

- *Activity Attribute Updates*

Any time attributes such as the resource requirements or start and finish dates are changed for an activity, these need to be updated in the activity attributes documentation.

- *Project Calendar Updates*

The project calendar is brought into this process as an input, and any schedule changes that change project calendars, such as resource usage that would affect a resource calendar, should be updated back to that calendar.

- *Requested Changes*

Change requests are a normal output of Schedule Development. For instance, if the project team determined that it was impossible to perform the scope of work within the allotted time, a change request to eliminate certain unnecessary components of the scope may be entered.

- *Project Management Plan Updates*

The project management plan contains the information on how the project will be managed. Of specific interest here is the schedule management plan, which details how the schedule will be managed and how changes to the schedule will be managed.

Schedule Control

WHAT IT IS:

As you can tell from its name, Schedule Control is a controlling process. The concept behind controlling processes in general is to compare the work results to the plan and ensure that they line up. In this process, the schedule is controlled to make sure that time-related performance on the project is in line with the plan.

One of the more important concepts to master with Schedule Control (and most controlling processes in general) is that schedule changes are not only reacted to, but the schedule is controlled proactively.

That is, the project manager should be out in front of the project, influencing changes before they affect the project. Of course, at times, changes to the schedule may occur, and the project manager will have to react to them, but the project manager should be proactive whenever possible.

WHY IT IS IMPORTANT:

Any time the schedule changes or a change request that affects the schedule occurs, the change should be evaluated and planned. The schedule should be monitored continuously against the actual work performed to ensure that things stay on target.

WHEN IT IS PERFORMED:

The process of Schedule Control is performed throughout the life of the project.

HOW IT WORKS:

- Inputs
  - *Schedule Management Plan*

The schedule management plan is an essential input into Schedule Control, since this is the portion of the project management plan that defines how the schedule will be managed and changed, and Schedule Control is the process that processes and manages those changes.

  - *Schedule Baseline*

The schedule baseline is another essential input into this process. It is the latest approved version of the project schedule. As changes are made or change requests are approved, the schedule baseline is updated to reflect those changes.

- *Performance Reports*

For the process of Schedule Control, we are most interested in the performance reports that relate to the schedule. This would include information such as which activities and milestones had been met and which were slipping.

- *Approved Change Requests*

Requested changes can, and usually do, affect the project schedule. They are factored into this controlling process, since Schedule Control is where the schedule changes are analyzed and managed.

- Tools

- *Progress Reporting*

The project schedule shows planned start and finish dates for each activity; however, it is the actual dates that determine when the project is delivered. Progress reports are used to show how the project is progressing against the schedule.

- *Schedule Change Control System*

The schedule change control system is part of the integrated change control system. It includes all of the procedures that are used to approve or reject a change.

- *Performance Measurement*

The performance measurement tool uses actual schedule data to compare against planned schedule data. Any differences between the plan and reality must be carefully analyzed and understood.

- *Project Management Software*

Software simply automates and simplifies the analysis used in the other tools and techniques described for Schedule Control.

- *Variance Analysis*

Variance analysis looks at the difference between what was scheduled and what was executed in order to understand any differences. These differences are analyzed to determine whether corrective action is required.

- *Schedule Comparison Bar Charts*

Bar charts, also known as Gantt charts, are an effective way to communicate the schedule and the actual results. With this tool, one bar shows the plan for each schedule activity and the other shows the actual data to date.

- Outputs

- *Schedule Model Data Updates*

The schedule model data is used to build the schedule. As the underlying data are affected by change requests, slippage, etc., these updates need to be reflected in the schedule model data updates.

- *Schedule Baseline Updates*

A baseline (whether scope, schedule, or cost), is the original plan plus all approved changes. In this case, when a change to the schedule is approved, this change becomes part of the schedule baseline.

- *Performance Measurements*

The updated schedule performance index (SPI), schedule variance (SV) and other earned value measurements relevant to the schedule need to be calculated and communicated out.

- *Requested Changes*

As the schedule is analyzed in detail, it is normal for change requests to be entered. For instance, if it were determined that the project was substantially ahead of schedule, it might be appropriate to reconsider scope items that had been omitted earlier in the interest of time.

- *Recommended Corrective Action*

Corrective action is not about fixing past mistakes; they are considered to be water under the bridge.

Instead, it is defined as anything done to bring future results in line with the plan. In the process of Schedule Control, after variance analysis has taken place, the recommended corrective action would be to evaluate ways that future performance and the schedule baseline could be made to align.

- *Organizational Process Assets Updates - See Chapter 2, Common Outputs*
- *Activity List Updates - See this chapter, Activity Definition, Outputs*
- *Activity Attribute Updates - See this chapter, Activity Definition, Outputs*
- *Project Management Plan Updates - See Chapter 2, Common Outputs*

As a general note regarding the four preceding outputs, updates to the organizational process assets, the activity list, the activity attributes, and the project management plan are normal results of the process of Schedule Control.

Any time one of these is added to or changes in any way, these updates need to be formally captured.

Special Focus: Critical Path Method

One very important tool used in Schedule Development is a mathematical analysis technique called critical path method. You need to be very comfortable with this technique for the PMP exam, so a good portion of the rest of this chapter is dedicated to the subject.

NETWORK PATHS

The term "network path" refers to a sequence of events that affect each other on the project from start to finish.

These activities form a path through the project. Paths are important because they illustrate the different sets of sequences in which activities must be performed, and they are used to identify areas of high risk on the project.

In a real project plan, there will usually be numerous paths through the network diagram, and software is typically used to represent and calculate them.

To understand the paths, refer to the network diagram and corresponding tables that follow.

Figure 6-1: Network Logic Diagram -Activity on Node

Duration units shown in weeks

Activity	Duration (weeks)
A	2
B	1
C	4
D	6
E	3
F	3

Table 6-2: Individual activities

In the diagram and tables above, each node has a corresponding duration as listed in Table 6-2. Each possible path through the network is determined by following the arrows in the diagram in Figure 6-1.

The arrows represent activity sequence, and in this example there are three paths through the system as illustrated in the next Table 6-3.

Path	Activities	Path Duration (weeks)
Start-A-B-C-Finish	2+1+4	7
Start-D-C-Finish	6+4	10
Start-D-E-F-Finish	6+3+3	12

Table 6-3: Network Paths

CRITICAL PATHS

Determining the critical path through the network is a tool used heavily in creating the schedule. Critical path calculations show you where most of the schedule risk will occur. The critical path is made up of activities that cannot be delayed without delaying the finish of the project.

By following the steps taken in creating the network diagram, the critical path is determined simply by identifying the longest path through the system. In the previous example, it is Start- D-E-F-Finish, because activities D (6) + E (3)+ F (3) will take 12 weeks, and this is the longest path through the system.

Keep in mind that it is not unusual to have more than one critical path on a project.

This occurs when two or more paths tie for the longest path. In this event, schedule risk is increased because there are an increased number of ways the project could be delayed. Consider the next network diagram.

This example is an activity-on-arrow diagram, so we calculate the activities and paths slightly differently.

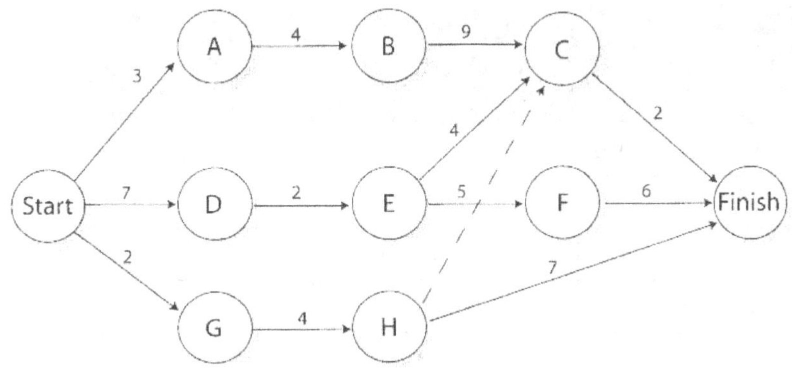

Figure 6-4: Network Logic Diagram -Activity on Arrow

Duration units shown in weeks

Activity	Duration (weeks)
Start-A	3
Start-D	7
Start-G	2
A-B	4
B-C	9
C-Finish	2
D-E	2
E-C	4
E-F	5
F-Finish	6
G-H	4
H-C	0
H-Finish	7

Table 6-5: Activity Durations

Path	Activities	Path Duration (weeks)
Start-A-B-C-Finish	3+4+9+2	18
Start-D-E-C-Finish	7+2+4+2	15
Start-D-E-F-Finish	7+2+5+6	20
Start-G-H-C-Finish	2+4+0+2	8
Start-G-H-Finish	2+4+7	13

Network paths

There are two key things to note about the previous example using activity on arrow.

1. The nodes in Table 6-5 are not listed individually, but in pairs. This is activity on arrow, so each arrow (activity) connects two nodes.

2. There is a "dummy activity," as indicated by the dotted line between nodes H and C. A dummy is technically not an activity, but is a way of representing a dependency between nodes; however, for the exam, it is easiest to think of a dummy activity as exactly the same as a normal activity, but with a duration of 0.

FLOAT

After you have mastered the concepts for activity on node and activity on arrow, calculating float (sometimes referred to as "slack") is also easy. Before attempting any exercises, it is critical to understand the following:

1. Float is simply how much time an activity can slip before it changes the critical path. Another way of thinking about float is that it is the maximum amount of time an activity can slip without pushing out the finish date of the project.

2. If an item is on the critical path, it has zero float. Although there are technical cases

where this might not be true, you should not encounter any such examples on the exam.

Keeping the preceding two items in mind, let us revisit the activity on node network diagram in figure 6-1. Using this example:

Question: What is the float for activity D?

Answer: 0 weeks. Activity D cannot slip without affecting the finish date because it is on the critical path.

Question: What is the slack for activity C?

Answer: 2 weeks. If activity C slips by more than 2 weeks, then the path Start-D-C-Finish would delay the finish of the project.

There is a simple way to do this that will allow you to breeze through the questions on the PMP exam. The method is a brute force approach. If the exam asks you to calculate the slack for an activity, simply look at the project network diagram and find the duration you can substitute for that activity that will put it on the critical path. You should be able to zero in on the float right away. If you have created a path chart, such as the one in Table 6-3, this method is a snap.

Path	Activities	Path Duration (weeks)
Start-A-B-C-Finish	2+1+4	7
Start-D-C-Finish	6+4	10
Start-D-E-F-Finish	6+3+3	12

In this example:

- Activity A has float of 5 weeks, because this activity may be delayed up to 5 weeks without delaying the finish of the project.

- Activity B also has float of 5 weeks, because paths with activity B could slip up to 5 weeks without changing the project's finish date.
- Activity C is slightly trickier. It has a float of only 2 weeks. Even though it could slip 5 weeks in the first path listed in Table 6-3, if it slipped from 4 weeks to 6 weeks, then path Start-D-C-Finish would be on the critical path. Therefore, we take the smallest slippage possible and that becomes the float for this activity.

After a little practice, this will become the easiest and quickest way to calculate float. There are detailed exercises at the end of this chapter to help cement your understanding of this.

Note that on the exam you may see the terms forward pass or backward pass related to the critical path. These methods of determining early or late start or early or late finish are presented below.

EARLY START

The early start date for an activity is simply the earliest date it can start when you factor in the other dependencies. This is the date the activity will start if everything takes as long as it was estimated. Consider the figure that follows (this diagram is the same as Figure 6-1, referenced earlier).

Question: Given the following diagram, what is the early start for activity B?

Units shown in weeks

The answer is week 3. The reason is that Activity A is scheduled to take 2 weeks. It begins on week 1, and finishes on week 2. Activity B could then begin on the first day of week 3.

This technique is called a "forward pass" because we have moved "forward" through the diagram, starting from the start date to perform our calculations.

EARLY FINISH

The early finish date is the early start date plus the duration estimate minus 1 unit. In the example above, the early finish date for activity 'A' would be 2 weeks after the early start date. Since the early start date is week 1, the early finish would be week 2. This makes sense since the activity takes all of week 1 and all of week 2, finishing at the end of week 2.

Here we also performed a "forward pass," as again we based our calculations on the start date.

LATE START

The late start date for activity turns the problem around and looks at it from the other end of the network diagram. It asks, "What is the latest this activity could start and not delay the project's finish date?"

Although there are many methods that may be used to calculate late start, the easiest way is to simply add the float to the early start. This will give us the absolute latest date that the activity can start and not impact the finish date, assuming that the activity takes as long as was estimated. Given the example in the table above, what is the late start for activity B?

Answer: Week 8. The reason is that this activity has 5 weeks of float, and if the early start is week 3, the late start would be week 3 + 5 weeks of float = week 8.

Remember that for activities with zero float or slack, the late start and the early start will be the same.

This technique is called a "backward pass." The reason is that you must begin at the end and work your way backward, evaluating how close activities may slide toward the finish without moving the finish.

LATE FINISH

An activity's late finish will be the late start plus the activity's duration estimate minus 1 unit.

In the example above, the late start is week 8, and the duration is 1 week, so the late finish date would also be week 8. If the fact that both the start and finish are shown as week 8 confuses you, keep in mind that the start represents is the first day of week 8, and the finish represents the last day of week 8.

Remember that for activities with zero float or slack, the late finish and the early finish will be the same.

Calculating the late finish date by adding the activity's duration to the late start date is another example of a "backward pass."

ES, EF, LS AND LF - Often a node on the project network diagram may be represented like this, with the early start in the upper left corner (represented by ES), the early finish in the upper right (EF), the late start in the bottom left (LS) and the late finish in the bottom right (LF). By dividing each node on a project network diagram into quadrants, all four of these may be represented for every node in the diagram

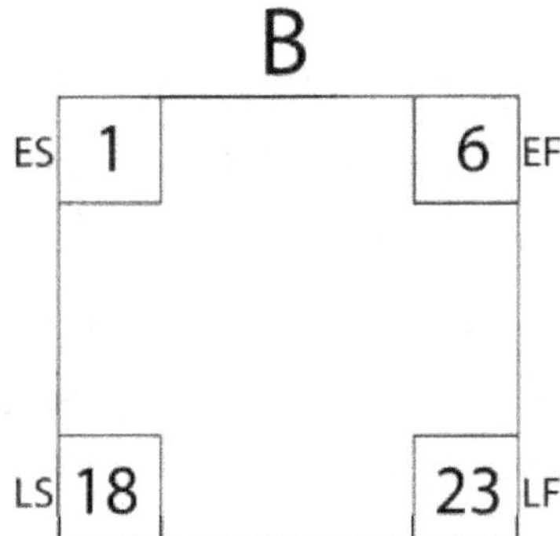

FREE FLOAT - Also known as "free slack," free float is the amount of time an activity can be delayed without affecting the early start date of subsequent dependent activities. If this leaves you scratching your head, consider the diagram below:

The free float of activity E would be determined by calculating how long it could slip before it impacted the early start of activity C. In this example, activity E has 4 units of free float. Note that this is different from its float, which is 19 units.

NEGATIVE FLOAT - Negative float is a situation that occurs when an activity's start date occurs before a preceding activity's finish date. For instance, suppose that there is a constraint for an activity (Activity D) that involves a 1 day final inspection of a building that is to begin and be completed on July 15th however, a preceding activity (Activity C) that encompasses testing the electrical components of the project does not finish until July 25th. If work is scheduled 7 days per week, then it could be stated that Activity D has a negative float of 10 days.

Technically, negative float exists when an activity's finish date happens before its start date. In the example above, Activity D is supposed to finish at the end of the 15th however, it cannot even begin until 10 days later. There Activity D has 10 days of negative float.

Negative float for an activity tells you that your schedule has problems. It most often occurs when immovable constraints or milestones are imposed by forces outside the project, causing an impossible situation.

Negative float may be resolved by several methods, such as reworking the logic of the schedule, crashing, or fast-tracking.

Summary of Key Terms

You must understand each of the following key terms. You do not have to memorize every term word for word, but you should be able to recall the general definition and apply them on the exam.

ACTIVITY ON ARROW DIAGRAM - A type of network diagram where the activities are represented by the arrows. The nodes (usually circles) are used to connect or show dependencies. Activity on arrow diagrams are always shown "finish to start," where one activity is finished before the next one begins. This is the diagramming method that uses dummy activities (usually represented by dashed lines).

ACTIVITY DECOMPOSITION - Activity decomposition is similar to scope decomposition (remember the asteroids metaphor), except that the final result here is an activity list instead of the WBS.

ACTIVITY LIST - A list of every activity that will be performed on the project.

ACTIVITY DURATION ESTIMATES - Probable number of periods (weeks, hours, days, months, etc.) this activity should take with the probable range of results. Example:

Activity Duration Estimate	Explanation
1 week +/- 3 days	The activity should take between 2 and 8 days, assuming a 5 day work week
1 month + 20% probability it will be accomplished later	There is an 80% likelihood that the activity will be completed within a month and a 20% chance that it will exceed a month.

ANALOGOUS ESTIMATING – A form of expert judgment often used early on when there is little information available. Example: "This project is similar to one we did last year, and it took three months." It is performed from the top down, focusing on the big picture.

BACKWARD PASS - The method for calculating late start and late finish dates for an activity (see explanation of float, early and late starts and finishes earlier in this chapter for a detailed explanation).

CRITICAL PATH - The paths through the network diagram that show which activities, if delayed, will affect the project finish date. For schedule, the critical path represents the highest risk path in the project.

DEPENDENCIES - Activities that must be completed before other activities are either started or completed.

Dependencies	Description	Example
Mandatory	Also called "hard logic," these activities must be followed in sequence.	Clearing the lot on a construction site before pouring the foundation.
Discretionary	Also called "soft logic." Expert judgment and best practices often dictate that particular activities are performed in a particular order. The dependencies are discretionary because they are based on expert opinion rather than mandatory or hard logic.	Painting the interior before putting down carpet.
External	Dependencies relying on factors outside of the project	Zoning approval for a new building. Weather for a rocket launch.

DUMMY ACTIVITY - An activity on a network diagram that does not have any time associated with it. It is only included to show a relationship, and is usually represented as a dotted or dashed line. Dummy activities only exist in activity on arrow diagrams.

DURATION COMPRESSION – This technique is primarily made up of two means of compressing the schedule: crashing and fast-tracking, described in the table below.

Technique	Description	Example
Crashing	Applying more resources to reduce duration. Crashing the	If setting up a computer network takes one person 6 weeks, three resources may be able to do it in two

	schedule usually increases cost.	weeks. Note – Crashing usually does not reduce the schedule by a linear amount.
Fast tracking	Perfoming activities in parallel that would normally be done in sequence. Fast tracking activities usually increases project risk, and these activities have a higher probability of rework.	Example: In XYX Corp, no coding on software modules is allowed until after the database design is complete, but when fast tracking, the activities could be done in parallel if it is not a mandatory dependency.

DELPHI TECHNIQUE - A means of gathering expert judgment where the participants do not know who the others are and therefore are not able to influence each other's opinion. The Delphi technique is designed to prevent groupthink and to find out a participant's real opinion.

EXPERT JUDGMENT - A method of estimating in which experts are asked to provide input into the schedule. Combining expert judgment with other tools and methods can significantly improve the accuracy of time estimates and reduce risk.

FLOAT - How much time an activity can be delayed without affecting the project's finish date. Also known as "slack."

FORWARD PASS - The method for calculating early start and early finish dates for an activity (see explanation of float and early and late starts and finishes earlier in this chapter).

FREE FLOAT - Also known as "free slack." How much time an activity can be delayed without affecting the early start date of subsequent dependent activities.

HEURISTICS - Rules for which no formula exists. Usually derived through trial and error.

LAG - The delay between the activity and the next one dependent upon it. For example, if you are pouring concrete, you may have a 3 day lag after you have poured the concrete before your subsequent activities of building upon it can begin.

LEAD - Activities with finish to start relationships cannot start until their predecessors have been finished; however, if you have 5 days of lead time on an activity, it may begin 5 days before its predecessor activity has finished. Think of it as getting a head start, like runners in a relay race. Lead time lets the subsequent task begin before its predecessor has finished.

MATHEMATICAL ANALYSIS - A technique to show scheduling possibilities where early and late start and finish dates are calculated for every activity without looking at resource estimates.

MILESTONES - High level points in the schedule used to track and report progress. Milestones usually have no time associated with them.

MONTE CARLO ANALYSIS - Computer simulation that throws a high number of "what if" scenarios at the project schedule to determine probable results.

NETWORK DIAGRAM - (also called network logic diagram or project network diagram). A method of diagramming project activities to show sequence and dependencies.

NEGATIVE FLOAT - Occurs when an activity's start date occurs before a preceding activity's finish date. Technically, negative float exists when an activity's finish date happens before its start date. Negative float for an activity indicates a schedule with problems. Reworking the logic of the schedule, crashing, or fast-tracking are potential solutions.

PRECEDENCE DIAGRAMMING METHOD - (also called Activity on Node). A type of network diagram where the boxes are activities, and the arrows are used to show dependencies between the activities.

RESERVE TIME (CONTINGENCY) - A schedule buffer used to reduce schedule risk. The chart below represents the most common types of reserve for a project.

Contingency	Example
Project %	Add 15% to the entire project schedule
Project lump sum	Add 2 months calendar schedule to the project
Activity %	Add 10% to each activity (or to key, high-risk activities)
Activitiy lump sum	Add 1 week to each activity (or to key, high-risk activities)

SCHEDULE BASELINE - The approved schedule that is used as a basis for measuring and reporting. It includes the originally approved project schedule plus approved updates.

SLACK - See "Float."

VARIANCE ANALYSIS - Comparing planned versus actual schedule dates.

Exercises

In order to test yourself on time management, complete the following questions based on Figure 6-7 below.

Figure 6-7: Project network diagram.

Durations shown in days

1. List all the paths through the network logic diagram as illustrated in Figure 6-7.
2. What is the critical path through the network diagram shown in Figure 6-7?
3. List the slack for each activity in the network.
4. Provided the table below, how many weeks long is the critical path?

Activity	Preceding Activity	Duration (in weeks)
Start		0
A	Start	6
B	A, E	2
C	B	2
D	C	3
E	Start	1
F	A, E	1
G	F, B	7
Finish	D, G	0

5. Given the table in question 4 above, describe the effect of activity D taking twice as long as planned.

6. An activity has 3 estimates, provided below:

Optimistic	=	*10 days*
Pessimistic	=	*25 days*
Realistic	=	*15 days*

What is the three point estimate for this activity?

7. What is the standard deviation for the activity in question 6?

8. Given the project network node below, fill in the value in the missing quadrant, and calculate the duration and the float for the activity.

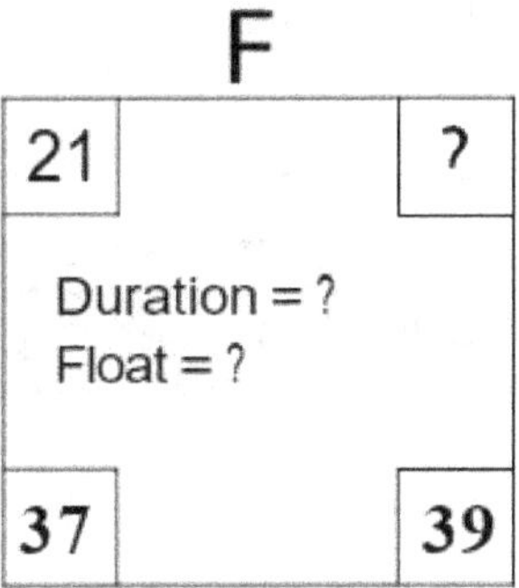

Answers to Exercises

1. There are six possible paths through the system, listed in the table below:

Path
Start-A-B-C-D-Finish
Start-E-F-G-D-Finish
Start-E-F-G-H-Finish
Start-I-J-G-D-Finish
Start-I-J-G-H-Finish
Start-I-J-K-Finish

2. Because two paths tie for the longest duration, there are two critical paths:

- Start-A-B-C-D-Finish - duration 29 days
- Start-I-J-G-H-Finish - duration 29 days

Path	Durations	Total
Start-A-B-C-D-Finish	16+3+7+3	29 days
Start-E-F-G-D-Finish	2+5+9+3	19 days
Start-E-F-G-H-Finish	2+5+9+10	26 days
Start-I-J-G-D-Finish	6+4+9+3	22 days
Start-I-J-G-H-Finish	6+4+9+10	29 days
Start-I-J-K-Finish	6+4+1	11 days

3. This task is easier than it first appears since over 70% of the activities were on the critical path. Those activities automatically have zero slack; thus, no calculations are necessary for most of the activities. The way to solve for these is to take each path listed above and go through them one at a time. If the path is a critical path, or if the activity is found on the critical path, simply skip it. If the path is not the critical path, then take the sum of the items on that path and subtract it from the total critical path. For activity E, it is found on 2 paths above. One of them totals 19, and the other 26. We always use the larger one and subtract it from the critical path total of 29. That leaves a slack of 29-26, or 3 for activity E.

Activity	On Critical Path?	Slack (Float) in days
A	Y	0
B	Y	0
C	Y	0
D	Y	0
E	N	3
F	N	3
G	Y	0
H	Y	0
I	Y	0
J	Y	0
K	N	18

4. Your first step in solving this problem is to draw out a network logic diagram. Your diagram should look similar to the one shown below:

The next step is to list out all paths through the network:

Path	Durations	Total
Start-A-B-C-D-Finish	6+2+2+3	13
Start-A-B-G-Finish	6+2+7	15
Start-A-F-G-Finish	6+1+7	14
Start-E-B-C-D-Finish	1+2+2+3	8
Start-E-B-G-Finish	1+2+7	10
Start-E-F-G-Finish	1+1+7	9

The answer is 15 weeks, based on the fact that Start-A-B-G-Finish has a duration of 15 weeks, and that is the longest duration of any path.

5. If activity D were to take twice as long as planned, that would change its duration from 3 to 6 weeks. This would have two effects:

- The critical path would change. The new critical path would be Start - A-B-C-D-Finish.

- The project finish date would be extended by 1 additional week, meaning the overall project would take 16 weeks.

To arrive at this solution, reconstruct the table as follows:

Path	Durations	Total
Start-A-B-C-D-Finish	6+2+2+6	16
Start-A-B-G-Finish	6+2+7	15
Start-A-F-G-Finish	6+1+7	14
Start-E-B-C-D-Finish	1+2+2+6	11
Start-E-B-G-Finish	1+2+7	10
Start-E-F-G-Finish	1+1+7	9

Note how the longest path has changed to the first one in the table above.

6. The formula for three-point estimates is:

$$\frac{\text{Pessimistic} + 4 * \text{Realistic} + \text{Optimistic}}{6}$$

or..

$$\frac{24 + 4 * 15 + 10}{6}$$

This yields a three-point duration estimate of 15.83.

7. The formula for standard deviation is:

$$\frac{\text{Pessimistic} - \text{Optimistic}}{6}$$

or...

$$\frac{25 - 10}{6}$$

This yields a standard deviation of 2.5.

8. For this exercise, the early finish (top right quadrant), the duration, and the float were in question.

   Given the example, it is easiest to begin with the duration. This may be calculated easily by subtracting the late start (bottom left) from the late finish (bottom right) and adding one. 39 – 37 + 1 yields an activity duration of 3 (if this confuses you, consider that days 37, 38, and 39 are all working days).

   Now, simply count 3 days from the early start (top left), using day 21 as a working day, and it gets you to an early finish of day 23 (day 21, 22, and 23). The float is the late start - the early start, represented as 37 - 21 = 16. Even though this is the only node we have, we could further deduce that the early start or early finish date could slip 16 days without affecting the critical path and endangering the finish date.

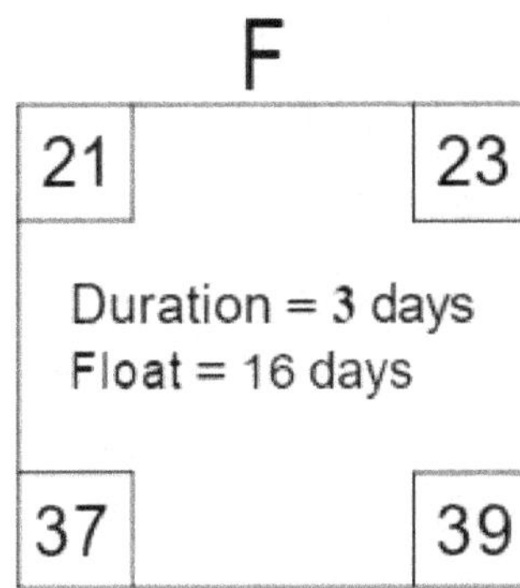

Time Management Questions

1. You are the project manager for the construction of a commercial office building that has very similar characteristics to a construction project performed by your company two years ago. As you enter Activity Definition, what is the BEST approach?

    A. Use the activity list from the previous project as your activity list.

    B. Generate your activity list without looking at the previous project's list and compare when your project's list is complete.

    C. Use the gap analysis technique to identify any differences between your project and the previous project.

    D. Use the previous activity list to help construct your list.

2. The customer has called a project team member to request a change in the project's schedule. The team member asks you what the procedure is for handling schedule changes. Where should you refer the team member to help him understand the procedure?

   A. The project office.

   B. The change control board.

   C. The schedule management plan.

   D. Inform the team member that the customer is always right.

3. If you were creating duration estimates for a schedule activity, which of the following tools or techniques would NOT be appropriate to use?

   A. Expert judgment.

   B. Reserve analysis

   C. Three-point estimating.

   D. Least-squares estimating.

4. Senior management has called you in for a meeting to review the progress of your project. You have been allocated 15 minutes to report progress and discuss critical issues. Which of the following would be BEST to carry with you in this case?

   A. Milestone chart.

   B. The project network diagram.

C. An expert from each functional area of the project so that all questions may be answered.

D. Project status reports from your team members.

5. How does activity on node differ from activity on arrow?

A. Activity on arrow is superior to activity on node.

B. Activity on node is superior to activity on arrow.

C. Activity on arrow may have dummy activities.

D. Activity on node may have dummy activities.

6. The amount of time that an activity may be delayed without extending the critical path is:

A. Lag.

B. Grace period.

C. Free factor.

D. Slack.

7. Crashing differs from fast tracking because crashing:

A. Usually increases value.

B. Usually increases the cost.

C. Usually saves more time.

D. Usually saves more money.

8. If senior management tells you "The last project we did like this cost us almost five million dollars," what estimating method is being used?

   A. Delphi technique.

   B. Principle of equivalency of activities.

   C. Analogous estimating.

   D. Bottom-up estimating.

9. You are advising a project manager who is behind schedule on his project. The sponsor on his project is very unhappy with the way things have progressed and is threatening to cancel the project. The sponsor has accepted a revised due date from the project manager but did not allow any increased spending. Which of the following would represent the BEST advice for the project manager in this case?

   A. Fast track the schedule.

   B. Ask senior management for a new sponsor within the organization.

   C. Crash the schedule.

   D. Talk with the customer to see if budget may be increased without the sponsor's involvement.

10. Which Schedule Development tool inserts non-working buffer time between schedule activities?

    A. Critical chain method.

    B. Critical path method.

    C. Resource leveling.

    D. Schedule modeling.

11. What is the BEST tool to use to calculate the critical path on a project?

    A. Work breakdown structure.

    B. GERT diagram.

    C. Gantt chart.

    D. Project network diagram.

12. Consider the table at right:

What is the critical path?

A. Start-A-E-H-Finish

B. Start-C-E-H-Finish

C. Start-B-D-I-Finish

D. Start-B-D-G-Finish

Activity	Duration	Dependent on
Start	0	
A	3	Start
B	4	Start
C	2	Start
D	2	B
E	5	A, C
F	1	B
G	6	D, F
H	11	E
I	8	D, F
Finish	0	G, H, I

13. Using the project network diagram you created for question 12, what is the float for activity D?

A. 0 days.

B. 3 days.

C. 5 days.

D. 7 days.

14. An activity has a duration estimate that is best case = 30 days, most likely = 44 days, and worst case = 62 days. What is the three-point estimate for this activity?

A. 44.67 days

B. 34.67 days

C. 5.33 days

D. 59.33 days

15. Which of the following choices best fits the description of a project manager applying the technique of What-if Scenario Analysis?

    A. Using project management software to build three versions of the project schedule.

    B. Using Monte Carlo analysis to identify what would happen if schedule delays occurred.

    C. Using critical path method to analyze what would happen if the critical path actually occurred.

    D. Discussing with the functional managers what they would do if certain project team members quit the project early.

16. How do the activity list and activity attributes relate to each other?

    A. The activity list focuses on schedule activities, while the activity attributes apply to WBS activities.

    B. The activity attributes are created prior to the activity list.

    C. The activity list may be substituted for the activity attributes in most processes.

    D. Activity attributes provide additional information for each activity on the activity list.

17. Which of the following is the BEST description of the critical path?

A. The activities that represent critical functionality.

B. The activities that represent the largest portions of the work packages.

C. The activities that represent the highest schedule risk on the project.

D. The activities that represent the optimal path through the network.

18. Consider the table at right.

Activity	Duration
Start - A	6
Start - E	11
A-B	7
B-C	8
C-D	14
D-Finish	8
E-F	21
F-C	0
F-H	6
E-G	8
G-H	7
H-D	7

What is the length of the critical path?

A. 43

B. 45

C. 53

D. 54

19. Consider the table in question 18. What is the slack for Activity H-D?

    A. 0

    B. 1

    C. 5

    D. 13

20. Your project schedule has just been developed, approved, and distributed to the stakeholders and presented to senior management when one of the resources assigned to an activity approaches you and tells you that her activity cannot be performed within the allotted time due to several necessary pieces that were left out of planning. Her revised estimate would change the schedule but would not affect the critical path. What would be the BEST way for the project manager to handle this situation?

    A. Stick with the published schedule and allow for any deviation by using schedule reserve.

    B. Go back to Activity Duration Estimating and update the schedule and other plans to reflect the new estimate.

    C. Hire an independent consultant to validate her claim.

    D. Replace the resource with someone who says they can meet the published schedule.

Answers to Time Management Questions

1. D. The previous activity list would make an excellent tool to help you ensure that you are considering all activities. Any historical information such as this is thought of as an organizational process asset. 'A' is incorrect because you cannot simply substitute something as intricate as a complete activity list. 'B' is incorrect because the other activity list would provide a good starting point and should be considered before you create your activity list. 'C' (gap analysis) is a tool that is used in the real world that is not defined by PMI, nor is it used in activity list definition.

2. C. The schedule management plan, which becomes part of the project plan, would be the best source of information on how changes to the schedule are to be handled. 'A' is incorrect because the project office's job is to define standards - not to make decisions on tactical items such as this. 'B' is incorrect because the change control board may or may not even exist, and if it does exist, it usually approves or rejects scope changes. Answer 'D' would be the worst choice. The customer is not always right when it comes to requesting changes. Procedures should be defined and followed in order to improve the project's chances of success.

3. D. Since we are creating activity duration estimates, we are performing the process of Activity Duration Estimating. Answers 'A', 'B' and 'C' are all tools used in Activity Duration Estimating, but 'D' is a made up term.

4. A. Milestone charts show the high level status, which would be appropriate given the audience and time allocated for this update.

5. C. Activity on arrow project network diagrams may have dummy activities, while activity on node diagrams cannot show dummy activities. While some debate over which method is superior may exist, answer 'C' is the most appropriate choice here.

6. D. The slack (or float) is the amount of time an activity may be delayed without affecting the critical path.

7. B. Crashing adds more resources to an activity. This usually increases the cost due to the law of diminishing returns which predicts that 10 people usually cannot complete an activity in half the time that 5 people can. The savings from crashing are rarely linear. 'A' is incorrect because crashing does not directly affect the project's value. 'C' is incorrect because crashing may or may not save more time than fast tracking - depending on the situation. 'D' is incorrect because crashing usually costs more money than fast tracking.

8. C. In this example management is providing you with analogous estimates. These estimates use actual costs from previous projects (historical information or organizational process assets) to produce estimates for a similar project.

9. A. In this case, you must compress the schedule without increasing the costs. Fast tracking does not directly add cost to the project and is the best choice in this case. 'B' is incorrect.

   The sponsor is paying for the project. Do this, and your sponsor will probably be asking for a new project manager instead! 'C' is incorrect because crashing usually adds cost to the project, and that is not allowed in this scenario. 'D' is incorrect because the sponsor authorizes budget. Doing an end-run around the sponsor and going to the customer would be very inappropriate.

10. A. The critical chain method uses buffers, which are non-working times, to help prevent the activities themselves from slipping.

11. D. The project network diagram shows duration and dependencies which would help you calculate the critical path. 'A' is incorrect because the WBS does not show durations or activity dependencies. 'B' is incorrect because GERT is most helpful for showing conditions and branches. 'C' is incorrect because a Gantt chart is very useful for showing percentage complete on activities but is not the best tool for showing activity dependencies or calculating the critical path.

12. A. The critical path is determined in 3 steps. The first step is to draw out the project network diagram. Yours should look similar to the one depicted below (Note that activities B and C were moved to make the diagram neater - don't worry if your diagram does not look this neat):

The next step is to list out all of the paths through the network. The six paths are:

Start-A-E-H-Finish

Start-C-E-H-Finish

Start-B-F-I-Finish

Start-B-F-G-Finish

Start-B-D-I-Finish

Start-B-D-G-Finish

The last step is to add up all of the values associated with each path as is done below:

Start-A-E-H-Finish = 0+3+5+11+0 = 19

Start-C-E-H-Finish = 0+2+5+11+0 = 18

Start-B-F-I-Finish = 0+4+1+8+0 = 13

Start-B-F-G-Finish =0+4+1+6+0 = 11

Start-B-D-I-Finish =0+4+2+8+0 = 14

Start-B-D-G-Finish = 0+4+2+6+0 = 12

The critical path emerges as Start-A-E-H-Finish because the path adds up to 19, which is longer than any of the other paths. If any of the activities in this path are delayed, the finish of the project will be delayed.

13. C. The float (or slack) of an activity is the amount of time it can slip without moving the critical path. In this case, we must calculate the float of activity 'D'. If activity 'D' was on the critical path, we would immediately know that the float was 0, but in this case it is not.

    To solve this problem, we must first list out all of the paths. We will use the list from the previous question.

    Start-A-E-H-Finish = 0+3+5+11+0 = 19

    Start-C-E-H-Finish = 0+2+5+11+0 = 18

    Start-B-F-I-Finish = 0+4+1+8+0 = 13

    Start-B-F-G-Finish = 0+4+1+6+0 = 11

    Start-B-D-I-Finish = 0+4+2+8+0 = 14

    Start-B-D-G-Finish = 0+4+2+6+0 = 12

    The next step is to identify the ones that have activity 'D' in them. They are:

    Start-B-D-I-Finish = 0+4+2+8+0 = 14

    Start-B-D-G-Finish = 0+4+2+6+0 = 12

    Now the task is simple. We simply subtract the path sums from the length of the critical path for each (19-14 = 5, and 19-12 = 7), and finally we take the smaller of those two values which is 5. Therefore, the float for activity D is 5.

14. A. The formula for a three-point estimate, also called a PERT estimate, is (Pessimistic + 4x Realistic + Optimistic) + 6. In this example, the terms were switched around slightly, but it equates to (62 + 4x44 + 30) / 6 = 268 / 6 = 44.67.

15. B. What-if analysis can take on many forms, but the form you are most likely to see on the exam is Monte Carlo analysis, which throws a large number of scenarios at the schedule to see what would happen if one or more bad scenarios occurred.

16. D. The activity attributes simply expand on the information for each activity. 'A' is incorrect since the activity attributes tie back to the activity list and not the WBS. 'B' is incorrect since the activity attributes may be created at the same time or after the activity list, but not before. 'C' is incorrect since the activity attributes may never be substituted for the activity list. Instead the activity attributes accompany the activity list, providing additional information on each activity.

17. C. This one may have been difficult for you, because it is a non- traditional definition of the critical path. The critical path is the series of activities, which if delayed, will delay the project. This makes these activities the highest schedule risk on the project. 'A' is incorrect because the critical path has no relationship with functionality 'B' is incorrect because the size of the work packages does not directly correlate to the critical path. 'D' is incorrect because the critical path does not represent the optimal path through the network.

18. D. Another project network diagram; however, this is an activity on arrow diagram. It is solved in the same 3 steps. First, draw the project network diagram. Yours should look similar to the one depicted below:

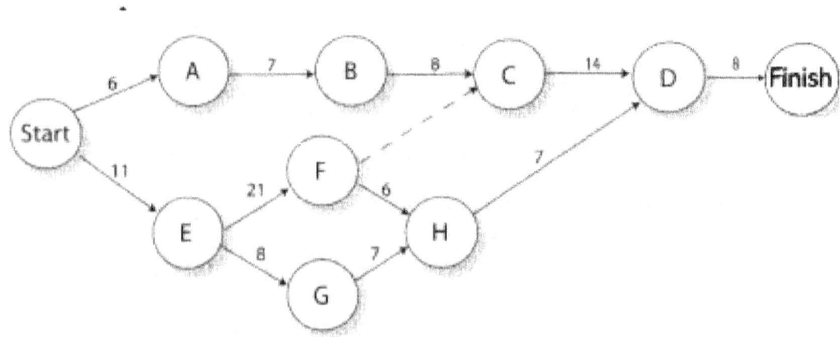

The second step in the process is to list all of the paths through the network. Don't forget that the "dummy activity" of F to C creates a relationship and thus another path through the network. The paths are:

Start-A-B-C-D-Finish

Start-E-F-C-D-Finish

Start-E-F-H-D-Finish

Start-E-G-H-D-Finish

Now add up all of the values associated with the activities on arrow as follows:

Start-A-B-C-D-Finish = 6+7+8+14+8 = 43

Start-E-F-C-D-Finish = 11+21+0+14+8 = 54

Start-E-F-H-D-Finish = 11+21+6+7+8 = 53

Start-E-G-H-D-Finish = 11+8+7+7+8 = 41

19. B. The slack, or float, for activity H - D is simple to calculate after the work we have already done on problem 18. The way to calculate this is to begin with the list we did for the critical path on the previous question.

Start-A-B-C-D-Finish = 6+7+8+14+8 = 43

Start-E-F-C-D-Finish = 11+21+0+14+8 = 54

Start-E-F-H-D-Finish = 11+21+6+7+8 = 53

Start-E-G-H-D-Finish = 11+8+7+7+8 = 41

Now, strike off every path that does not have activity H-D, as follows:

Start-A-B-C-D-Finish = 6+7+8+14+8 = 43

Start-E-F-C-D-Finish = 11+21+0+14+8 = 54

Start-E-F-**H-D-**Finish = 11+21+6+7+8 = 53

Start-E-G-**H-D**-Finish = 11+8+7+7+8 = 41

Finally, we take the critical path and subtract the sum from the remaining paths as follows:

54-53 = 1

54-41 = 13

The smaller of those two values (1) represents the slack, or float for activity H-D. It indicates how long activity H-D may slip before it moves the critical path. In this case, if it moves more than 1 unit, the critical path moves as well, and the project will be delayed.

20. B. Changes happen. Some of them are submitted as change requests, and some of them come out of nowhere. In this case, you would want to return to planning and update the plans. The project will not be delayed, and the resource has given a good reason why the dates need to be revisited (a common occurrence in the real world). 'A' is incorrect, because the plan should reflect reality - not an unrealistic estimate. 'C' is incorrect, because you cannot possibly get an outside opinion every time a resource needs to change a date. 'D' is incorrect, because the resource gave a good reason for the adjustment. It was not that she was lacking in training or ability, but that pieces were left out of planning. Therefore, 'B' represents the all-around best answer.

Cost Management

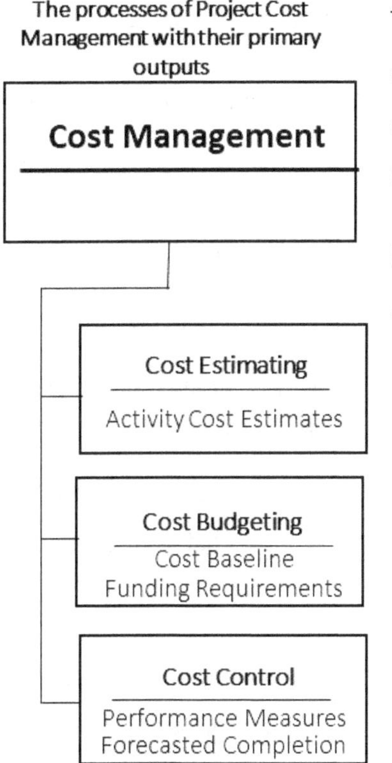

The topic of Cost Management, like time management, has critical formulas that must be learned and understood. This chapter will explain these formulas clearly and provide methods and exercises for quick retention.

Most of the principles and techniques explained here, such as earned value, did not originate with PMI but were derived from long-standing practices in the fields of cost accounting, managerial accounting, and finance.

From the indicators at the top of the page, you can tell that most people find this chapter to be one of the more difficult ones. In order to help you prepare for this topic, this book has clearly broken down the practices and outlined the techniques and formulas needed to ace the questions on the exam.

Philosophy

While the actual tools and techniques behind Cost Management may be different from time management, the driving philosophy has several similarities. Costs should be planned, quantified, and measured. The project manager should tie costs to activities and resources and build the estimates from the bottom up.

It is common practice for high-level budgets to be determined prior to knowing costs. The reason for this is that many companies use fiscal year planning cycles that must be done far in advance of their project planning.

Budgetary constraints are a fact of life, but instead of blindly accepting whatever budget is specified by management, the project manager carefully reviews the scope of work and the duration estimates and then reconciles them to the scope and projected costs.

Adjustments to the project scope, the budget, or the schedule are much easier to justify by working up from a detailed level instead of from the top down. Although summary budgets are often the first thing created in the real world, when it comes to detailed planning, the overall approach advocated here is scope first, schedule second, and budget third.

Throughout this book, you will see that estimates should be built from the bottom up. At the point in the process where budgets are created, you should have a well-defined work breakdown structure, an activity list with resource and duration estimates for each activity, and a schedule. Now budgeting becomes a task of applying rates against those resources and activities to create activity cost estimates, a cost baseline, and a cost management plan.

It is the project manager's job to constantly monitor and control cost against time, scope, quality and risk to ensure that all projections remain realistic and clearly defined.

Importance

The topic of Cost Management is of high importance for the exam, both in the understanding of PMI's processes and in the application of key formulas, which play a part here as well.

Preparation

Learning the 12 key formulas for Cost Management is a must. Learning to apply them is equally as important. The good news is that none of the formulas are overly difficult, and there are plenty of explanations and examples in this book to help cement the concepts.

As mentioned briefly above, memorization is important; however, understanding the formulas is even more important. Once you grasp the formulas and concepts, the memorization will be a snap. In fact, some people studying for the exam only memorize the concepts and reconstruct the formulas as needed. This is possible because each formula does make sense, so read and reread this chapter until they are clear to you.

Important Concepts

Process Group	Cost Management Process
Initiating	(none)
Planning	Cost Estimating, Cost Budgeting
Executing	(none)
Controlling	Cost Control
Closing	(none)

In the knowledge area of Cost Management, it is also essential that you know the main outputs that are produced during each process. The different tasks that are performed in each process are summarized in the chart that follows.

Process	Key Output (s)
Cost Estimating	Activity cost estimates
Cost Budgeting	Cost baseline, Funding requirements
Cost Control	Performance measurements, Forecasted completion, Recommended corrective actions

LIFE-CYCLE COSTING

Instead of simply asking "How much will this product cost to develop?" life-cycle costing looks at the total cost of ownership from purchase or creation, through operations, and finally to disposal. It is a practice that encourages making decisions based on the bigger picture of ownership costs. For instance, it may be less expensive for the project to use generic computer servers to develop a software product; however, if the organization will have to maintain those servers, and if that organization has expertise and service contracts with IBM, then the project has made a short-sighted decision that will have adverse effects downstream.

VALUE ENGINEERING

Value engineering is the practice of trying to get more out of the project in every possible way. It tries to increase the bottom line, decrease costs, improve quality, shorten the schedule, and generally squeeze more benefit and value out of each aspect of the project. The key to value engineering is that the scope of work is not reduced by these other efforts.

Cost Estimating

WHAT IT IS:

One of the biggest improvements evident in this edition of the PMBOK Guide is that most of the processes are intuitively named, making it that much easier to understand them. Cost Estimating is an example of that.

Its name indicates exactly what it does. In Cost Estimating, each activity is analyzed to evaluate the activity time estimates and the resource estimates associated with them, and a cost estimate is produced.

WHY IT IS IMPORTANT:

If you've heard the old saying that the devil is in the details, then you know why Cost Estimating is important. In this process, you gain a detailed understanding of the costs involved in performing a project.

WHEN IT IS PERFORMED:

It may be misleading to suggest that Cost Estimating is only performed once. This process, like many others, may be performed over and again throughout the project; however, there are a few essential predecessor processes that must be completed before it can be performed adequately.

It is normal to perform Activity Definition, Activity Resource Estimating, and Activity Duration Estimating before estimating the costs.

The reason for this is that costs are estimated against schedule activities. It is also helpful to know the number of resources and how long they are expected to work on the project.

The order of certain processes from Scope Definition through Cost Estimating

HOW IT WORKS:

Cost estimates, prepared for each activity, are thought of in terms of their accuracy. In other words, how much leeway are you giving yourself with your estimating?

When it comes to estimates that result from this process, there are many options. Consider the table that follows:

Estimate type	Range
Order of Magnitude Estimate	-50% to +100%
Conceptual Estimate	-30% to +50%
Preliminary Estimate	-20% to +30%
Definitive Estimate	-15% to +20%
Control Estimate	-10% to +15%

These five different types of estimating are used, depending on the need. For instance, in the initiation of a project, an order of magnitude estimate may suffice, while later in the project, a definitive estimate may be in order. For activities with relatively few unknowns, a control estimate may be appropriate.

Typically, the closer in time you get to actually spending money for an activity, the more precise you want that activity's estimate to be.

- Inputs

  - *Enterprise Environmental Factors - See Chapter 2, Common Inputs*

  - *Organizational Process AssetsSee Chapter 2, Common Inputs*

  - *Project Scope Statement*

The project scope statement, created earlier in Scope Definition, defines the scope and ties each element of the scope back to the underlying need it was designed to address. Also, the project scope statement provides information on constraints and assumptions related to the scope, and these can dramatically affect the cost estimates.

All of this information found in the project scope statement should be carefully considered as the cost estimates are being created.

  - *Work Breakdown Structure*

The work breakdown structure contains everything in scope and the how these deliverables are organized. Costs are mapped back to work packages on the WBS.

  - *WBS Dictionary*

Any time you see the WBS, you are likely to see the WBS dictionary. It provides additional attributes and expanded information about the WBS that may prove helpful.

  - *Project Management Plan*

The project management plan provides the plan for how the project will be executed and controlled. Specifically, the schedule, cost, and staffing management plans, which are all part of the project plan, should be considered. Also, the risk register should be carefully considered since risks and cost estimates are typically tightly linked.

- Tools

    - *Analogous Estimating*

The tool of analogous estimating uses the actual results of projects that have been performed by your organization as the estimates for your activities. Analogous estimates are typically easier to use, and their accuracy depends on how similar the two projects actually are.

    - *Determine Resource Cost Rates*

The resource cost rates are applied to the activity resource estimates, determined earlier. For instance, if it were determined that a senior Java programmer would be needed for 120 hours on an activity, then the hourly rate for that resource would be needed in order to know the full cost of those 120 hours. These resource cost rates may be actual or estimated.

    - *Bottom-Up Estimating*

The technique of bottom-up estimating produces a separate estimate for each schedule activity. These individual estimates are then aggregated up to summary nodes on the WBS.

Bottom-up estimating is considered to be highly accurate; however, it can also be time consuming and labor-intensive.

    - *Parametric Estimating*

Parametric estimating is a tool often used on projects with a high degree of historical information, and it works best for linear, scalable projects. For instance, if knew that it cost $4,000,000 to build a mile of roadway, then you could estimate that it would cost $32,000,000 to build 8 miles of road.

- *Project Management Software*

The tool of project management software is most useful in facilitating the other tools and techniques, performing the routine calculations, and organizing and storing the large amounts of information used to build the cost estimates.

- *Vendor Bid Analysis*

Bids should be analyzed, and most specifically the winning bid, if for no other reason than to improve the project team's understanding of costs.

- *Reserve Analysis*

It is normal for the cost estimates that will be produced as a result of this process to include reserve amounts, also called contingency. This is simply a buffer against slippage on the project.

The reserve amounts should be analyzed as part of the Cost Estimating process simply to ensure that the amount of reserve being planned properly reflects the risk associated with the project.

- *Cost of Quality*

The technique of evaluating the cost of quality, often abbreviated as COQ, looks at all of the costs that will be realized in order to achieve quality. This tool is also used in the Quality Planning process. The costs of items that are not conformant to quality standards are known as "cost of poor quality," often abbreviated as COpQ.

- **Outputs**

  - *Activity Cost Estimates*

The activity cost estimates are the primary output of this process. These estimates address how much it would cost to complete each schedule activity on the project.

- *Activity Cost Estimate Supporting Detail*

When it comes to the PMBOK Guide, you can never have too much supporting detail. Here, especially, it is important to include enough information on how you derived the activity cost estimates.

- *Requested Changes*

When costs are estimated, it is normal that change requests to the project scope or schedule would be made. For instance, if the cost estimates came in higher than the allowable cost constraint, then the project team might elect to review elements of the scope to determine whether there were any non-essential elements that could be cut in order to get costs back on track.

- *Cost Management Plan Updates*

The cost management plan (discussed in the Integration chapter), details how the project costs will be managed and how change or requested changes to the project costs will be managed. As the activity cost estimates are created, it is normal that this plan would be updated, sometimes significantly.

Cost Budgeting

WHAT IT IS:

In order to understand this process, it is necessary to understand what a budget is. A budget, also known as the cost baseline, takes the estimated project expenditures and maps them back to dates on the calendar. In other words, the Cost Budgeting process time phases the costs so that the performing organization will know how to plan for cash flow and likely expenditures.

WHY IT IS IMPORTANT:

A good cost baseline will help the organization plan their expenditures appropriately and will prevent the organization from tying up too much money throughout the life of the project. For instance, a construction project may have relatively low costs early on, but these costs may rise dramatically in the construction phase. The cost baseline will reflect this, helping the organization to plan accordingly.

Although a high-level budget, similar to a cost constraint, may be determined early in the project, this cost baseline describes a detailed budget that shows costs and timelines for each work package or activity.

WHEN IT IS PERFORMED:

Because the budget typically maps back to schedule activities, it should be performed after Activity Definition, Activity Duration Estimating, and Activity Resource Estimating have been performed. Additionally, because it is time-phased, it should be performed after Schedule Development, since that is where the project's timeline is determined, and since it is based on the activity costs estimates, it should be performed after the process of Cost Estimating.

HOW IT WORKS:

- Inputs

  - *Project Scope Statement*

The project scope statement contains information on why the scope was set where it was, what its limits are, and what other scope-related constraints exist. For instance, certain elements of scope may be non- negotiable since they are required by contract, while other requirements may be easily changed. As the project's budget is being created, the scope should be carefully considered.

  - *Work Breakdown Structure*

Most often, the budget is not only mapped back to the schedule (that is, time-phased), but costs are also tied back to a node on the work breakdown structure.

  - *WBS Dictionary*

The WBS dictionary is almost inseparable from the WBS. It provides expanded attributes, information, and details on the work packages.

  - *Activity Cost Estimates*

The activity cost estimates are a primary input into this process. They provide details on what each schedule activity is estimated to cost.

Because every schedule activity maps back to a single work package, these activity cost estimates are added together to get the cost for their parent work package.

- *Activity Cost Estimate Supporting Detail*

As in other areas, you can never have too much supporting detail. The supporting detail for the activity cost estimates describes how you derived the cost estimates you are using here.

- *Project Schedule*

The project's budget is time-phased, meaning that it shows what costs will be incurred and when they will be incurred. The schedule helps tie these costs back to periods of time for planning purposes.

- *Resource Calendars*

Resource calendars show when resources are going to be utilized and when they will be available. This, along with the project schedule, will help plan for when costs will be incurred.

- *Contract*

The contract may provide information on what costs the project is contractually obliged to incur. For instance, the contract may specify that only a specific brand of computer server may be used in a data center, or that the project is obligated to use at least five of these servers. Any contractual information that affects the cost or expenditures should be factored into the Cost Budgeting process.

- *Cost Management Plan*

The cost management plan is created during Develop Project Management Plan, and it describes how the cost baseline will be created and managed and how changes to the cost baseline will be managed.

- Tools

  - *Cost Aggregation*

Even though costs are estimated at an activity level, these cost estimates should be aggregated to the work package level where they will be measured, managed, and controlled during the project.

Individual cost estimates aggregating back to a single work package

  - *Reserve Analysis*

The tool of reserve analysis is related to risk. Almost all projects maintain a financial reserve to protect them against cost overrun. How much they keep, and how they track it varies from project to project. These buffers go by various names such as management reserve and contingency reserve.

For instance, one project may decide to pad each activity cost estimate by 10% across the board, while another project may only pad the activities that are considered to be at highest risk for cost overrun. Still another project may add a total cost buffer of 20% as a lump sum to the entire project cost baseline. There is no one prescribed way to perform reserve analysis; however, the reserve amount should be in keeping with the risk levels and tolerances on the project.

- ***Parametric Estimating***

Parametric estimating typically uses simple formulas to estimate costs. It may be used for an individual activity, as in Cost Estimating, or for the entire cost baseline, as is the case here. An example of parametric estimating would be basing the cost of the development of a software application on the estimated number of lines of code or screens involved or the cost of an airport runway based on its planned length.

Parametric estimates work best when the project being undertaken is highly similar to previous projects and there is significant historical information available within your organization or industry.

- ***Funding Limit Reconciliation***

Because companies typically operate on fiscal years, they are required to budget for projects long before the actual scope is known. Because of that, it is normal for a project to receive a funding limit or constraint as the project is begun.

It is important for the project to reconcile planned spending with these funding limits. For instance, the organization may specify that they will only be able to provide $200,000 in the first month of a project, but $450,000 in the second month. The project's cost baseline ultimately needs to be compatible with these limitations.

- Outputs

    - *Cost Baseline*

The cost baseline is another term for the project's budget, and you should expect to see questions related to it on the exam. It not only specifies what costs will be incurred, but when they will be incurred. Larger projects may be divided into multiple cost baselines. For instance, one cost baseline may track domestic labor costs, while a second cost baseline is used to track international labor costs.

- *Project Funding Requirements*

It would be impractical for most projects to petition management for authorization on each individual cost, so instead the project determines funding requirements. The cost baseline is used to determine the project's funding requirements.

The funding requirements are almost always related to the planned expenditures, but they are not identical to them. For instance, a project may request $20,000 per month throughout the life of the project, or they may require a larger portion early on in order to purchase equipment or incur other fixed costs. The funding requirements should also include any planned contingency or reserve funds, since these must be available to the project as soon as they are needed.

- *Cost Management Plan Updates*

Often times performing the process of Cost Budgeting results in changes to the project's scope, schedule, or costs. If any of these changes affect the way in which costs are managed (or the way change requests to the costs are managed) then the cost management plan should be updated to reflect this.

- *Requested Changes*

As with most processes, performing this process may result in changes that should be channeled back through the appropriate change control process.

Cost Control

WHAT IT IS:

Cost Control, in many ways, is the quintessential monitoring and controlling process. There are two important things to keep in mind about controlling processes:

1. They are proactive. They do not merely wait for changes to occur. Instead, they try to influence the factors that lead to change.

2. Controlling processes measure what was executed against what was planned. If the results of what was executed do not match the cost baseline, then appropriate steps are taken to bring the two back in line. This could either mean changing future plans or changing the way the work is being performed.

Cost Control is primarily concerned with cost variance. In project management, cost variances are described as either being positive (good) or negative (bad). It is important to understand that even a positive cost variance needs to be understood and the plan must be adjusted. Accurate planning is paramount.

WHY IT IS IMPORTANT:

Cost Control is an essential process for ensuring that costs are carefully monitored and controlled. It ensures that the costs stay on track and that change is detected whenever it occurs.

WHEN IT IS PERFORMED:

Cost Control is not a process that is performed only once. Instead, it is performed regularly throughout the project, typically beginning as soon as project costs are incurred. The activities associated with Cost Control are usually performed with more frequency as project costs increase. For instance, many projects will perform Cost Control monthly during planning phases and weekly (or even more frequently) during construction phases, where costs typically peak.

HOW IT WORKS:

- Inputs
  - *Cost Baseline*

The cost baseline is a primary input into the process of Cost Control. A baseline is the original plan plus all approved changes. The cost baseline (also known as the budget) shows what costs are projected and when they are projected to be incurred. The cost baseline is the plan against which the actual costs are measured.

- *Project Funding Requirements*

Like the preceding input of the cost baseline, the project funding requirements are also part of the plan against which the actual funding is measured. In this case, positive or negative variances from the planned funding requirements will be evaluated so that corrective action may be taken if necessary.

- *Performance Reports*

The performance reports provide a summary of how costs are progressing against the plan for work completed. Performance reports are covered in greater detail in Chapter 10 - Communication Management.

- *Work Performance Information - See Chapter 2, Common Inputs*

- *Approved Change Requests*

Any change requests that are approved should be evaluated for their impact on the cost baseline.

- *Project Management Plan*

The project management plan contains the cost management plan, which guides the Cost Control process. The cost management plan tells how costs will be managed and how changes or variances to the costs will be managed.

- Tools

  - *Cost Change Control System*

The cost change control system describes how the cost baseline may be changed. Like any other system described here, it can include people, departments, systems, policies, forms, etc.

- *Performance Measurement Analysis*

There are numerous performance measurement techniques used within project management, most of them originating in the field of cost accounting. These fall under the heading of Earned Value, discussed later in this chapter, and include Earned Value (EV),

Planned Value (PV), Actual Cost (AC), Estimate to Complete (ETC), Estimate At Complete (EAC), Cost Variance (CV), Schedule Variance (SV), Cost Performance Index (CPI), Schedule Performance Index (SPI), and the Cumulative CPI (CPI^C). Each of these formulas and concepts are covered in detail later in this chapter.

- *Forecasting*

The technique of forecasting uses current and previous cost information to predict future costs. This focuses on the concepts of Estimate At Completion (EAC) and Estimate To Completion (ETC), covered later in this chapter under the heading of Earned Value.

- *Project Performance Reviews*

Periodic reviews of the project performance are meetings held to measure actual performance against the plan. As part of Cost Control, particular attention is paid to cost performance.

- *Project Management Software*

Because the calculations involved in Cost Control (especially the earned value calculations) can be tedious and complex, project management software is typically used to calculate actual values and assist with "what if" analysis.

- *Variance Management*

How you will handle variances is determined in the cost management plan (part of the overall project management plan). Keep in mind that positive variance is considered to be good, while negative variance is bad, but even positive variances need to be managed, and the cost baseline should be adjusted to accurately reflect anticipated costs.

- Outputs

- *Cost Estimate Updates*

As previous cost estimate changes are detected and understood, it may be appropriate to update future cost estimates. For instance, if a contractor has been producing every deliverable at half the estimated cost, and there is reason to believe that will continue, then the project manager may decide to adjust future cost estimates to reflect this performance increase.

- *Cost Baseline Updates*

It is a common misconception that baselines may not be updated, but they can. Once a change becomes approved, it becomes part of the baseline. You should understand that if a change was not approved, it does not become part of the baseline.

- *Performance Measurements*

Performance measurements show how the project is performing against the plan. For the process of Cost Control, the performance measurements of CV, SV, CPI, SPI, and CPIC are especially applicable, since they help show variances and trends on how the project is performing against the plan.

- *Forecasted Completion*

The two values that relate most closely to completion are EAC and ETC. These numbers are used to help forecast a likely completion date for the project.

- *Requested Changes*

As Cost Control is performed, requested changes are a normal output. For instance, if the process of Cost Control showed that the project was going to cost significantly less or more than the cost baseline, then certain changes would likely result to bring the project back in line. These changes could take the form of reducing the scope, increasing the budget, or changing factors related to execution.

- *Recommended Corrective Actions*

Corrective action is anything done to bring future results in line with the plan. For the process of Cost Control, if the actual costs did not line up with the cost baseline and the activity cost estimates, then corrective action could mean changing the plan or changing something with the execution (e.g. training or changing vendors) to help control costs.

- *Organizational Process Assets Updates*

Every mistake is potentially an asset if it is documented in the forms of lessons learned, but lessons learned can document more than mistakes. Organizational process assets could also include things done well that resulted in costs falling into line.

- *Project Management Plan Updates*

Anything that changes the project plan, such as changes to the cost baseline, changes to the cost management plan, or changes to any other parts of the project management plan, should be updated back into the project management plan.

Special Focus: Earned Value

In a double entry accounting system, for every debit to one account, there is a corresponding credit to another account. Earned value is similar in that if you spend a dollar on labor for your project, that dollar doesn't just evaporate into thin air. You are "earning" a dollar's value back into your project. If you buy bricks or computers, write code or documentation, or perform any work on the project, those activities earn value back into your project. There are 12 key formulas associated with earned value management that often appear on the test, and they require both memorization and understanding. Following is a chart that presents a summary of the key terms used in earned value calculations. Note that for Planned Value, Earned Value, and Actual Cost, there are older, equivalent terms that still show up on the exam. These older terms and their associated abbreviations are shown along with the current terms on the chart on the following page, and you must be able to recognize and apply either one on the exam.

Term	Abbreviation	Description	Formula
Budgeted At Completion	BAC	How much was originally planned for this project to cost.	No one formula exists. BAC is derived by looking at the total budgeted cost for the project.
Planned Value (also known as Budgeted Cost of Work Scheduled)	PV (or BCWS)	How much work should have been completed at a point in time based on the plan. (Derived by measuring where you had planned to be in terms of work completed at a point in the schedule).	Planned % Complete X BAC
Earned Value (also known as Budgeted Cost of Work Performed)	EV (or BCWP)	How much work was actually completed at a point in time. Derived by measuring where you actually are in terms of work completed at a point in the schedule.	Actual % Complete X BAC
Actual Cost (also known as Actual Cost of Work Performed)	AC (or ACWP)	The money spent at a point in time.	Sum of all costs for the given period of time.
Cost Variance	CV	The difference between what we expected to spend and what was actually spent at a point in time.	EV-AC
Schedule Variance	SV	The difference between where we planned to be in the schedule and where we are in the schedule.	EV-PV
Cost Performance Index	CPI	The rate at which the project performance is meeting cost expectations.	EV ÷ AC
Cumulative CPI	CPI^C	Forecasts project costs at completion.	$CPI^C = EV^C/AC^C$
Schedule Performance Index	SPI	The rate at which the project performance is meeting schedule expectations.	EV ÷ PV
Estimate At Completion	EAC	Projecting the total cost at completion based on project performance at a point in time.	BAC ÷ CPI (Note that there are over 25 ways to calculate the EAC, but this one should be sufficient for the exam.)
Estimate To Completion	ETC	Projecting how much will be spent on the project, based on past performance.	EAC - AC
Variance At Completion	VAC	The difference between what was budgeted and what will actually be spent.	BAC - EAC

EVM EXAMPLE

Consider the following example:

You are the project manager for the construction of 20 miles of sidewalk. According to your plan, the cost of construction will be $15,000 per mile and will take 8 weeks to complete. 2 weeks into the project, you have spent $55,000 and completed 4 miles of sidewalk, and you want to report performance and determine how much time and cost remain.

Below, we will walk through each calculation to show how we arrive at the correct answers.

BUDGETED AT COMPLETION

In the approach outlined by this book, we will always begin by calculating BAC. Budgeted at completion simply means, "how much we originally expected this project to cost". It is typically very easy to calculate. In our example, we take 20 miles of sidewalk * $15,000/mile. That equates to a BAC of $300,000.

BAC = $300,000

PLANNED VALUE

The planned value is how much work was planned for this point in time. The value is expressed in dollars.

Planned Value = Planned % complete * BAC

We do this by taking the BAC ($300,000) and multiplying it by our % complete. In this case, we are 2 weeks complete on an 8 week schedule, which equates to 25%. $300,000 * .25 = $75,000. Therefore, we had planned to spend $75,000 after two weeks.

PV = $75,000

EARNED VALUE

If you have been intimidated by the concept of earned value, relax. Earned value is based on the assumption that as you complete work on the project, you are adding value to the project. Therefore, it is simply a matter of calculating how much value you have "earned" on the project.

Planned value is what was planned, but earned value is what actually happened.

EV = Actual % Complete * BAC

In this case, we have completed 4 miles of the 20 mile project, which equates to 20%. We multiply that percentage by the BAC to get EV. It is $300,000 *20%= $60,000. This tells us that we have completed $60,000 worth of work, or more accurately, we have earned $60,000 of value for the project.

EV = $60,000

ACTUAL COST

Building on the above illustration, we will calculate our actual costs. Actual cost is the amount of cost you have incurred at this point, and we are told in the example that we have spent $55,000 to date. In this example, no calculation is needed.

AC = Actual Cost

AC = $55,000

COST VARIANCE

Cost variance (CV) is how much actual costs differ from planned costs. We derive this by calculating the difference between EV and AC. In this example, it is EV of $60,000 - AC of $55,000. A positive variance (as in this case), reflects that the project is doing better on cost than expected.

For those who are curious, the reason we use EV in this formula instead of PV is that we are calculating how much the actual costs have varied. If we used PV, it would give us the variance from our plan, but the cost variance measures actual cost variance, and EV is based on actual performance, whereas PV is based on planned performance.

A positive CV is a good thing. It indicates that we are doing better on costs than we had planned. Conversely, a negative CV indicates that costs are running higher than planned.

CV = EV-AC

CV = $5,000

SCHEDULE VARIANCE

Schedule variance (SV) is how much our schedule differs from our plan. Where people often get confused here is that this concept is expressed in dollars. SV is derived by calculating the difference between EV and PV. In this example, the schedule variance is EV of $60,000 - PV of $75,000.

A negative variance (as in this case) reflects that we are not performing as well as we had hoped in terms of schedule. A positive SV would indicate that the project is ahead of schedule.

SV = EV-PV

SV = -$15,000

COST PERFORMANCE LNDEX

The cost performance index gives us an indicator as to how much we are getting for every dollar. It is derived by dividing Earned Value by the Actual Cost. In this example, Earned Value = $60,000, and our Actual Cost = $55,000. $60,000 / $55,000 = 1.09.

This figure tells us that we are getting $1.09 worth of performance for every $1.00 we expected. A CPI of 1 indicates that the project is exactly on track. A closer look at the formula reveals that values of 1 or greater are good, and values less than 1 are undesirable.

CPI = EV / AC

CPI = 1.09

SCHEDULE PERFORMANCE LNDEX

A corollary to the cost performance index is the schedule performance index, or SPI. The schedule performance index tells us how fast the project is progressing compared to the project plan. It is derived by dividing earned value by the planned value. In this example, earned value= $60,000, and our planned value = $75,000. $60,000 / $75,000 = 0.8. This tells us that the project is progressing at 80% of the pace that we expected it to, and when we look at the example, this conclusion makes sense.

We had expected to lay 20 miles of sidewalk in 8 weeks. At that rate, after 2 weeks, we should have constructed 5 miles, but instead the example tells us that we had only constructed 4 miles. That equates to 4/5 performance, which is 80%. Like the cost performance index, values of 1 or greater are good, and values that are less than 1 are undesirable.

SPI = EV+ PV

SPI = 0.8

A common way for the cost performance and schedule performance index to be used is to track them over time. This is often displayed in the form of a graph as illustrated below. This graph may be easily interpreted if you consider that a value of 1 indicates that the index is exactly on plan.

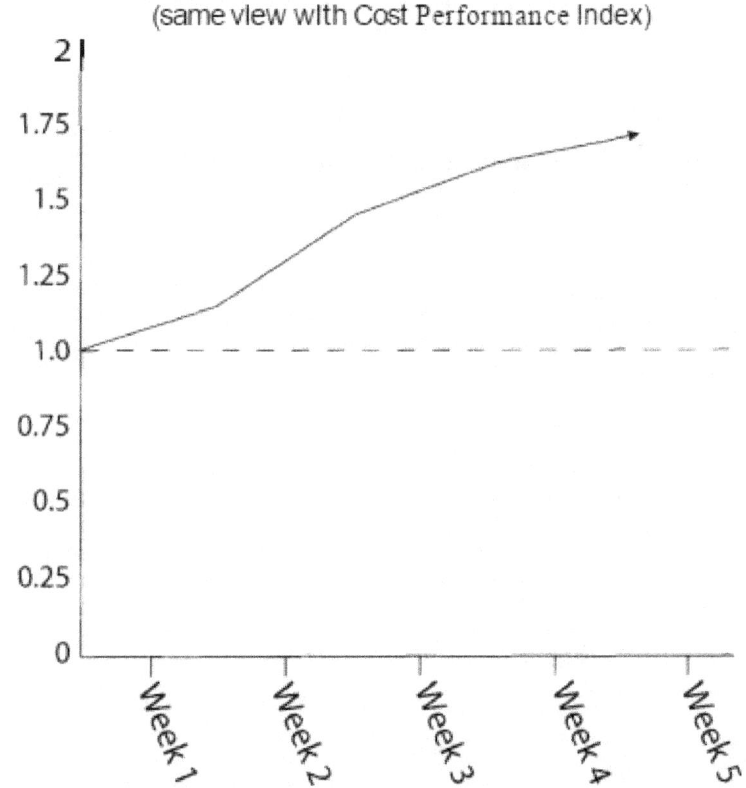

ESTIMATE AT COMPLETION

Estimate at completion is the amount we expect the project to cost, based on where we are relative to cost and schedule. If that sounds confusing, think of it this way. If you know you are half way through the project, and you are currently 20% over budget, then the estimate at completion factors that variance out to the end of the project. There are many ways to calculate EAC; for the exam, the most straightforward way to calculate it is to take the BAC and divide it by our cost performance index. In this example, we expected to spend (BAC) $300,000 and our CPI is 1.09.

$300,000 / 1.09 = $275,229.36. This should make sense. We are doing better on costs than we had originally planned, and this value reflects that.

EAC = BAC / CPI

EAC = $275,229.36

ESTIMATE TO COMPLETION

Estimate to completion is simply how much more we expect to spend from this point forward based on what we've done so far. It can be easily backed into by taking our estimate at complete (what we expect to spend) and subtracting what we have spent so far (Actual Cost). Given the numbers above, it would be EAC of $275,229.36 - AC of $55,000 = $220,229.36. This tells us that we expect to spend $220,229.36 more, given our performance thus far.

ETC = EAC - AC

ETC = $220,229.36

VARIANCE AT COMPLETION

Variance at completion is the difference between what we originally budgeted and what we expect to spend. A positive variance indicates that we are doing better than projected, and a negative variance indicates that we expect the project to run over on costs.

In this example, our BAC was $300,000; however, our EAC is now $275,229.36.

$300,000 - $275,229.36 = $24,770.64.

VAC = BAC – EAC

VAC = $24,770.64

CUMULATIVE CPI (CPI^C)

Once you understand the concept of earned value, the cumulative cost performance index (also expressed as CPI^C) is not as intimidating as it may first look. Recall that the regular CPI is simply EV/AC. The CPI^C is simply all of the earned value calculations added together and divided by all of the actual cost calculations added together.

For example, if a company took earned value measurements at monthly intervals for the past three months and they were:

	EV	AC
Month 1	$22,000	13,700
Month 2	$51,000	37,900
Month 3	$107,000	98,400

Given the preceding table, the CPI^C would be calculated by adding up all of the earned value figures ($180,000) and dividing by the sum of the actual costs ($150,000). This yields a cumulative CPI of 1.2.

This number is useful because the regular CPI is only a snapshot of your earned value at a point in time, but the cumulative CPI can show a number that factors in previous trends, and the cumulative CPI has been shown to be a good predictor of performance at completion, even when used very early in the project. Because this is a relatively new topic, it is highly likely that you will see questions related to the cumulative CPI on the exam.

Types of Cost

Several types of questions regarding cost may appear on the exam. It is important to understand the difference between the different types of cost presented below:

Cost type	Explanation
Fixed	Costs that stay the same throughout the life of a project. An example is a piece of heavy equipment, such as a bulldozer.
Variable	Costs that vary on a project. Examples are hourly labor, and fuel for the bulldozer.
Direct	Expenses that are billed directly to the project. An example is the materials used to construct a building.
Indirect	Costs that are shared and allocated among several or all projects. An example could be a manager's salary. His people might be direct costs on a project, but his salary is overhead and would be considered an indirect cost.
Sunk	Costs that have been invested into or expended upon the project. Sunk costs are like spilt milk. If they are unrecoverable, they are to be treated as if they are irrelevant! This is difficult for many people to understand, but the statement "we've spent over 10 million

	dollars on this project, and we're not turning back now" is not good decision-making if the costs are unrecoverable, or "sunk."

Exercises

EXAMPLE 1

You are constructing 6 additional rooms on an office building. Each of the six rooms is identical, and the projected cost for the project is $100,000 and is expected to take 5 weeks.

At the end of the 2nd week, you have spent $17,500 per room and have finished 2 rooms; you are ready to begin on the 3rd

1. Based on the information provided in the example above, fill in the values for the following table:

	Value
Budgeted At Completion	
Planned Value	
Earned Value	
Actual Cost	
Cost Variance	
Schedule Variance	
Cost Performance Index	
Schedule Performance Index	

Estimated At Completion	
Estimated To Completion	
Variance At Completion	

2. Is the project ahead of or behind schedule?

3. Is the project going to be completed over or under budget?

EXAMPLE 2

Here is another example to test your understanding of these concepts.

You have planned for a project to write a software application to take 1 year. The costs to this project are budgeted at $12,500 per month. Six months into the project, you find that the software application is 50% completed, and you have spent $70,000.

4. Based on the information provided, fill in the values for the following table:

	Value
Budgeted At Completion	
Planned Value	
Earned Value	
Actual Cost	
Cost Variance	
Schedule Variance	
Cost Performance Index	

Schedule Performance Index	
Estimated At Completion	
Estimated To Completion	
Variance At Completion	

5. Is the project ahead of or behind schedule?

6. Is the project going to be completed over or under budget?

7. Mark one value in each column that shows the most desirable value given the information provided. (Note that some of these attributes are covered in Chapter 4 - Project Scope Management)

IRR	SPI	CPI	NPV	Payback Period	BCR	ROI
22%	1	.5	$25,000	16 mos	2	9%
0%	0	1	$95,000	2 yrs	1.5	12%
12%	.8	1.2	$50,000	16 wks	1	-2%
-3%	1.2	1.15	$71,000	25 mos	.2	3%

8. Project X was projected to take four months and cost $70,000 per month. At the end of month one, the project was 20% complete, and had spent $89,000. At the end of month two, it was 40% complete and had spent $151,000. What is the cumulative CPI for Project X at the end of month two?

Answers to Exercises

1. Your answers should look like these:

	Value
Budgeted At Completion	$100,000
Planned Value	$40,000.00
Earned Value	$33,333.33
Actual Cost	$35,000.00
Cost Variance	-$1,666,67
Schedule Variance	-$6,666.67
Cost Performance Index	0.95
Schedule Performance Index	0.83
Estimated At Completion	$105,263.15
Estimated To Completion	$70,263.15
Variance At Completion	-$5,263.15

BAC = $100,000.00

PV = 2 weeks / 5 weeks = 40% complete * 100,000 = $40,000

EV = 2 rooms / 6 rooms = 33.3% complete * 100,000 = $33,333.33

AC = $17,500 per rooms * 2 rooms = $35,000

CV = (EV) $33,333.33- (AC) $35,000.00 = -$1,666.67

SV = (EV) $33,333.33 - (PV) $40,000.00 = -$6,666.67

CPI = (EV) $33,333.33 / (AC) $35,000.00 = 0.95

SPI = (EV) $33,333.33 / (PV) $40,000.00 = 0.83

EAC = (BAC) $100,000.00 / (CPI) 0.95 = $105,263.15

ETC = (EAC) $105,263.15- (AC) $35,000.00 = $70,263.15

VAC = (BAC) $100,000.00- (EAC) $105,263.15 = -$5,263.15

2. Is the project ahead of or behind schedule?

   The project is behind schedule. The easiest way to determine this is by looking at the SPI. Since it is less than 1, we can determine that the project is not doing well in terms of the schedule.

3. Is the project going to be completed over or under budget?

   There are two ways to see that the project is going to run over budget. First, the CPI is less than 1. Second, the VAC is negative.

4. Check your answers against these:

	Value
Budgeted At Completion	$150,000.00
Planned Value	$75,000.00
Earned Value	$75,000.00
Actual Cost	$70,000.00
Cost Variance	$5,000.00
Schedule Variance	$0
Cost Performance Index	1.07

Schedule Performance Index	1
Estimated At Completion	$140,186.91
Estimated To Completion	$70,186.91
Variance At Completion	$9,813.09

BAC = $150,000.00

PV = 6 months / 12 months = 50% complete * 150,000 = $75,000

EV = Project is 50% complete * 150,000.00 = $75,000.00

AC = $70,000.00

CV = (EV) $75,000.00 - (AC) $70,000.00 = $5,000.00

SV = (EV) $75,000.00 - (PV) $75,000.00 = $0

CPI = (EV) $75,000.00 / (AC) $70,000.00 = 1.07

SPI = (EV) $75,000.00 / (PV) $75,000.00 = 1

EAC = (BAC) $150,000.00 / (CPI) 1.07 = $140,186.91

ETC = (EAC) $140,186.91- (AC) $70,000.00 = $70,186.91

VAC = (BAC) $150,000.00 - (EAC) $140,186.91 = $9,813.09

5. Is the project ahead of or behind schedule?

   This would be classified as a "trick" question, as neither answer is correct. Since the SPI is 1, we can see that the project is exactly on schedule.

6. Is the project going to be completed over or under budget?

The project is projected to finish ahead of budget, due to the cost performance index being greater than 1.

7. The most desirable project attributes for each column are shaded in the chart below. Note that some of these formulas came from Chapter 4, Scope Management:

IRR	SPI	CPI	NPV	Payback Period	BCR	ROI
22%	1	.5	$25,000	16 mos	2	9%
0%	0	1	$95,000	2 yrs	1.5	12%
12%	.8	1.2	$50,000	16 wks	1	-2%
-3%	1.2	1.15	$71,000	25 mos	.2	3%

Did you notice that for most of these measurements, the bigger value is the best one? That is true for all except for the payback period, where you want the shortest time to recoup project costs.

8. If you got this one, give yourself a big pat on the back! In order to answer it, you first had to calculate the earned value for months one and two.

Month 1 EV = 0.2 * $280,000 = $56,000

Month 1 AC = $89,000

Month 2 EV = 0.4 * $280,000 = $112,000

Month 1 AC = $151,000

After getting these values, the CPI^C is a snap. It is simply the sum of the earned value numbers divided by the sum of the actual costs. This yields $168,000 / $240,000 = a cumulative CPI of 0.7.

Cost Management Questions

Note: Some of these questions are based on material covered in chapter 2 as well as this chapter.

1. Your schedule projected that you would reach 50% completion today on a road construction project that is paving 32 miles of new highway. Every 4 miles is scheduled to cost $5,000,000. Today, in your status meeting, you announced that you had completed 20 miles of the highway at a cost of $18,000,000. What is your Planned Value?

   A. $12,800,000.

   B. $18,000,000.

   C. $20,000,000.

   D. $40,000,000.

2. If the CPI is 0.1, this indicates:

   A. The project is performing extremely poorly on cost.

   B. The project is costing 10% over what was expected.

   C. The project is only costing 90% of what was expected.

   D. The project is performing extremely well on cost.

3. Activity cost estimates are used as an input into which process?

   A. Cost Estimating.

B. Cost Budgeting.

C. Cost Analysis.

D. Cost Control.

4. Based on the following Benefit Cost Ratios, which project would be the best one to select?

A. BCR = -1.

B. BCR = 0.

C. BCR = 1.

D. BCR = 2.

5. The difference between present value and net present value is:

A. Present value is expressed as an interest rate, while net present value is expressed as a dollar figure.

B. Present value is a measure of the actual present value, while net present value measures expected present value.

C. Present value does not factor in costs.

D. Present value is more accurate.

6. Your best cost estimate for an activity is $200,000, but the estimate you document has a range of $100,000 to $400,000. This ranged estimate represents a(n):

A. Cost estimate.

B. Budgeted estimate.

C. Order of magnitude estimate.

D. Definitive estimate.

7. Which of the following process sequences is correct?

   A. Create WBS, then Cost Budgeting, then Cost Estimating.

   B. Create WBS, then Cost Estimating, then Cost Budgeting.

   C. Cost Budgeting, then Cost Estimating, then Create WBS.

   D. Cost Estimating, then Cost Budgeting, then Create WBS.

8. One of your team members makes a change to the budget with your approval. In what process is he engaged?

   A. Cost Planning.

   B. Cost Estimating.

   C. Project Cost Management.

   D. Cost Control.

9. After measuring expected project benefits, management has four projects from which to choose. Project 1 has a net present value of $100,000 and will cost $50,000. Project 2 has a net present value of $200,000 and will cost $75,000. Project 3 has a net present value of $500,000 and will cost $400,000. Project 4 has a net present value of $125,000 and will cost $25,000. Which project would be BEST?

   A. Project 1.

   B. Project 2.

   C. Project 3.

   D. Project 4.

10. Your project office has purchased a site license for a computerized tool that assists in the task of cost estimating on a very large construction project for a downtown skyscraper. This tool asks you for specific characteristics about the project and then provides estimating guidance based on materials, construction techniques, historical information, and industry practices. This tool is an example of:

   A. Bottom-up estimating.

   B. Parametric modeling.

   C. Analogous estimating.

   D. Activity duration estimating.

11. You are managing a project that is part of a large construction program. During the execution of your project, you are alerted that the construction of a foundation is expected to experience a serious cost overrun. What would be your FIRST course of action?

    A. Evaluate the cause and size of the overrun.

    B. Stop execution until the problem is solved.

    C. Contact the program manager to see if additional funds may be released.

    D. Determine if you have sufficient budget reserves to cover the cost overrun.

12. If earned value = $10,000, planned value = $8,000, and actual cost = $3,000, what is the schedule variance?

    A. -$2,000

    B. $2,000

    C. $5,000

    D. -$5,000

13. Estimate to complete indicates:

    A. The total projected amount that will be spent, based on past performance.

    B. The projected remaining amount that will be spent, based on past performance.

    C. The difference between what was budgeted and what is expected to be spent.

D. The original planned completion cost minus the costs incurred to date.

14. If a project has a CPI of .95 and an SPI of 1.01, this indicates:

    A. The project is progressing slower and costing more than planned.

    B. The project is progressing slower and costing less than planned.

    C. The project is progressing faster and costing more than planned.

    D. The project is progressing faster and costing less than planned.

15. Project A would yield $100,000 in benefit. Project B would yield $250,000 in benefit. Because of limited resources, your company can perform only one of these. They elect to perform Project B because of the higher benefit. What is the opportunity cost of performing Project B?

    A. -$150,000.

    B. $150,000.

    C. -$100,000.

    D. $100,000.

16. As a project manager, your BEST use of the project cost baseline would be to:

    A. Measure and monitor cost performance on the project.

    B. Track approved changes.

    C. Calculate team performance bonuses.

D. Measure and report on variable project costs.

17. The value of all work that has been completed so far is:

A. Earned value.

B. Estimate at complete.

C. Actual cost.

D. Planned value.

18. If you have a schedule variance of $500, this would indicate:

A. Planned value is less than earned value.

B. Earned value is less than the estimate at complete.

C. Actual cost is less than earned value.

D. The ratio of earned value to planned value is 5:1.

19. If budgeted at complete = $500, estimate to complete = $400, earned value = $100, and actual cost = $100, what is the estimate at complete?

A. $0.

B. $150.

C. $350.

D. $500.

20. You have spent $322,168 on your project to date. The program manager wants to know why costs have been running so high. You explain that the resource cost has been greater than expected and should level out over the next six months. What does the $322,168 represent to the program manager?

   A. Earned value

   B. Actual cost

   C. Planned value

   D. Cost performance index

Answers to Cost Management Questions

1. C. Planned value is calculated by multiplying the Budgeted At Completion by planned % complete. Our cost per mile is planned at $1,250,000 ($5,000,000 / 4 miles), and our Budgeted At Completion is 32 miles * 1,250,000/mile = $40,000,000. We planned to be 50% complete. Therefore, $40,000,000 * .50 = $20,000,000.

2. A. Understanding the concepts behind the earned value calculations is important for the exam and will help you with questions like this one. In this question, the terrible cost performance index indicates that we are getting ten cents of value for every dollar we spent; thus the project is doing very poorly on cost performance.

3. B. Cost Budgeting takes the activity cost estimates and uses them (along with other inputs) to create a budget. 'A' is incorrect because Cost Estimating is the process that creates the activity cost estimates, so it stands to reason that they would be an output and not an input. 'C' sounds like a decent guess, but it is not a real process. 'D' is incorrect because the process of Cost Control is not concerned with just the individual activity cost estimates. Instead, it uses inputs of the cost baseline.

4. D. With Benefit Cost Ratios, the bigger the better! BCR is calculated as benefit/cost, so the more benefit, and the less cost, the higher the number.

5. C. There is a difference between present value and net present value. Present value tells the expected value of the project in today's dollars. Net present value is the same thing, but it subtracts the costs after calculating the present value.

6. C. Order of magnitude estimates are -50% to +100%. In this example, $100,000 and $400,000 are -50% to +100% of $200,000.

7. B. This question may not look like it is about inputs and outputs, but it actually is. Create WBS is performed first out of the three processes, and the output is the Work Breakdown Structure (WBS). The WBS is used as an input for the next process of the three, Cost Estimating, where the costs of the activities are estimated and aggregated back to the WBS. Finally, the output of that process, the Activity Cost Estimates, is used as an input into Cost Budgeting, which occurs last out of the three processes listed. By understanding how the outputs of one process become the inputs into another, it becomes simpler to understand the logical order of many of these processes.

8. D. The main clue here is "change." If they are making approved changes, they are in a control process. 'A' is not a real process. 'B' is incorrect since Cost Estimating is the process where the original estimates are developed and not where they are updated. 'C' is the knowledge area (careful not to get these confused with processes).

9. C. This one was very tricky! Net present value already has costs factored in, so they can be ignored here. The net present value is the only value you need to consider, and bigger is better!

10. B. This is an example of parametric modeling. Parametric modeling is common in some industries, where you can describe the project in detail, and the modeling tool will help provide estimates based on historical information, industry standards, etc.

11. A. This illustrates one of PMI's biggest biases on these questions. Your job as a project manager is almost always to evaluate and understand first. Know what you are dealing with before you take action, and don't just accept anyone's word for it - verify the information yourself!

12. B. Schedule Variance is calculated as EV-PV. In this example, \$10,000 - \$8,000 = \$2,000.

13. B. The estimate to complete is what we expect to spend from this point forward, based on our performance thus far.

14. C. Did the wording trip you up on this one? Make sure you read the questions and answers carefully since things were switched around on this one. A schedule performance index greater than 1 means that the project is progressing faster than planned. A cost performance index that is less than 1 means that the project is costing more than planned. Therefore choice 'C' is the only one that fits.

15. D. Opportunity cost is simply how much cost you are passing up. In this case, by choosing project B, you are forgoing \$100,000 in expected benefit from project A, and that \$100,000 represents the opportunity cost.

16. A. The cost baseline is used to track cost performance based on the original plan plus approved changes.

17. A. Earned value is defined as the value of all work completed to this point.

18. A. This is another tricky question because of the way it is worded. Schedule variance is calculated as earned value- planned value. In this case, schedule variance could only be positive if earned value is greater than planned value (or stated otherwise, if planned value is less than earned value). 'A' is the only choice that has to be true.

19. D. The estimate at complete is what we expect to have spent at the end of the project. It is calculated by taking our budgeted at complete and dividing it by our cost performance index. Step 1 is to calculate our cost performance index. It is earned value / actual cost, and in this case, it equals 1. Budgeted at complete is $500, and $500/ 1 = $500. Therefore, 'D' is the correct answer, indicating that we are progressing exactly as planned.

20. B. Look at the first sentence "You have spent $322,168.. ."Actual cost is what you have spent to date on the project.

Quality Management

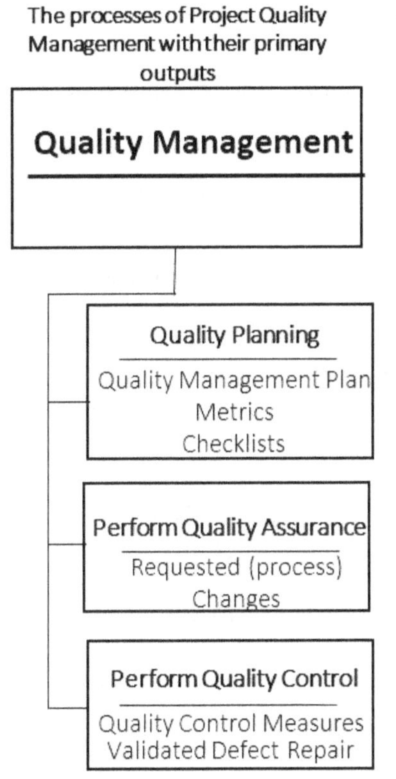

The topics on the PMP borrow from numerous business disciplines such as psychology, math, accounting, and even law. Previous formal study of these disciplines is beneficial but not necessary to pass the exam. This is especially true for the topic of quality management.

The study of quality and its effect on business and projects has been the focus of considerable research, especially since the end of World War II.

Volumes of research, usually focused on a few central theories, have been conducted and documented.

Quality borrows heavily from the field of statistics, and a high-level explanation of some of the statistical tools and techniques is provided here.

The quality processes presented in the PMBOK Guide and on the PMP Exam come largely from the theories of Deming, Crosby, and Juran.

There are numerous theories on quality, how it should be implemented, how it should be measured, and what levels of quality should be attained. You need to have a solid grasp of the theories presented in this chapter before taking the exam.

Please note that the statistics and statistical examples here are greatly simplified, but they are adequate to get you through the exam.

Topics on the PMP Exam, such as sampling, distribution, and deviations from the mean are built on underlying assumptions that you do not need to know in order to pass, but you do need to know how to apply them.

Philosophy

PMI's philosophy of quality is derived from several leading quality theories, including TQM, ISO-9000, Six Sigma, and others. The foundational work in this field performed by PMI looks at each of these theories in terms of the tools and techniques they provide.

PMI's philosophy of quality is also a very proactive approach. Whereas early theories on quality relied heavily on inspection, current thinking is focused on prevention over inspection. This evolution of thought is based on the fact that it costs more to fix an error than it does to prevent one.

The responsibility of quality in PMI's philosophy falls heavily on the project manager. Everyone on the team has an important contribution to make to project quality; however, it is management's responsibility to provide the resources to make quality happen, and the project manager is ultimately responsible and accountable for the quality of the project.

The PMI process, as it relates to quality, is perhaps more important here than most other places. Quality Planning, Perform Quality Assurance, and Perform Quality Control map closely to the Plan-Do-Check-Act cycle as described by Deming, and several questions on the exam rely heavily on your understanding of how quality activities flow and connect.

It is also important to understand that some of the investment in quality is usually borne by the organization, since it would be far too expensive for each project to have its own quality program.

An example would be a company investing in a site license for a software testing product that can be used across numerous projects.

Importance

Project quality management is one of the slimmer chapters both in this book and in the PMBOK Guide, but it is of high importance on the exam. You should expect to see several exam questions that will relate directly to this chapter, so it will be necessary to become acquainted with the terms and theories as prescribed below. Then reread this chapter to ensure that you have mastered the topic.

In real world practice, quality formulas abound; however, the PMP does not currently require you to memorize or apply them. This chapter focuses, instead, on the processes, concepts and terms.

Some parts of this topic will be revisited in later chapters in order to show how quality fits into the overall project management context.

Preparation

The quality processes, tools and techniques, and outputs found in this chapter must be learned and understood. You should expect to see questions on the PMP Exam that relate directly to these concepts. Special attention will be paid to the key quality theories that show up on the exam, as well as terminology that you need to know. Pay careful attention to the differences between Quality Planning, Perform Quality Assurance, and Perform Quality Control. This distinction is a tricky area for many people on the exam.

Quality Management Processes

DEFINITION OF QUALITY

The definition of quality you should know for the exam is "the degree to which a set of inherent characteristics fulfill requirements." It is also important to realize that the requirements or needs of the project may be stated or implied.

QUALITY PROCESSES

There are only three processes within project quality management, as listed in the table below. In the PMI framework, these processes touch three process groups: planning (Quality Planning), executing (Perform Quality Assurance), and controlling (Perform Quality Control).

Process Group	Quality Management Process
Initiating	(none)
Planning	Quality Planning
Executing	Perform Quality Assurance
Controlling	Perform Quality Control
Closing	(none)

The primary outputs associated with the three quality management processes are shown in the table below.

Process	Primary Outputs
Quality Planning	• Quality management plan • Quality baseline
Perform Quality Assurance	• Requested changes • Recommended corrective action
Perform Quality Control	• Quality control measurements

	• Defect repair

QUALITY TERMS AND PHILOSOPHIES

A. Total Quality Management (TQM)

A quality theory popularized after World War II. that states that everyone in the company is responsible for quality and is able to make a difference in the ultimate quality of the product. TQM applies to improvements in processes and in results. TQM also includes statistical process control.

TQM shifts the primary quality focus away from the product that is produced and looks instead at the underlying process of how it was produced.

B. Continuous Improvement

Also known as "Kaizen," from the Japanese Management term. A philosophy that stresses constant process improvement, in the form of small changes in products or services.

C. Kaizen

(See Continuous Improvement)

D. Just-In-Time (JIT)

A manufacturing method that brings inventory down to zero (or near zero) levels. It forces a focus on quality, since there is no excess inventory on hand to waste.

E. ISO 9000

Part of the International Standards Organization to ensure that companies document what they do and do what they document. ISO 9000 is not directly attributable to higher quality, but may be an important component of Perform Quality Assurance, since it ensures that an organization follows their processes.

F. Statistical Independence

When the outcomes of two processes are not linked together or dependent upon each other, they are statistically independent. Rolling a six on a die the first time neither increases nor decreases the chance that you will roll a six the second time. Therefore, the two rolls would be statistically independent.

G. Mutually Exclusive

A statistical term that states that one choice excludes the others. For example, painting a house yellow and painting it blue or white are mutually exclusive events.

H. Standard Deviation

The concept of standard deviation is an important one to understand for the exam. Standard deviation is a statistical calculation used to measure and describe how data is organized. The following graphic of a standard bell curve illustrates standard deviation:

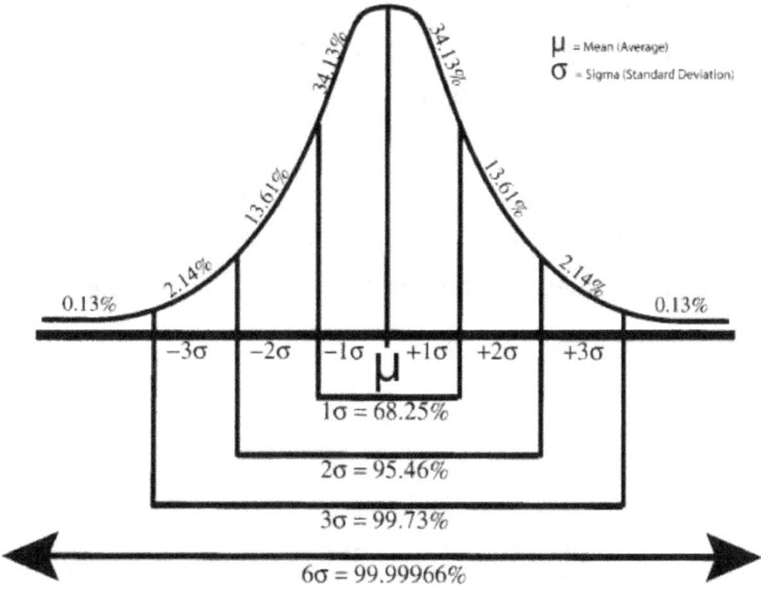

The standard deviation, represented by the Greek symbol σ, is calculated first by averaging all data points to get the mean, then calculating the difference between each data point and the mean, squaring each of the differences, and dividing the sum of the squared differences by the number of data points minus one. Finally, take the square root of that number, and you have the standard deviation of the data set. If the data set is "normally distributed," as it is in the preceding chart, the following statistics will be true:

- 68.25% of the data points (or values) will fall within 1 σ from the mean.
- 95.46% of the values will fall within 2 σ from the mean.
- 99.73% of the values will fall within σ from the mean.
- 99.99966% of the values will fall within 6 σ from the mean.

The standard deviation may be used in a few different ways in quality. For instance, the higher your standard deviation, the more diverse your data points are. It is also used to set quality levels (see the Six Sigma topic later in this chapter), and to set control limits to determine if a process is in control (see the control charts topic later in this chapter).

Even though you should not expect to have to perform standard deviation calculations for the exam, you will likely see questions related to the application of the standard deviation. The more you understand about this concept, the more prepared you will be for the exam.

I. Six Sigma

Six sigma is a popular philosophy of quality management that focuses on achieving very high levels of quality by controlling the process and reducing defects (a defect is defined as anything that does not meet the customer's quality standards).

As you will remember from the section on standard deviation, a σ (sigma) is defined as 1 standard deviation from the mean. At the level of one sigma quality, 68.25% of all outputs will meet quality standards.

At the three sigma quality level, that number jumps to 99.73% of all outputs that meet quality standards. At the six sigma level, the number is 99.99966% of all outputs that meet quality standards. This means that when quality reaches six sigma standards, the results will be such that only 3.4 out of every 1,000,000 outputs do not meet quality standards. Six sigma quality strives to make the overwhelming majority of the bell curve fall within customer quality limits.

Six sigma puts a primary focus on quantifying, measuring, and controlling the quality of products, services, and results. It is based on the underlying theory that anything will vary if measured to a fine enough level. The goal is to refine the process so that human error and outside influence no longer exist, and these variations are completely random.

If done properly, the statistical outcome should follow the bell curve illustrated previously under the topic of standard deviation. The goal is to make six standard deviations (sigmas) of the outputs fall within the customer's quality limits.

If this seems like a lot of information, the most important things to know for the exam are that six sigma is a quality management philosophy that sets very high standards for quality, and that one sigma quality is the lowest quality level, allowing 317,500 defects per 1,000,000 outputs, three sigma quality is higher, allowing 2,700 defects per 1,000,000, and six sigma quality is the highest of these, allowing only 3.4 defects per 1,000,000.

Also know that six sigma quality levels may not be high enough for all projects or all industries. For instance, the pharmaceutical industry, the airline industry, and power utilities typically strive for higher levels of quality than six sigma would specify in some areas of their operations.

J. Prevention vs. Inspection

Prevention is simply keeping defects from occurring, while inspection is about catching the errors that have occurred before they impact others outside the project.

K. Attribute Sampling vs. Variable Sampling

Attribute sampling is binary; either a work result conforms to quality or it does not. Variable sampling, on the other hand, measures how well something conforms to quality. Consider, for example, a production facility making prescription drugs. Using attribute sampling, they would define tolerances, a batch of product would be tested, and it would either pass or fail that inspection. Using variable sampling, however, the batch of product would be rated on a continuous scale (perhaps on parts per million) that showed how well the batch conformed to ideal quality.

L. Special Causes vs. Common Causes

Within the topic of statistical process control (see control charts later in this chapter), there is a concept of special causes and common causes. Special causes are typically considered unusual and preventable, whereas common causes are normal. For instance, if your manufacturing process produced 250 defects per 1,000,000 due to assembly errors, that might be considered a special cause, whereas if your manufacturing process produced one defect in a million due to bad raw materials, that might be considered a common cause. In general, special causes are considered preventable by process improvement, while common causes are generally accepted.

M. Tolerances vs. Control Limits

Tolerances deal with the limits your project has set for product acceptance. For instance, you may specify that any product will be accepted if it weighs between 12 and 15 grams. Those weights would represent your tolerances for weight. Control limits, on the other hand, are a more complex concept. Typically, control limits are set at three standard deviations above and below the mean. As long as your results fall within the control limits, your process is considered to be in control. Control limits are discussed further under the topic of control charts, covered later in this chapter. If that explanation still leaves you scratching your head, consider that tolerances focus on whether the product is acceptable, while control limits focus on whether the process itself is acceptable.

Quality Planning

WHAT IT IS:

Quality Planning is named appropriately. It is the process where the project team identifies what the quality specifications are for this project and how these specifications will be met.

WHY IT IS IMPORTANT:

As stated earlier in this chapter, quality is planned in from the start, and not inspected in after the product has been constructed. Quality Planning is the process where this planning is performed to make sure that the resulting product is of acceptable quality.

WHEN IT IS PERFORMED:

Quality Planning begins early in the project planning phase of the project. In fact, it typically is performed concurrently with other planning processes and the development of the project management plan.

The reason Quality Planning is performed early in the project planning is that decisions made about quality can have a significant impact on other decisions such as scope, time, cost, and risk.

HOW IT WORKS:

- Inputs
  - *Enterprise Environmental Factors - See Chapter 2, Common Inputs*
  - *Organizational Process Assets*

Assets such as quality management plans from previous projects should be considered and used as part of the project plan. Also, the performing organization's quality policy should be brought into this process. The quality policy is usually brief and defines the performing organization's attitudes about quality across all projects. It must be considered if it exists. If it does not exist, the project team should write one for this project.

  - *Project Scope Statement*

Many project management practitioners view scope and quality as inseparable. The PMBOK Guide also views them as tightly linked, and this is why the project scope statement is a primary input into Quality Planning. The project scope statement defines the requirements, and it also specifies the acceptance criteria for these requirements.

  - *Project Management Plan - See Chapter 2, Common Inputs*

- Tools

  - *Cost-Benefit Analysis*

Quality can be expensive to achieve; however, there is a golden rule in quality that all benefits of quality activities must outweigh the costs. No activities should be performed that cost more (or even the same) as the expected benefits. These benefits potentially include acceptance, less rework, and overall lower costs.

  - *Benchmarking*

The technique of benchmarking is where a project's quality standards are compared to those of other projects which will serve as a basis for comparison. These other projects may be from the industry, such as an automaker setting quality standards based on those of other automobiles in their class, or it may be based on projects previously executed by the performing organization.

  - *Design of Experiments*

The technique of design of experiments, often abbreviated as DOE, is an important technique in Quality Planning. It uses data analysis to determine optimal conditions. For instance, rather than conducting a series of individual trials, an information technology project may use design of experiments analysis and optimize hardware and bandwidth needed to run an Internet-based application, finding the optimal match of system response time and overall cost.

- *Cost of Quality*

The technique of evaluating the cost of quality (often abbreviated as COQ) looks at all of the costs that will be realized in order to achieve quality. The costs of items that do not conform to quality are known as "cost of poor quality" often abbreviated as COpQ.

- *Additional Quality Planning Tools*

There are nearly countless additional Quality Planning tools that are designed by specific industries or for use with specific types of manufacturing processes. For the exam, focus energy on the tools presented here.

- Outputs

  - *Quality Management Plan*

The quality management plan is a key output of the Quality Planning process. It describes how the quality policy will be met.

- *Quality Metrics*

Quality metrics specifically define how quality will be measured. For instance, it is not adequate for the team to say that the system needs to have a rapid response time. Instead, a quality metric might specify that a system must respond within two seconds to 99% of all requests up to 1,000 simultaneous users. Quality metrics may relate to any quality measure.

- *Quality Checklists*

A checklist is a Quality Planning output created to ensure that all steps were performed, and that they were performed in the proper sequence. They are created here and used in the process of Perform Quality Control.

- *Process Improvement Plan*

The concept of process improvement is tightly linked to quality, and the process improvement plan (part of the project management plan), deals with how quality activities will be streamlined and improved.

- *Quality Baseline*

A baseline is the original plan plus approved changes. The quality baseline is the quality objectives and the plan for achieving those objectives. The quality baseline, like other baselines, may be changed through the change control process.

- *Project Management Plan Updates*

As Quality Planning is performed, it is normal to update the project management plan (specifically the process improvement plan and the quality management plan).

Perform Quality Assurance

WHAT IT IS:

For many people, Perform Quality Assurance can be a tricky process to understand. One of the most common mistakes exam takers make is to confuse Perform Quality Assurance with Perform Quality Control (covered next), and the difference between the two can appear subtle.

Perform Quality Assurance is an executing process, and the most important thing for you to remember about this process is that it is primarily concerned with overall process improvement. It is not about inspecting the product for quality or measuring defects. Instead, Perform Quality Assurance is focused on steadily improving the activities and processes undertaken to achieve quality. Numerous questions on the exam are missed because of a misunderstanding of this key distinction!

WHY IT IS IMPORTANT:

Too often, people think of quality as simply measuring, testing, and inspecting the final product; however, quality management should be about improving the process as well as the product. Perform Quality Assurance is important because if the quality of the process and activities are improved, then quality of the product should also improve, along with an overall reduction of cost.

WHEN IT IS PERFORMED:

The process of Perform Quality Assurance is performed as an ongoing activity in the project management lifecycle. It typically begins early and continues throughout the life of the project. Because Perform Quality Assurance uses many of the outputs of Quality Planning, it is not undertaken until after Quality Planning has been performed.

HOW IT WORKS:

- Inputs

    - *Quality Management Plan*

The quality management plan gives guidance as to how the process of Perform Quality Assurance will be executed. As changes are made to the way quality is managed, they should be documented back into the quality management plan.

- *Quality Metrics*

The quality metrics created earlier in Quality Planning are used here. Since Perform Quality Assurance is primarily concerned with process improvement, the metrics provide an objective means of measurement.

- *Process Improvement Plan*

The process improvement plan is one of the primary inputs into the process of Perform Quality Assurance, as it defines much of what this process is about. It describes how the quality process will be improved and how these improvements will be measured and managed.

- *Work Performance Information*

Since Perform Quality Assurance is primarily about process improvement, the work performance information is a key input. It will help identify areas where process improvement is needed, as well as areas where process improvements have worked.

- *Approved Change Requests*

Any change requests should be evaluated for their impact on the quality of the project (among other things). A change in scope or schedule can potentially have a significant effect on the quality of the project.

- *Quality Control Measurements*

The quality control measurements can be thought of as a feedback loop. As changes are evaluated here in the process of Perform Quality Assurance, they are measured in Perform Quality Control and fed back into this process for evaluation.

- *Implemented Change Requests; Implemented Corrective Actions;*
- *Implemented Defect Repair; Implemented Preventive Actions*

This applies to the four preceding inputs. As a change or correction is implemented, it should be brought back into Perform Quality Assurance. The reason for this is that Perform Quality Assurance is constantly monitoring the quality process to ensure that it is working, and any change, positive or negative, needs to be a part of that evaluation.

- Tools

  - *Quality Planning Tools and Techniques*

The same tools and techniques used for Quality Planning are used in Perform Quality Assurance.

- *Quality Audits*

Quality audits are the key tool in Perform Quality Assurance. The reason for this is that audits review the project to evaluate which activities taking place on the project should be improved and which meet quality standards. The goal of the audits is both to improve acceptance of the product and the overall cost of quality.

- *Process Analysis*

The process analysis carefully reviews the quality process to ensure that it is working efficiently and effectively.

- Outputs

- *Requested Changes*

When this much analysis and evaluation is taking place, requested changes are a normal and expected outcome of this process.

- *Recommended Corrective Actions*

Corrective action is anything done to bring future results in line with the plan. In Perform Quality Assurance, the recommended corrective actions are going to be focused on adjusting the use of the tools in this process.

- *Organizational Process Assets Updates*

These assets are anything that can be borrowed, built-upon, and reused for future projects within this organization. Any such assets should be updated as new practices are implemented or new information is learned.

- *Project Management Plan Updates*

As the process of Perform Quality Assurance changes the way in which the project is managed, the project management plan (and specifically the quality management plan) should be updated.

Perform Quality Control

WHAT IT IS:

Perform Quality Control looks at specific results to determine if they conform to the quality standards. It involves both product and project deliverables, and it is done throughout the project - not just at the end.

Perform Quality Control typically uses statistical sampling rather than looking at each and every output. Many volumes have been written about sampling techniques, and the practice is often very complex and is highly tailored to industry. This process uses the tool of inspection to make sure the results of the work are what they are supposed to be. Any time you find a part being inspected for quality, you can be sure that you are in the control process.

WHY IT IS IMPORTANT:

Perform Quality Control is the process where each deliverable is inspected, measured, and tested. This process makes sure everything produced meets quality standards.

WHEN IT IS PERFORMED:

This process typically takes place throughout much of the project. It is performed beginning with the production of the first product deliverable and continues until all of the deliverables have been accepted.

HOW IT WORKS:

- Inputs
    - *Quality Management Plan*

The quality management plan, a component of the project management plan, is a primary input into Perform Quality Control. This provides the plan for how Perform Quality Assurance and Perform Quality Control will be carried out.

- *Quality Metrics*

The quality metrics will be used to measure whether the work results meet quality specifications.

- *Quality Checklists*

The quality checklists show the steps that were taken to achieve quality on the product. Now that the product is in Perform Quality Control, the checklists will assist in assessing its conformance to quality.

- *Organizational Process Assets Updates*

Any informational assets (or other types of assets) should be brought into this process. For instance, if a company invested in a license for a tool that assisted in software testing, that would be appropriate to bring into this process.

- *Work Performance Information*

The work performance information can shed light on the current state of the deliverable and where it should be.

For example, if a software project is scheduled to have the database completed by this date, and the work performance information shows that it is still in design, that will affect the Perform Quality Control process. In this example, you would likely use the work performance information to alter your testing scenarios to reflect the database's unfinished state.

- *Approved Change Requests*

Change requests can not only change the product, but also the way the product is constructed, and either of these types of changes will affect quality management.

- *Deliverables*

The project's deliverables are a primary input into this process. The deliverables are inspected and measured to ensure that they conform to quality standards.

- Tools

  - *Cause and Effect Diagram*

Cause and effect diagrams are also known as Ishikawa diagrams or fishbone diagrams. In Quality Control, these are used to show how different factors relate together and might be tied to potential problems.

In Quality Control, cause and effect diagrams are used as part of an approach to improve quality by identifying quality problems and trying to uncover the underlying cause.

An example of a Cause and Effect (or Ishikawa or Fishbone) Diagram

Cause and Effect Diagrams, or Ishikawa Diagrams, are also known as fishbone diagrams due to the way the chart appears like a fish's skeleton.

- ***Control Charts***

Control charts are part of a set of quality practices known as Statistical Process Control. If a process is statistically "in control," it does not need to be corrected. If it is "out of control," then there are sufficient variations in results that must be brought back statistically in line. A control chart is one way of depicting variations and determining whether or not the process is in control. Control charts graph the results of a process to show whether or not they are in control. The mean of all of the data points is represented by a line drawn through the average of all data points on the chart. The upper and lower control limits are set at three standard deviations above and below mean.

If measurements fall outside of the control limits, then the process is said to be out of control. The assignable cause should then be determined. An interesting rule that is used with control charts is known as the rule of seven. It states that if seven or more consecutive data points fall on one side of the mean, they should be investigated. This is true even if the seven data points are within the control limits. Some control charts, especially those used in a manufacturing environment, represent the upper and lower limits of a customer's specification for quality as lines on the control chart. Everything between those lines would be considered within the customer's quality specification.

An example of a control chart

- ***Flowcharting***

Flowcharts show how various components relate in a system. Flowcharting can be used to predict where quality problems may happen. In addition to traditional flow charts, Cause and Effect Diagrams are another type of flowcharting.

- ***Histogram***

A histogram is another word for a column chart. Histograms show how often something occurs, or its frequency, A Pareto chart (see next tool) is one example of a type of histogram.

- ***Pareto Chart***

Another type of chart used as a tool of Perform Quality Control is the Pareto diagram. This is a histogram showing defects ranked from greatest to least. It is used to focus energy on the problems most likely to change results.

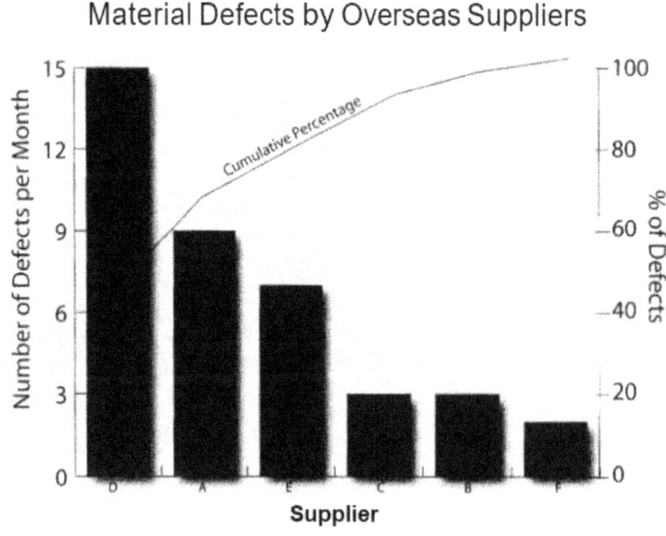

Sample Pareto chart showing a ranking of problems by supplier

- ***Pareto's Law***

Pareto diagrams are based on Pareto's Law, which is also known as the 80/20 rule. This rule states that 80% of the problems come from 20% of the causes, but there are variations on this theme. For the exam, know that Pareto's Law is also known as the 80/20 rule and that a Pareto chart is used to help determine the few root causes behind the majority of the problems on a project.

- *Run Chart*

When you measure quality, it is somewhat like taking a snapshot of the way things are at a point in time. This can be useful as a stand-alone tool, but when those snapshots are compiled over time, they can provide excellent information on trends. Using the tool of a run chart, such as the one pictured here, the project manager can analyze how quality is trending over time.

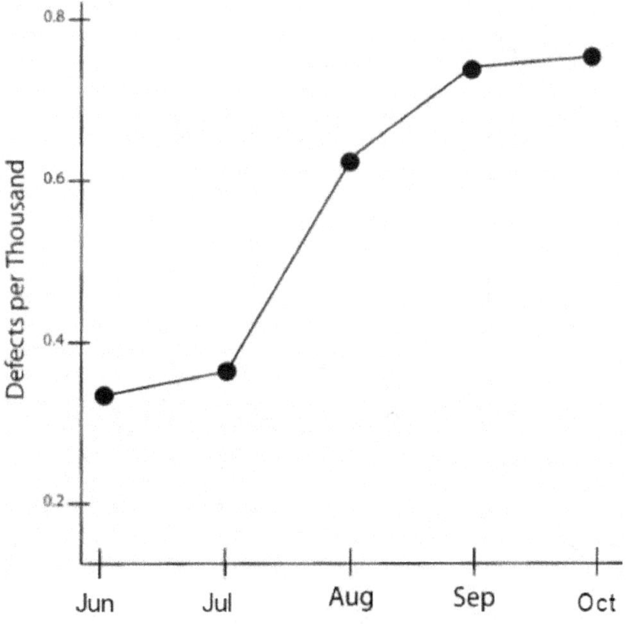

- *Scatter Diagram*

Scatter diagrams are a particularly powerful tool for spotting trends in data. Consider the following three scatter diagrams:

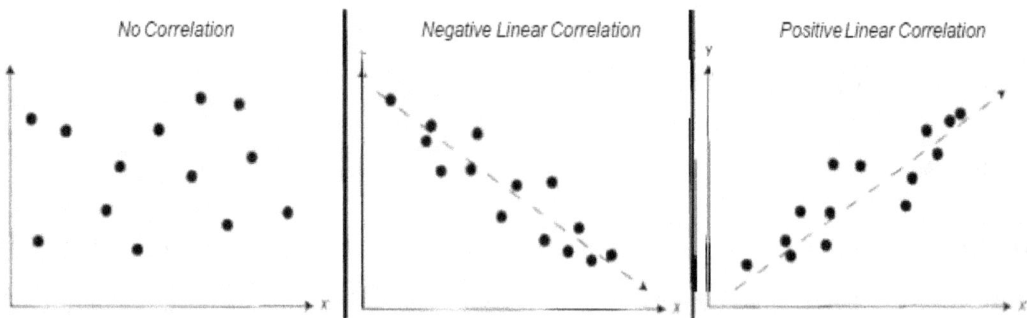

Scatter diagrams are made using two variables (a dependent variable and an independent variable). What you are looking for is correlation between the two variables. Considering the previous examples, suppose that the horizontal, or X axis, represented hours of study, which is your independent variable.

The vertical, or Y axis, represented your score on the PMP Exam, which is the dependent variable. In that case, the third graph would make sense, since the more people studied, the higher their scores tended to be. When you see a graph that looks like the second graph where the more the person studied, the lower their score, you might deduce that the book they are reading is actually having a negative effect.

The first example graph below might lead you to deduce that the study material being used has no effect at all and therefore there is no correlation.

- *Statistical Sampling*

Statistical sampling is a powerful tool where a random sample is selected instead of measuring the entire population. By sampling properly, you can dramatically cut down on the number of measurements you need to take. Note that you will not need to know how to statistically sample for the PMP Exam. Opinion on statistical sampling varies widely among experts and industries.

- *Inspection*

The tool of inspection is exactly what it sounds like. Inspection may be testing a module of a software application or performing a walkthrough on a building.

- *Defect Repair Review*

When defects are noted, they must be fixed. The tool of the defect repair review is used to ensure that the defects were fixed and meet quality specifications.

- Outputs

  - *Quality Control Measurements*

As the process of Perform Quality Control is performed, important measurements of quality levels and compliance are a normal output and should be documented.

- *Validated Defect Repair*

Defects need to be corrected, and as they are corrected, they need to be inspected again to validate that they now meet quality standards.

- *Quality Baseline Updates*

The quality baseline, like other baselines, may be changed. This quality baseline is the quality management plan and standards that must be achieved, and as either the quality limits or the plan on how to achieve or manage that quality changes, the baseline should be officially modified.

- *Validated Deliverables*

The validated deliverables are the key output of Perform Quality Control.

- *Recommended Corrective Actions - See Chapter 2, Common Outputs*
- *Recommended Preventive Actions - See Chapter 2, Common Outputs*
- *Requested Changes - See Chapter 2, Common Outputs*
- *Recommended Defect Repair - See Chapter 4, Monitor and Control Project Work, Outputs*
- *Organizational Process Assets Updates - See Chapter 2, Common Outputs*
- *Project Management Plan Updates See Chapter 2, Common Outputs*

Quality Management Questions

1. You are a project manager, and your manager wants to meet with you to evaluate your project's performance in order to see how it is meeting the quality standards supplied by the company. In what process is your boss engaged?

   A. Total Quality Management.

   B. Perform Quality Control.

   C. Quality Planning.

   D. Perform Quality Assurance.

2. If you were using a fishbone diagram to determine root causes of problems, you would you be involved in:

A. Quality Inspection.

B. Quality Prevention.

C. Quality Control.

D. Quality Audits.

3. Quality Planning includes all of the following outputs EXCEPT:

   A. Quality management plan.

   B. Acceptance.

   C. Operational definitions.

   D. Checklists.

4. In a control chart, the mean is represented as a horizontal line. This represents:

   A. The average of the control limits.

   B. The average of all data points.

   C. The average of all data points that are within control limits.

   D. A means of identifying assignable cause.

5. Quality audits are an important part of quality management because:

   A. They allow for quantification of the risk.

   B. They randomly audit product results to see if they are meeting quality standards.

C. They check to see if the quality process is being followed.

D. They are conducted without prior notice and do not allow team members time to cover up defects.

6. If the results of activity A have no bearing on the results of activity B, the two activities would be considered:

   A. Statistically unique.

   B. Statistically independent.

   C. Correlated, but not causal.

   D. Mutually exclusive.

7. The BEST tool to use to look for results that are out of control is:

   A. Pareto chart.

   B. Control chart.

   C. Ishikawa diagram.

   D. Statistical sampling.

8. You are a project manager with limited resources on the project. Several quality defects have been discovered, causing the stakeholders concern. You wish to begin by attacking the causes that have the highest number of defects associated with them. Which tool shows defects by volume from greatest to least?

   A. Pareto chart.

B. Control chart.

C. Ishikawa diagram.

D. Cause and effect diagram.

9. In the process of managing a construction project, you discover a very serious defect in the way one particular section has been built. Your engineers analyze the section of the building and decide that the problem is actually relatively minor. In which process are you involved?

   A. Quality Planning.

   B. Perform Quality Assurance.

   C. Perform Quality Control.

   D. Quality Management.

10. You are performing a project that has a lot in common with a project completed by your company two years ago. You want to use the previous project to help you determine quality standards for your project. Which of the following tools would be the BEST one to help you with this?

   A. Benchmarking.

   B. Control chart.

   C. ISO 9000.

   D. Total Quality Management.

11. Which of the following is most representative of the Total Quality Management philosophy?

    A. Decreasing inventory to zero or near zero levels.

    B. Everyone can contribute to quality.

    C. Zero defects.

    D. Continuous improvement is preferred over disruptive change.

12. Which of these quality standards is the highest?

    A. It is impossible to determine without further information.

    B. 99% quality.

    C. Three sigma quality.

    D. Six sigma quality.

13. Which quality process is performed first?

    A. Quality Planning.

    B. Perform Quality Assurance.

    C. Perform Quality Control.

    D. Quality Definition.

14. A project team is having their first quality meeting and plans to review the organization's quality policy when it is discovered that the company has never developed an organizational quality policy. The project manager is very concerned about this discovery What would be the BEST course of action?

    A. Document the absence of a quality policy in the quality management plan and take corrective action.

    B. Write a quality policy just for this project.

    C. Substitute benchmark data for the quality policy.

    D. Suspend execution until the organization provides a quality policy.

15. On a control chart, the customer's acceptable quality limits are represented as:

    A. Control limits.

    B. Mean.

    C. Specification.

    D. Normal distribution.

16. A customer is concerned that the quality process is not being followed as laid out in the quality management plan. The best way to see if this claim is accurate is:

    A. Random sampling.

    B. Kaizen.

    C. Personally participate in the quality inspections.

D. Audits.

17. Your organization practices just-in-time management. Which of the following would be the highest concern for a project manager operating in this company?

A. Absenteeism.

B. Lower quality of parts.

C. Conflicting quality processes.

D. Inventory arriving late.

18. Reduced quality on a project would MOST likely lead to which of the following?

A. Rework and increased cost risk.

B. Absenteeism and decreased cost.

C. Increased inspections and decreased cost.

D. Reduced quality limits.

19. Which of the following is NOT a part of the Quality Planning process?

A. Benchmarking

B. Audits.

C. Flowcharting.

D. Design of experiments.

20. A project manager wants to perform a code review, but over two million lines of code have already been written for this project, and more are being produced every day. Rather than reviewing each line of code, the manager should consider:

    A. Automated testing tools.

    B. Trend analysis.

    C. Statistical sampling.

    D. Regression analysis.

Answers to Quality Management Questions

1. D. In this example your boss is auditing you to see if you are following the process. Remember that audits are a tool of Perform Quality Assurance.

2. C. Quality Control is the correct answer here.

3. B. Acceptance is not an output of Quality Planning. It is an output of Perform Quality Control.

4. B. The mean represents the average of all of the data points shown on the chart, calculated simply by adding the values together and dividing by the number of values. 'C' is not correct because the mean includes everything. If only the values that were within the control limits were used, it could make the mean look better than it should.

5. C. Audits are a tool of Perform Quality Assurance that checks to see if the process is being followed. Choices 'A' and 'D' are incorrect, and choice 'B' is referring to inspection, which is a tool of Perform Quality Control.

6. B. If two events have no bearing on each other, they are statistically independent.

Choice 'D' is when two events cannot both happen at the same time.

7. B. That is how a control chart is used. It visually depicts whether a process is in or out of control. Choice 'A' is used in Perform Quality Control to rank problems by frequency. 'C' is used in Quality Planning to anticipate problems in advance. Choice 'D' is used in Perform Quality Control to pick random samples to inspect.

8. A. Pareto charts rank defects from greatest to least, showing you what should get the most attention.

9. C. Perform Quality Control is the best choice here. Your clue here was the fact that your engineers had inspected something specific. This wasn't related to planning or process- it was a physical inspection of a work result, and that is what happens during the Perform Quality Control process.

10. A. Benchmarking takes results from previous projects and uses them to help measure quality on your project. Benchmarks give you something against which you can measure.

11. B. Total Quality Management stresses, among other things, that everyone contributes to the quality of the product and process.

12. D. Six sigma represents 99.99966% of all work results that will be of acceptable quality in the manufacturing process. This is higher than 99% or 3 sigma, which represents a 99.73% quality rate.

13. A. Quality Planning should always happen first. Perform Quality Assurance and Perform Quality Control would come after the quality management plan is in place. 'D' is not a PMI process. Keep in mind that the quality processes do run in a cycle, but planning should always happen first.

14. B. If no organizational quality policy exists, you should develop one for this project. 'A' and 'C' are incorrect since you should not proceed without a quality policy. 'D' would be a good way to lose your job! You should try to fix this problem yourself rather than force your organization to write a quality policy.

15. C. The quality specification is the customer's quality requirements. 'A' represents the limits for what is in and out of statistical control, typically set at three standard deviations from the mean. 'B' is the average of all of the data points. 'D' is a statistical term relating to the way the data points are scattered.

16. D. Audits, part of the Perform Quality Assurance process, review the process and make sure that the process is being followed.

17. D. The PMP Exam has several questions structured like this one. You could have any of the problems listed here, but the one you would be most concerned with is parts arriving late. An organization that practices just-in-time (JIT)does not keep spare inventory on hand. Instead, the inventory is ordered so that the parts arrive only slightly before they are needed.

18. A. Rework and increased cost are likely outcomes of low quality. 'B' and 'C' are incorrect since decreased cost is not related to low quality. 'D' is incorrect, because you would not lower the quality limits or specifications just because the quality is bad.

19. B. Audits are part of Perform Quality Assurance. Choices 'A', 'C', and 'D' are all part of Quality Planning.

20. C. Statistical sampling (may appear as random sampling) is the best choice. If the overall population is too large, accurate sampling can give you the same statistical results as measuring the entire population. 'A' might be a good choice in the real world, but you should focus on the inputs, tools, and outputs contained in the PMBOK Guide for the exam, and automated testing tools is not a part of quality management.

Human Resource

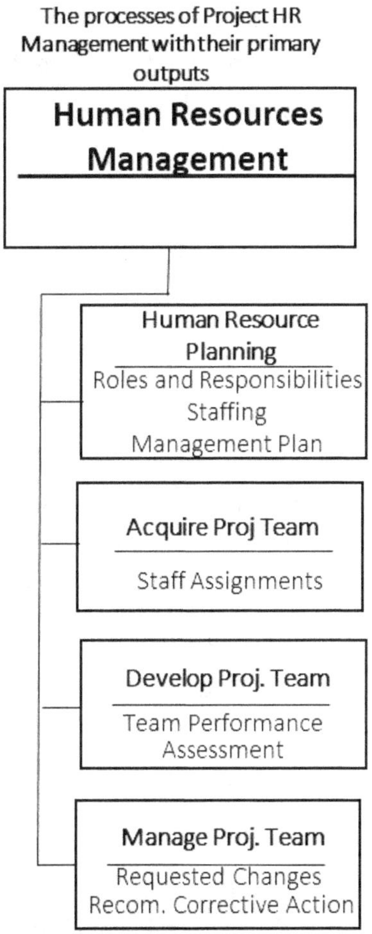

The processes of Project HR Management with their primary outputs

Human Resources Management

Human Resource Planning
Roles and Responsibilities
Staffing
Management Plan

Acquire Proj Team
Staff Assignments

Develop Proj. Team
Team Performance Assessment

Manage Proj. Team
Requested Changes
Recom. Corrective Action

Although the PMBOK Guide treats the subject of human resources management lightly, you should expect several questions on the exam that cover a variety of theories, including leadership, motivation, conflict resolution, and roles within a project.

The content for this area of the test is drawn from basic management theory, organizational behavior, psychology and of course, the field of human resources. If you have ever formally studied these subjects, you have probably been exposed to many of the theories in this chapter.

In managing a project, the project manager must also lead people. Some project managers may excel at organizing tasks and planning activities and be dismal at motivating other people.

This chapter covers the processes, inputs, tools, and outputs you will need to know in order to pass the exam.

Philosophy

PMI's approach to the area of human resource management is to define a role for everyone on the project and to define the responsibilities for each of these roles.

Many people make the mistake of only understanding the role of the project manager, and never understanding the proper role of senior management, the sponsor, or the team. Instead of merely looking at their own role on the project, project managers must help define the roles and influence everyone who has a role on the project.

PMI's philosophy of leadership and power are based on the realization that the project manager is rarely given complete and unquestioned authority on a project. Instead, he must be able to motivate and persuade people to act in the best interest of the project and must be able to build a team and lead members to give their best effort to the project.

Importance

After professional responsibility, human resource management questions are considered by many people to be the easiest questions on the exam. This is because this is one of the few sections where many of the questions can often be answered by using common sense.

Unless you find these questions particularly difficult, it is a good idea to get comfortable with the information in this chapter and focus your attention on more challenging subject matter. A careful review of the theories and content prior to taking the exam should be adequate to help you through this section.

Preparation

The focus of this chapter will be on roles and responsibilities, motivational theories, forms of power, and leadership styles. While this knowledge area is considered to be one of the easiest ones, take the time to learn it. Questions from this section will appear on the exam.

This chapter, in particular, contains a significant amount of content that is not found in the PMBOK Guide.

Human Resource Management Processes

There are four processes within project human resource management. In the PMI framework, these processes touch three process groups: planning (Human Resource Planning), executing (Acquire Project Team and Develop Project Team), and controlling (Manage Project Team).

Process Group	Human Resource Management Process
Initiating	(none)
Planning	Human Resource Planning
Executing	Acquire Project Team, Develop Project Team
Controlling	Manage Project Team
Closing	(none)

The primary outputs associated with the four human resource management processes are shown in the table below.

Process	Primary Outputs
Human Resource Planning	• Roles and responsibilities • Project organizational chart • Staffing management plan
Acquire Project Team	• Staff assignments
Develop Project Team	• Team performance assessment
Manage Project Team	• Change requests • Recommended corrective actions

	• Recommended preventive actions

Human Resource Planning

WHAT IT IS:

It is a common pattern throughout the PMI processes to have a knowledge area that starts off with a planning process to set the tone for the remaining processes in that area.

This is the case with Human Resource Management, which begins with Human Resource Planning. This process gives guidance to the rest of the human resource processes, defining the roles and responsibilities and creating the staffing management plan.

WHY IT IS IMPORTANT:

This process lays out how you will staff, manage, team-build, assess, and improve the project team.

WHEN IT IS PERFORMED:

Human Resource Planning typically takes place very early on the project, and it may be performed iteratively. In other words, you may do some work on the scope, some work on the schedule, and then plan human resources, and then do more work on the scope, returning again to human resources, and so on. It should not be thought of as a process that is tackled only one time.

HOW IT WORKS:

- Inputs

- *Enterprise Environmental Factors - See Chapter 2, Common Inputs*
- *Organizational Process Assets - See Chapter 2, Common Inputs*
- *Project Management Plan - See Chapter 2, Common Inputs*

- Tools
  - *Organization Charts and Position Descriptions*

There are many way to represent who will be working on the project and what they will be responsible for doing. For the exam, you need to know about three primary formats:

A. Hierarchical

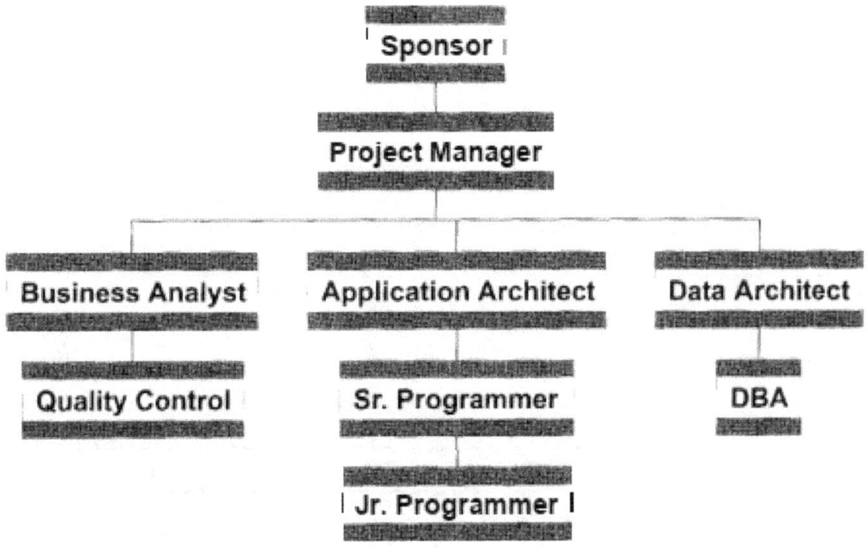

B. Matrix

Matrix charts are used to illustrate which roles on the project will be working with which work packages and what their responsibilities will be. One of the most popular categories is the Responsibility Assignment Matrix (RAM), which displays work packages in the rows and the roles in the columns. Each cell shows how that role will work on that particular work package.

One popular type of the RAM is known as the RACI chart (pronounced “ray-cee”). RACI charts, such as the one that follows, list each work package in the rows and list the roles in the columns. RACI charts derive their name from the way each cell is assigned either an 'R' for Responsible, 'A' for Accountable, 'C' for Consult, or 'I' for Input.

Generally, only one person is assigned accountability for a work package, but more than one person may be responsible for performing the work on a work package.

Work Package \ Role	Project Manager	Business Analyst	Data Architect	Application Architect	Jr. Programmer	Sr. Programmer	Quality Control
Document scope	C	R					A
Review scope	R	I	C	C	C	C	C
Approve scope	R	A	A	A	A	A	A
Create database	C	C	R	I			
Design application	I	I	I	R	I	A	C
Code application	I	I	I	I	A	R	C
Application testing	I	I	I	I	I	I	R

R = Responsible, A = Accountable, C = Consult, I = Inform

C. Text

Text formats basically follow the format of a position description, detailing out what responsibilities each position on the project will involve and what qualifications will be needed to fill these positions. This tool is particularly useful in recruiting.

- ***Networking***

Networking is the process of communicating with others within your "network" of contacts. By tapping into his network of contacts, a project manager can leverage his expertise on issues related to human resources on the project.

- ***Organizational Theory***

Groups behave differently than individuals, and it is important to understand how organizations and teams behave. Familiarizing yourself with the vast amount of work that has been done to understand organizational theory can pay dividends throughout the project.

- Outputs

  - ***Roles and Responsibilities***

Each role on the project should be defined with a title, its level of responsibility, authority, and the skill level or competency needed to be able to perform this role.

- ***Project Organization Charts***

A project organization chart shows how team members relate on a project. This is particularly important since reporting relationships that exist on a project will often be different than those which exist in the organization.

- ***Staffing Management Plan***

The staffing management plan, which is part of the project management plan, is the plan which details how and when the project will be staffed, how and when the staff will be released, and other key human resources components such as how they will be trained.

One common component of the staffing management plan is a resource histogram. A resource histogram (see example below) simply shows the usage for resources for a given period of time. On most projects, resources increase from planning through executing, falling off in control and closure.

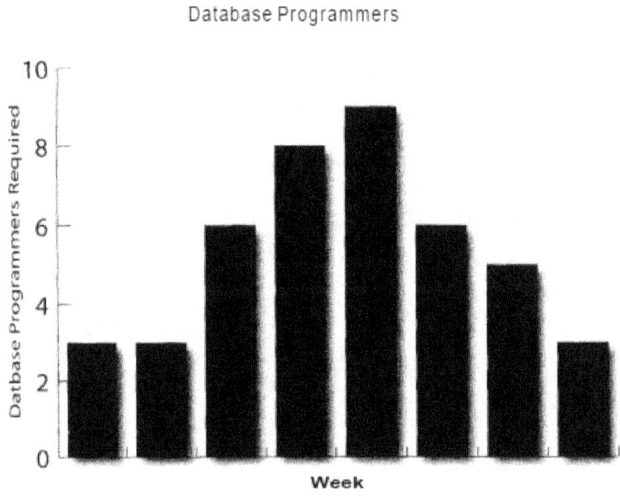

Sample Resource Histogram

Acquire Project Team

WHAT IT IS:

Acquire Project Team is another process that sounds exactly like what it is. This process focuses on staffing the project. Because it is an executing process, you can think of it as the process that carries out the staffing management plan.

WHY IT IS IMPORTANT:

This process gets the right people working on the project. Careful attention to this process should pay off in the form of the quality of staff you bring on.

WHEN IT IS PERFORMED:

Make sure that you understand that the process of Acquire Project Team is typically performed throughout the project, as you may need different skill sets throughout the life of the project. For instance, you may need business analysts early on in the life of the project, while you may need more quality engineers later in the project. Acquire Project Team would be performed as long as the project was adding new staff.

HOW IT WORKS:

- Inputs
  - *Enterprise Environmental Factors - See Chapter 2, Common Inputs*
  - *Organizational Process Assets - See Chapter 2, Common Inputs*
  - *Roles and Responsibilities*

As you are staffing the project, the roles and responsibilities you defined earlier in Human Resource Planning (usually in the form of position descriptions) will be indispensable here.

  - *Project Organization Charts*

The organization charts show the number of people, how they will relate to each other on the project, and how they fit into the overall project organization.

  - *Staffing Management Plan*

The staffing management plan details when resources will be needed and how long they will be needed. Other human resources information that would be useful to functional managers and the team may be included, such as whether the resources will be internal or external, how they will be released, etc.

- Tools

  - *Pre-Assignment*

It is normal on project for the roles to be defined first. Later, resources are assigned to perform those roles and fulfill the responsibilities; however, occasionally specific resources will be pre-assigned to fill a role. This may occur before the staffing management plan has been created and even before the project formally begins.

  - *Negotiation*

Negotiating is an important skill for project managers to cultivate. Project managers often have to negotiate for resources, both inside and outside the organization.

  - *Acquisition*

The tool of acquisition, as used here, can be a bit misleading, since the overall process is "Acquire Project Team." The tool of acquisition refers to looking outside the organization for resources when they cannot be provided by your organization.

  - *Virtual Teams*

Virtual teams have become much more popular over recent years. A virtual team is a group of individuals who may or may not see each other in person. Instead, they typically use communication tools to meet online, share information, and collaborate on deliverables.

- Outputs

- *Project Staff Assignments*

The assigned staff is the primary output of this process. Each role that was defined should have a resource assigned to it. Understand that these assignments may happen several times throughout the process as resources are needed. For instance, it would typically be difficult to assign a particular person to a role that will not be needed for a year.

- *Resource Availability*

As resources are assigned to the project, the time they are assigned to work on activities should be documented. Each resource's forecasted time on the project should be documented.

- *Staffing Management Plan Updates*

It is normal for the staffing management plan to undergo revisions and updates as the staff are assigned. Specifically, as things work or do not work when staffing a project, elements of the staffing management plan may be updated to reflect re-planning and corrective action.

Develop Project Team

WHAT IT IS:

Where the exam is concerned, Develop Project Team is the most important process in Project Human Resource Management. It is an executing process that focuses on building a sense of team and improving their performance.

WHY IT IS IMPORTANT:

A team performs better than a group of disconnected individuals. This is true not only for sports teams, but also for project teams.

WHEN IT IS PERFORMED:

The process of Develop Project Team is performed throughout the project. In other words, as long as there is a team on the project, you should perform this process, but it is considered to be most effective when done early.

HOW IT WORKS:

- Inputs
  - *Project Staff Assignments*

The staff assignments, which were made as part of the Acquire Project Team process, are brought into this. The rationale behind this is that the staff assignments contain a list of all team members for the project.

  - *Stuffing Management Plan*

The staffing management plan provides the guidance for this entire process. It outlines how project team members will be trained and how Develop Project Team will be conducted. This may include such things as offering flex time, bonus pay, or options of telecommuting.

  - *Resource Availability*

The resource availability specifies, for each resource, when project team members would be available to take part in team building activities.

- Tools

- *General Management Skills*

General management skills cover a whole host of topics, but are particularly focused on producing key results by driving and managing tasks and motivating team members.

There is a difference between leading and managing on a project. Managing has been defined as being about producing key results, while leading involves establishing direction, aligning people to that direction, and motivating and inspiring.

There are several different styles of leading that are recognized throughout the field of project management. The following graphic may prove very helpful for the exam. In the early phases of the project, the project manager should take a very active role in the leadership of the project, usually directing the activities and providing significant leadership. As the project progresses, however, other styles of leading may be more appropriate. These styles of leading are less heavy-handed. Of course, the particular style of leading needed will vary from one project to the next.

- *Forms of Power*

Project managers, especially those in matrix and functional organizations, are often tasked with responsibility for the project without much formal authority in the organization.

Understanding the forms of power can help the project manager maximize his ability to influence and manage the team.

A. Reward Power

Reward power is the ability to give rewards and recognition. Examples include a pay raise, time off, or any other type of reward that would motivate a team member.

B. Expert Power

Expert power exists when the manager is an expert on the subject. For instance, the person who architected a part of a software system would probably have significant expert power on a project that used that system. People would listen to the architect because he had credibility. A subject matter expert usually has significant power to influence and control behavior.

C. Legitimate

Also known as formal power, legitimate power is the power that the manager has because of his position. This type of power comes from being formally in charge of the project and the people and has the backing of the organization.

Strong, broad-based, formal authority for a project manager is unusual and would typically indicate a projectized organizational structure.

D. Referent

Referent power is a form of power that is based on respect or the charismatic personality of the manager. It is ultimately rooted in a persuasive ability with people. Another usage of referent power is when a less powerful person allies with a more powerful person and leverages some of the superior's power. For instance, if the project manager is very close to the CEO of the company, his power will probably be higher because of that alliance.

E. Punishment

Also known as coercive power, this type of influence is the ability to punish an employee if a goal is not met. "If this module does not pass quality control by the end of next week, you are all fired," would be an example of a manager using punishment power.

F. Best Forms of Power

In addition to being able to identify the different types of power a project manager can use, you should also know that PMI considers reward and expert the most effective forms of power and punishment the least effective.

- *Training*

Training can include a wide range of activities, but it may be thought of as any instruction or acquisition of skills that increases the ability of the team or individuals to perform their jobs.

If a team member does not have the skills needed to carry out their responsibilities, then training may be a good option. In general, training is highly favored for the exam, but in most cases it should be paid for by the performing organization or the functional manager and not by the customer or the project.

- *Team-Building Activities*

A team-building activity is any activity that enhances or develops the cohesiveness of the team. This usually occurs by focusing on building bonds and relationships among team members. It is important to keep in mind that although team-building may be treated as a special event, it can occur while performing regular project responsibilities, and it actually becomes more important as the project progresses.

Team-building cannot be forced. It should be modeled by the project manager, who should work to include all members of the team and create a shared goal.

- *Ground Rules*

A project's ground rules are the formal or informal rules that define the boundaries of behavior on the project. For instance, a project may lay down ground rules that say that everyone on the project shares responsibility for protecting the security of project data. In this case, everyone on the project would be expected to treat information with the same general caution. It is important that ground rules be defined and communicated to the team members.

- *Co-Location*

Co-location is the act of physically locating team members in the same general space. The most common example of this is to create a war room where all the team members work, or to co-locate the project team at the customer's site.

- *Recognition and Rewards (Theories of Motivation)*

Recognition and reward systems are typically defined as part of the staffing management plan. In general, desirable behaviors should be rewarded and recognized. For the exam, you should focus on win-win rewards as the best choices for team building. Win-lose rewards, such as a contest where one team member wins and the others do not, can be detrimental to the sense of team.

There is a substantial body of knowledge on recognition and reward theories. These theories, also known as Theories of Motivation, regularly appear on the PMP exam. A thorough understanding of several theories is needed to be fully prepared to pass the exam, including Maslow's Hierarchy of Needs, McGregor's Theory X and Y, Contingency Theory, Herzberg's Motivation-Hygiene Theory, Expectancy Theory, and Achievement Theory.

A. Maslow's Hierarchy of Needs

Maslow's Hierarchy of Needs is a basic theory of human motivation that project managers should understand. Abraham Maslow grouped human needs into five basic categories as illustrated in the following diagram.

Maslow's theory states that these needs form a hierarchy, since the needs at the bottom must be satisfied before the upper needs will be present. As an example, people cannot reach their full potential if they do not have sufficient food or safety.

Every project manager should understand the needs of the team members and how they interrelate so that he can help them to perform at their full potential.

B. McGregor's Theory X and Theory Y

McGregor's organizational theory states that there are two ways to categorize and understand people in the workplace.

Managers who ascribe to Theory X presume that people are only interested in their own selfish goals. They are unmotivated, they dislike work, and they must be forced to do productive work. Theory X managers believe that constant supervision is necessary to achieve desired results on a project.

Those who practice Theory Y assume that people are naturally motivated to do good work. "Y managers" believe that their team members need very little external motivation and can be trusted to work toward the organization's or project's goals.

An assembly line organization may treat everyone as an "X Person," monitoring and measuring every move, whereas an organization that encourages telecommuting might be more prone to treat employees as "Y People." However, it should be understood that it is the manager, not the organization, that ascribes to Theory X or Theory Y, and the style of management is not necessarily determined by the type of work being performed.

C. Contingency Theory

The Contingency Theory, developed by Fred E. Fiedler in the 1960s and 1970s, states that a leader's effectiveness is contingent upon two sets of factors. The first set of factors measures whether the leader is task-oriented or relationship-oriented. The second set evaluates situational factors in the workplace, such as how stressful the environment is.

The practical application of this theory suggests that in stressful times, a task-oriented leader will be more effective, while in relatively calm times, a relationship-oriented leader will function more effectively. The inverse is also true. What makes a leader effective in one setting may actually work against him in another.

D. Herzberg's Motivation-Hygiene Theory

This theory has nothing to do with personal hygiene as the name might incorrectly cause you to conclude. Instead, Herzberg conducted studies to quantify what factors influence satisfaction at work.

Similar to Maslow's theory, Herzberg's Motivational-Hygiene theory states that the presence of certain factors does not make someone satisfied, but their absence will make someone unsatisfied. In this case, hygiene factors must be present, but they do not motivate by themselves. Motivation factors will motivate, but they will not work without the hygiene factors in place.

Hygiene Factors	Motivation Factors
Company Policy	Achievement
Supervision	Recognition
Good relationship with boss	Work
Working conditions	Responsibility
Pay check	Advancement

Personal life	Growth
Status	
Security	
Relationship with co-workers	

Herzberg's theory is based on the assumption that workers want and expect to find meaning through their work.

E. Expectancy Theory

The expectancy theory states that the anticipation of a reward or good outcome is a motivation. This theory is often employed in training animals, and it relies on providing a reward for a job well done.

For instance, if you promise your team a day off if a certain milestone is achieved, that would possibly motivate them to work harder to reach that goal, but expectancy theory only works if the person or team believes the outcome is achievable. If a sales team is promised a $10,000 bonus if they reach a goal but none of them believe the goal is remotely possible, they will probably not be motivated by the reward and will not work harder to reach the goal.

F. Achievement Theory

Achievement theory states that people need three things: achievement, power, and affiliation. Achievement is the desire to accomplish something significant. Power is the desire to influence the behavior of others. Affiliation is the desire to belong to a group, or to fit in with your coworkers.

Achievement theory focuses on using these three human desires as the means to motivate employees.

- Outputs

    - *Team Performance Assessment*

The output of team performance assessment focuses on measuring and evaluating how the team is doing. For instance, you may measure the work performance, the experience the team is acquiring, the turnover rate, or even such factors as the schedule performance index and similar metrics.

Manage Project Team

WHAT IT IS:

Keep in mind that any time you encounter a controlling process, you are generally going to be making sure that the plan matches up with the results. That holds true here, where you are managing the team. You need to ensure that the staffing management plan is working and change it where it must be changed. You need to handle changes to your team, performance problems, and anywhere the plan does not match up with reality.

But keep in mind that even if the differences are positive, you still need to document the differences between the plan and the results. For instance, if you planned for 7% turnover and had no turnover on the project, you would want to document that so that a future team could use your lessons learned to build on your successes.

WHY IT IS IMPORTANT:

Out of all the areas on the project, the human resource side often has the most trouble matching the plan. People don't perform as expected. Some leave the project, teams experience unexpected conflict, individuals suffer from low morale, and all of these events directly affect objective measures such as the budget, the schedule, and quality.

This uncertainty becomes even more challenging when you consider that team members often report to a different functional manager and have "dotted line" responsibilities to the project manager.

In the Manage Project Team process, the project manager considers all of these factors and works to keep the team at their optimal performance.

WHEN IT IS PERFORMED:

Manage Project Team is performed as long as there is a team on the project.

HOW IT WORKS:

- Inputs

The first five inputs are basic ingredients for making this process work.

- *Organizational Process Assets - See Chapter 2, Common Inputs*
- *Project Staff Assignments - See Acquire Project Team, Outputs*
- *Roles and Responsibilities - See Human Resource Planning, Outputs*
- *Project Organization Charts - See Human Resource Planning, Outputs*
- *Staffing Management Plan - See Human Resource Planning, Outputs*
- *Team Performance Assessment*

The team performance assessment is an important and ongoing input into the Manage Project Team process. The project manager should regularly assess the performance of the team so that issues can be identified and managed. The format and frequency of the assessment may vary.

- *Work Performance Information*

While the project work is being performed, the project manager monitors the performance of team members. How the team members perform their job, how they handle their responsibilities, and how they meet their deadlines are all considered and brought into the process of Manage Project Team.

- *Performance Reports*

The performance reports are an objective measure of progress against the plan. Such items as scope, time, and cost are measured carefully, and any difference between these and the plan is documented.

- Tools and Techniques

  - *Observation and Conversation*

These informal tools are used to monitor team morale and identify problems, whether potential or real.

- *Project Performance Appraisals*

A project performance appraisal is where the project manager and other personnel managers on the project meet with the people who report to them on the project and provide feedback on their performance and how they are conducting their job.

In recent years, the tool of 360 degree feedback has gained in popularity. Using this tool, feedback is provided from all directions and often from individuals both internal and external to the project, and occasionally even vendors and external contractors.

- *Conflict Management*

Managing conflict in a constructive way helps improve team morale and performance. As discussed earlier in this chapter, conflict may occur between any individuals or groups on the project, but it most often occurs between project managers and functional managers. It is important to manage and resolve conflict; however, if conflict cannot be resolved among the parties involved, it should be escalated.

Conflict that hurts the project should be dealt with in progressively more official channels, such as escalation to the project manager, escalation to the functional managers, and ultimately escalation to the human resources department.

A. Methods of Conflict Management

Consider for a moment the problem of a door that is stuck shut. There are several ways to approach this problem:

- You may want to throw your weight against the door, pounding it with your shoulder.
- You might elect to try to go in the room from another point of entry.
- You could try to take the hinges off the door to make it come apart.
- You might choose to ignore the problem of the stuck door, avoiding it altogether, or hope someone else will take care of it.
- You could attempt to find out why the door was stuck in the first place and deal with that problem.

In the same way there are several ways to approach conflict resolution. Because conflict is inevitable, you should be aware of the common ways of handling it:

- **Problem- Solving**

Problem-solving involves confrontation, but it is confrontation of the problem and not the person! It means dealing with the problem head on. Using this technique, the project manager gets to the bottom of the problem and resolves the root causes of the conflict. One common term in problem-solving is "confrontation."

Although the word confrontation may have negative connotations, this type of conflict resolution is highly favored, as it is proactive, direct, and deals with the root of the problem. Consequently, it is most often the correct answer on the exam when questions of conflict resolution arise.

- **Compromise**

Compromise takes place when both parties sacrifice something for the sake of reaching an agreement. On the test, compromise may be presented as "lose - lose" since both parties give up something.

- **Forcing**

Forcing is exactly what the name implies. It is bringing to bear whatever force or power is necessary to get the door open. Although forcing may work well in the case of a stuck door, this is considered to be the worst way to resolve project conflict. Forcing doesn't help resolve the underlying problems, it reduces team morale, and it is almost never a good long-term solution.

- **Smoothing**

Using smoothing, the project manager plays down the problem and turns attention to what is going well. The statement "We shouldn't be arguing with each other. Look at how well we've done so far, and we're ahead of schedule," would be an example of smoothing.

Smoothing downplays conflict instead of dealing with it head on and does not produce a solution to the conflict. Instead, smoothing merely tries to diminish the problem.

- Withdrawal

Withdrawal is technically not a conflict resolution technique but a means of avoidance. A project manager practicing withdrawal is merely hoping the problem will go away by itself. Needless to say, PMI does not favor this method of conflict resolution because the conflict is never resolved.

- *Constructive and Destructive Team Roles*

Related to the area of conflict management is the project manager's ability to recognize and deal with constructive and destructive roles on his or her team.

Constructive Team Roles	
Initiators	Clarifiers
Information seekers	Harmonizers
Information givers	Summarizers
Encouragers	Gate Keepers

Destructive Team Roles	
Aggressors	Topic Jumper
Blockers	Dominator

Withdrawers	Devil's Advocate
Recognition Seekers	

- *Constructive Team Roles*

A. Initiators

An initiator is someone who actively initiates ideas and activities on a project. This role is considered positive because it is proactive and can be highly productive.

B. Information seekers

Information seekers are people on the team who actively seek to gain more knowledge and understanding related to the project. This is a positive role because fostering an understanding among the team is important, and open communication should be valued.

C. Information givers

An information giver, as its name implies, is someone who openly shares information with the team. Although not all information may be shared (for instance Classified or secret information must be kept confidential), the overarching principal is to foster good communication and a good flow of information on the project.

D. Encouragers

Encouragers maintain a positive and realistic attitude. On the project, they focus on what can be accomplished, not on what is impossible. This is a positive role because it contributes to team morale.

E. Clarifiers

A clarifier, as the name suggests, is someone who works to make certain that everyone's understanding of the project is the same. This is a positive role because it ensures that everyone has a common understanding of the project goals and details.

F. Harmonizers

In music, harmony is not the same as the melody, but it complements and enhances the melody. Similarly, a harmonizer on the project will enhance information in such a way that understanding is increased. This is a positive role because the overall understanding of the project and the project context, or the details surrounding it, is enhanced.

G. Summarizers

Summarizers take the details and restate them succinctly or relate them back to the big picture. This is a positive role because details on the project may become overwhelming, but the summarizer can keep things simple enough for everyone to understand the higher purpose of the tasks.

H. Gate Keepers

The term gate keeper has two possible uses in project management literature. The first definition is used differently in project management than it is in other business disciplines. A gate keeper is someone who draws others in. Someone who says, "We haven't heard from the other end of the table today," would be an example of a gate keeper. This is a very positive role because it encourages the entire team to participate on the project.

The other usage comes from someone who judges whether the project should continue at different stages (known as the stage-gate approach).

This gate keeper makes decisions about whether the project is still achieving the business need and if it is justified in continuing to a subsequent phase. Both of these usages of gate keeper are considered to be constructive roles on the project.

- *Destructive Team Roles*

The effective project manager will be able to identify destructive roles within the team and diminish or eliminate them.

A. Aggressors

An aggressor is someone who is openly hostile and opposed to the project. This is a negative role because it serves no productive purpose on the project.

B. Blockers

A blocker is someone who blocks access to information and tries to interrupt the flow of communication.

This is a negative role because of the disruptive effect poor communication can have on a project.

C. Withdrawers

A withdrawer does not participate in discussion, resolution, or even the fleshing out of ideas. Instead, he is more likely to sit quietly or not participate at all. This is a negative role because it usually produces a team member that does not buy in to the project and can have a negative effect on the overall team morale.

D. Recognition Seekers

A recognition seeker looks at the project to see what is in it for him. He is more interested in his own benefit rather than the project's success.

This is a negative role because of the damaging effect on team morale and because a recognition seeker may ultimately jeopardize the project if doing so somehow personally benefits him.

E. Topic Jumper

A topic jumper disrupts effective communication by constantly changing the subject and bringing up irrelevant facts. This is a negative role, because it prevents issues from being fully discussed and brought to closure.

F. Dominator

The dominator is someone who disrupts team participation and communication by presenting opinions forcefully and without considering the merit of others' contributions. He will likely talk more than the rest of the group and will bully his way through the project. This is a negative role because valid opinions are often quashed, and the project may take on a one-dimensional quality.

G. Devil's Advocate

A devil's advocate is someone who will automatically take a contrary view to most statements or suggestions that are made. This may be a positive or negative role on the project, but it is most often associated with a negative role since it often disrupts and frustrates communication, discourages people from participating, and stalls progress.

- *Sources of Conflict*

One last area of conflict management you should understand is based on research that suggests that the greatest project conflict occurs between project managers and functional managers. Most conflict on a project is the result of disagreements over schedules, priorities, and resources. This finding runs contrary to a commonly held belief that most conflicts are the result of personality differences.

- *Issue Log*

An issue may be thought of as anything that threatens project progress. It could be specific, such as a technical concern, or general, such as a personality conflict among team members. A documented log of issues is important, since it gives the project team a place to record issues that require resolution. Along with each issue, the person or people responsible for resolving the issue should be documented, as well as due dates for the desired resolution.

- **Outputs**

    - ***Requested Changes***

Requested changes to the staff could occur for a number of reasons. The team could be dysfunctional, the workers could be under-qualified or over-qualified, or you may need to augment the staff with more resources than previously planned. Any of these or other scenarios that change the staffing management plan should be documented and processed through the change control system.

- *Recommended Corrective Action*

Corrective action is anything done to bring future results in line with the plan. Where the Manage Project Team process is concerned, this would include anything that helps the team perform as planned. This might include incentives, changing staff, or getting them more training to augment their skills.

- *Recommended Preventive Action*

Preventive action is all about helping to prevent a problem from occurring. For instance, the development of attendance policies could be viewed as preventive action that can help prevent attendance problems.

- *Organizational Process Assets Updates - See Chapter 2, Common Inputs*
- *Input to Organizational Performance Appraisals*

The project manager's feedback on the performance of team members to their functional managers is valuable and will help in planning future projects.

- *Lessons Learned Documentation*

Lessons learned focus on variances between what was planned and what occurred. Here, your lessons learned would document things that you would do differently if you had the project to do again. The lessons learned can be a valuable input for future projects. It helps ensure that the organization keeps learning from historical information.

- *Project Management Plan Updates - See Chapter 2, Common Outputs*

Human Resource Management Questions

1. If you hear a project manager saying to a customer "We all agree that this project is important. Let's not fight over a few thousand dollars," what conflict resolution technique is the project manager trying to use?

   A. Smoothing.

   B. Problem Solving.

   C. Forcing.

   D. Compromising.

2. Who manages the resources in a matrix organization?

   A. Senior management.

   B. Functional managers.

   C. Project manager.

   D. Human resources.

3. What is considered the LEAST desirable form of power for a project manager to exercise?

   A. Formal.

   B. Referent.

   C. Punishment.

   D. Forcing.

4. Which statement below BEST matches a Theory X manager's beliefs:

   A. People want to be rewarded for their work.

   B. People have higher needs that will not emerge until the lower needs have been satisfied.

   C. People will contribute to work if left alone.

   D. People cannot be trusted.

5. The staffing management plan:

   A. Must be created by the human resources department.

   B. Is a part of the resource management plan.

   C. Is a tool of team development.

   D. Is an output of the Human Resource Planning process.

6. Which technique produces the most lasting results?

   A. Problem-solving.

   B. Smoothing.

   C. Compromising.

   D. Withdrawing.

7. The most important role of the project sponsor is to:

A. Manage and resolve conflicts between the team and upper management.

B. Provide and protect the project's financial resources.

C. Provide and protect the project's human resources.

D. Balance the project's constraints regarding time, scope, and cost.

8. Human resource management encompasses:

A. Organizational Planning, Staff Acquisition, Performance Reporting, and Manage Project Team.

B. Human Resource Planning, Staff Acquisition, Performance Reporting, and Develop Project Team.

C. Human Resource Planning, Staff Acquisition, Develop Project Team, and Release Project Team.

D. Human Resource Planning, Acquire Project Team, Develop Project Team, Manage Project Team.

9. Which of the following is NOT an input into Human Resource Planning?

A. Enterprise environmental factors.

B. Roles and responsibilities.

C. Organizational process assets.

D. Project management plan.

10. Which of the following is a constructive team role?

    A. Information seeker.

    B. Recognition seeker.

    C. Blocker.

    D. Devil's advocate.

11. Maslow's Hierarchy of Needs theory states that:

    A. The strongest motivation for work is to provide for physiological needs.

    B. Hygiene factors are those that provide physical safety and emotional security.

    C. Psychological needs for growth and fulfillment can be met only when lower-level physical or security needs have been fulfilled.

    D. The greater the financial reward, the more motivated the workers will be.

12. Which of the following is NOT true of team building?

    A. Team agreement should be obtained on all major actions.

    B. Team building requires role modeling on the part of the project manager.

    C. Team building becomes less important as the project progresses.

    D. Teamwork cannot be forced.

13. Team building is primarily the responsibility of:

A. The project team.

B. The project manager.

C. Senior management.

D. The project sponsor.

14. A war room is an example of:

A. Contract negotiation tactics.

B. Resource planning tools.

C. A functional organization.

D. Collocation.

15. Which of the following is NOT a process of human resources management?

A. Human Resource Planning.

B. Acquire Project Team.

C. Team Performance Reporting.

D. Develop Project Team.

16. A project coordinator is distinguished from a project manager in that:

A. A project coordinator has no decision-making power.

B. A project coordinator has less decision-making power.

C. A project coordinator has no authority to assign work.

D. A project coordinator takes on larger projects than a project manager.

17. Which of the following is not a tool used in Develop Project Team?

A. General Management Skills.

B. Training.

C. Ground Rules.

D. Encouragement.

18. One potential disadvantage of a matrix organization is:

A. Highly visible project objectives.

B. Rapid responses to contingencies.

C. Team members must report to more than one boss.

D. The matrix organization creates morale problems.

19. A project manager in Detroit is having difficulty getting the engineers in his company's Cleveland office to complete design documents for his project. He has sent numerous requests to the VP of Engineering (also in Cleveland) for assistance in getting the design documents, but so far his efforts have been unsuccessful. What kind of organization does this project manager work in?

A. Functional.

B. Hierarchical.

C. Strong matrix.

D. Projectized.

20. Which of the following is not true about a project's ground rules?

    A. Ground rules should be communicated to all team members.

    B. Ground rules should be consistent across projects in an organization.

    C. Ground rules should be clearly defined.

    D. Ground rules define behavioral boundaries on a project.

Answers to Human Resource Management Questions

1. A. Smoothing occurs when the person trying to resolve the conflict asks everyone to focus on what they agree upon and diminishes the items on which there is disagreement.

2. B. The functional manager has resource responsibilities in a matrix organization. In this type of organizational structure, the project manager must work with the functional managers to secure resources for a project. If you were tempted to choose 'C', keep in mind that the project manager primarily manages the project. The benefit of a matrix organization is that the project manager does not need to divert as much attention to managing the resources as he or she would in a projectized organization.

3. C. Punishment. 'D' is a problem solving technique - not a form of power.

4. D. Theory X managers believe that people cannot be trusted and must be watched and managed constantly.

5. D. The staffing management plan is created during the Human Resource Planning process.

6. A. Problem-solving (sometimes referred to as confrontation) is getting to the root of the problem and is the best way to produce a lasting result and a real solution.

7. B. This question comes from Chapter 2 - Foundational Terms and Concepts. The project sponsor provides the funds for the project. He may or may not take on other roles, but this is his defining role on the project.

8. D. The four processes are Human Resource Planning, Acquire Project Team, Develop Project Team, Manage Project Team, and yes, you do need to know all of them before taking the PMP Exam.

9. B. Role and responsibility assignments are the result, or output, of human resource planning. An even easier way to answer this question, however, was to realize that enterprise environmental factors, organizational process assets, and the project management plan are inputs into almost every planning process.

10. A. Information seeker. Recognition seekers are more concerned with getting in the spotlight than with facilitating communication. Blockers reject others' viewpoints and shut down discussion. Devil's advocate - bringing up alternative viewpoints - can be either positive or negative, but it is listed in most project management literature as a destructive team role because when it is negative it is very negative! Information seekers are constructive because they ask questions to gain information.

11. C. This question might have been difficult for you. 'A' is not necessarily true, because Maslow stated that any level of his "pyramid," provides the greatest level of motivation when the needs of the levels below have already been met. Thus physiological needs such as food and shelter will be the greatest motivator for workers to do a good job when those needs are unmet. But once the lower level needs are met, the needs of the next level become the greatest motivators.

12. C. Successful team building begins early in project development, but it is a continuous process throughout the life of the project.

13. B. Team building must be carried out under the direction of a strong leader. The project manager has the only project role that allows for regular, direct interaction with the team.

14. D. Collocation is the practice of locating all team members in a central location, or collocation. Another variation of a war room is a conference room devoted exclusively to use by a particular project team. It is a tool of Develop Project Team used in human resource management.

15. C. Team Performance Reporting is not a real process. It does sound similar to the Communications process of Performance Reporting, but this question asked which of the processes did not belong to the knowledge area of human resources management.

16. B. A project coordinator has some authority and some decision making power, but less than a project manager.

17. D. Encouragement may be a great idea, but it is not specified as a tool of Develop Project Team. 'A', 'B', and 'C' all are. If you missed this question keep in mind that you should favor the terms, vocabulary, and phrases you see here on the exam. Few people can commit all 592 inputs, tools and techniques, and outputs to memory, but you should learn to recognize them and pick out the ones that do not belong.

18. C. In a matrix organization, team members report to both the project manager and the functional manager. This can sometimes cause confusion and can lead to conflict on a project and within the organization.

19. A. The clue in the question that indicates a functional organization is the project manager's low authority; he must appeal to the head of the engineering department rather than making his request directly to the team members.

20. B. Ground rules may be unique to the project, and they certainly don't have to be the same across all projects in an organization. For instance, a project that has high security might have more stringent ground rules than a less secure one. 'A' and 'C' are incorrect, because clearly defining ground rules and communicating them to everyone helps to make sure they are understood and will be followed. 'D' is incorrect, because that is exactly what ground rules do - they define the boundaries of behavior that team members should respect.

Communications Management

The processes of Project Communications Management with their primary outputs

Communications Management

- **Communications Planning** — Communications Management Plan
- **Information Distribution** — Requested Changes
- **Performance Reporting** — Performance Reports, Requested Changes
- **Manage Stakeholders** — Resolved Issues, Approved Change Requests

Many consider communications management to be one of the easier knowledge areas to study for the PMP. Although project communications management can be very political and tricky in an organization, the volume of material for the test is moderate and not exceedingly difficult.

Many people who study this material are surprised that it is not related to the skill of communication through verbal and written media, in areas such as project writing styles, persuasion, and presentation methods. Rather, communications management covers all tasks related to producing, compiling, sending, storing, distributing and managing project records.

This knowledge area is made up of four processes to determine what to communicate, to whom, how often, and when to reevaluate the plan. It involves understanding who your stakeholders are and what they need to know.

Communications management also requires that you accurately report on the project status, performance, change, and earned value, and that you pay close attention to controlling the information to ensure that the communication management plan is working as intended.

Philosophy

There is an old joke in project management circles about "mushroom project management," in which you manage projects the same way you grow mushrooms - by keeping everyone buried in manure, leaving them in the dark, and checking back periodically to see what has popped up.

PMI's philosophy is, as you may have guessed, quite different. It focuses on keeping the stakeholders properly informed throughout the project.

Communication under PMI's philosophy may be a mixture of formal and informal, written and verbal, but it is always proactive and thorough. It is essential that the project manager distribute accurate project information in a timely manner and to the right audience.

Importance

Communications management is of medium importance on the exam, bordering on high. You may see several questions that relate directly to this chapter, so it will be necessary to become acquainted with the terms and theories presented here.

Preparation

Although the volume of material in communications management is smaller than most of the other areas, there are key concepts that must be learned. Be prepared for several questions on the test specifically related to the inputs, tools and techniques, and outputs for each process. The reason the focus is on these areas is that they are critical to the smooth operation of the processes, and most test takers do not find them intuitive.

Two other areas of key importance are the communications model and understanding channels of communication. You can expect to see questions about these on the exam.

There are not as many exam questions related to this section as there are to some of the others, and by carefully reviewing this chapter, and the exercises, you should be in good shape for the exam. There are four processes within project communications management. In the PMI framework, these processes touch three process groups: planning (Communications Planning), executing (Information Distribution), and monitoring and controlling (Performance Reporting, and Manage Stakeholders).

Process Group	Communications Management Process
Initiating	(none)
Planning	Communications Planning
Executing	Information Distribution
Controlling	Performance Reporting, Manage Stakeholders
Closing	(none)

The primary outputs associated with the four communications management processes are shown in the table below.

Process	Primary Outputs
Communications Planning	Communications management plan
Information Distribution	Requested changes
Performance Reporting	Performance reports Forecasts Requested changes
Manage Stakeholders	Resolved issues Approved change requests Approved corrective actions

Project Manager's Role in Communications

The project manager's most important skill set is that of communication. It is integral to everything the project manager does. You may see questions on your exam asking you what the project manager's most important job or most important skills are, or how most of the project manager's time is spent.

The answer is almost always related to communications. It is estimated that an effective project manager spends about 90% of his time communicating, and fully 50% of that time is spent communicating with the project team.

Also note that while communications take up a majority of the project manager's working day, one individual cannot control everything communicated on a project, nor should he try. Project managers who ask that every single e-mail or conversation be filtered through them first are generally demonstrating that they are not in control on the project. Instead, the project manager should be in control of the communications process. '

This is done by creating a strong communications management plan, adhering to it, and regularly monitoring and controlling.

Communications Planning

WHAT IT IS

For the exam, you should consider that Communications Planning is all about the communications management plan, its sole output. The communications management plan is, as you might guess, the plan that drives communication on the project. It defines:

- How often communications will be distributed and updated

- In what format the communications will be distributed (e.g. e-mail, printed copy, web site, etc.)

- What information will be included in the project communications

- Which project stakeholders will receive these communications

WHY IT IS IMPORTANT

This plan sets stakeholder expectations on the project, letting them know what information they will receive and when and how they will receive it.

If the project manager invests time in defining these lines of communication up front, conflict should be less than if it were undefined. Keep in mind that projects will vary greatly in how formally they define the communications management plan.

On a small project, it may not make sense for the project manager to go to great lengths to define an overly formal communications management plan.

WHEN IT IS PERFORMED

Like many planning processes, Communications Planning is typically performed early on the project, before regular project communications commence; however, it may be revisited as often as needed.

HOW IT WORKS

- Inputs

  - *Enterprise Environmental Factors - See Chapter 2, Common Inputs*

  - *Organizational Process Assets - See Chapter 2, Common Inputs*

- *Project Management Plan - See Chapter 2 Common Inputs*
- *Project Scope Statement*

One of the important components of the communications management plan is to communicate with the project's stakeholders. Since the stakeholders were identified during the Scope Definition process and documented in the project scope statement, the list of identified stakeholders should be brought into this process.

- Tools &Techniques
    - *Communications Requirements Analysis*

This one tool can cover quite a bit of ground. It is relatively simple to define, but sometimes quite tricky to perform on an actual project. The goal of this technique is to identify which stakeholders should receive project communications, what communications they should receive, how they should receive these communications, and how often they should receive them.

A. Communication Channels

A significant part of analyzing the project's communications requirements is determining the communication channels, or paths of communication, that exist within it. Expect at least a couple of questions on the exam to relate directly to this topic. Because the project manager needs to manage and be in control of project communications, it is important to understand that adding a single person on a project can have a significant impact on the number of paths or channels of communication that exist between people.

$$\text{Channels} = n(n-1)/2$$

(Where n = the number of people on the project)

The formula above for calculating communications channels looks complicated, but is actually a simple geometric expansion. Before memorizing the formula, refer to the two illustrations below. You can see from the drawing that four people produce six communication channels, as is confirmed by the formulas:

$$4^* (4-1) / 2 = 6$$

If there were five people, the formula would be applied as:

$$5^* (5-1) / 2 = 10$$

By understanding the drawings that follow and the way people interrelate to form communication channels, the concept should be easy to comprehend.

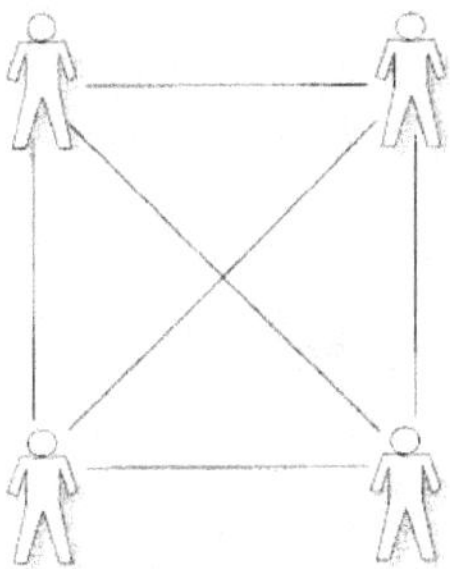

Four people create six communication channels as illustrated above.

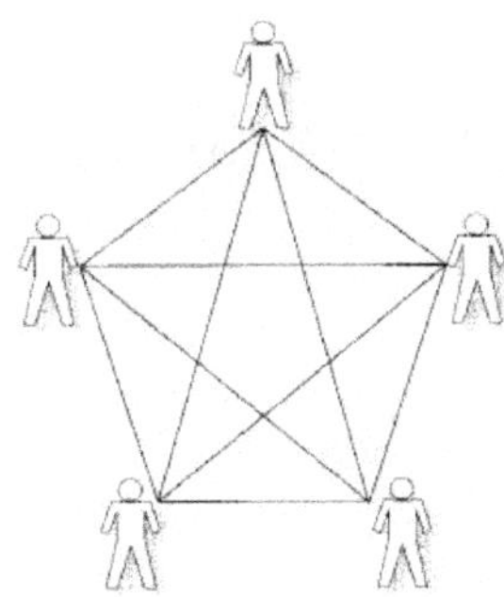

Five people create ten communication channels, or paths, as depicted in this illustration

B. Official Channels of Communication

The number of communication channels is of specific concern when analyzing the project's communication requirements. If there are a large number of channels of communication on the project, the project manager should work to define which communication channels are official. For example, it may be necessary to determine who can officially communicate with the customer or with key subcontractors.

- *Communications Technology*

Technology is a tool, and the right tool should be selected for a given communications need. Whereas formal face-to-face meetings may be needed for some projects, a project web site may be more appropriate for others. The technology should be tailored to the need.

- Outputs

  - *Communications Management Plan*

The communications management plan is part of the project management plan, and it defines the following:

- Who should receive project communications
- What communications they should receive
- Who should send the communication
- How the communication will be sent
- How often it will be updated

- Definitions so that everyone has a common understanding of terms

Information Distribution

WHAT IT IS

It is easiest to consider the process of Information Distribution as the execution of the communications management plan. In other words, the communications management plan lays out how communications will be handled, and the process of Information Distribution carries that out.

Keep in mind that while Information Distribution is performed according to the communications management plan, it must also be flexible so that unplanned information requests may be handled.

WHY IT IS IMPORTANT

Information Distribution is the process where the bulk of project communications takes place.

WHEN IT IS PERFORMED

Information Distribution generally updates stakeholders on the progress of the project according to the communications management plan. It may start quite early on the project, but typically elevates in importance and activity during the construction phase of the project.

HOW IT WORKS

- Inputs
  - *Communications Management Plan*

The communications management plan is the only input into this process. It is indispensable to this process since it defines how this process will be carried out.

- Tools &Techniques

    - *Communications Skills*

As part of the tool of communications skills, it is important to understand the communications model.

A. Communications Model

The communications model is a formal way of understanding how messages are sent and received. This model defines the responsibilities between the sender and the receiver.

The Communications Model

The sender's responsibilities are to:

1. Encode the message clearly
2. Select a communication method
3. Send the message

4. Confirm that the message was understood by the receiver

The receiver's responsibilities are to:

1. Decode the message
2. Confirm that the message was understood

Messages can be conveyed in verbal and nonverbal ways. Below are some terms related to different ways of communicating:

- *Active listening*

Active listening requires that the receiver takes active steps to ensure that the sender was understood. It is similar to effective listening (below).

- *Effective listening*

Effective listening requires the listener's full thought and attention. To be effective as a listener means to monitor non-verbal and physical communication and to provide feedback indicating whether the message has been clearly understood.

- *Feedback*

Feedback refers to the verbal and nonverbal cues a speaker must monitor to see whether the listener fully comprehends the message. Nodding and smiling might be considered positive feedback and indicate that the message is understood and received, whereas nodding and a blank stare might indicate that the message needs to be re-coded for better communication. Asking questions or repeating the speaker's words are also ways to give feedback.

- *Non-Verbal*

Non-verbal communication takes place through body language such as facial expressions, posture, hand motions, etc. In fact, most communication between the sender and receiver is non-verbal. Therefore, in order to understand the message, a good listener must carefully attend to non-verbal communication.

- *Paralingual*

Paralingual communication is vocal but not verbal - for example, tone of voice, volume, or pitch. A high-pitched squeal does not employ words, but it certainly communicates!

B. Communication Blockers

A communication blocker is anything that interferes with the sender encoding the message or the receiver decoding it. It can include anything that disrupts the communication channels.

C. Methods of Communication

Another topic related to the tool of communications skills in Information Distribution is that of methods of communication. There are four methods of communication covered on the exam, and it is very important to understand what they are and how and when they are used. Many people find the difference between formal and informal to be non-intuitive the first time they encounter it. The methods of communication are covered in the following table.

Method	Examples	When used
Informal written	E-mail messages, memorandums	Used frequently on the project to convey information and communicate
Formal written	Contracts, legal notices,	Used infrequently, but essential

	project documents (e.g. the Charter), important project communications	for prominent documents that go into the project record. The project plan is a formal written document.
Informal verbal	Meetings, discussions, phone calls, conversations	Used to communicate information quickly and efficiently.
Formal verbal	Speeches, mass communications, presentations	Used for public relations, special events, company-wide announcements, sales.

Be aware that not just the message but the medium determines whether a form of communication is formal or informal.

- ***Information Gathering and Retrieval Systems***

Regardless of what system you select and implement for information gathering and retrieval, it is important to be able to organize and access material quickly and easily.

- ***Information Distribution Methods***

Information on the project may be distributed to the stakeholders in many ways - oral or written, electronic or paper, live or taped. The important thing about these methods is that it they are used to distribute accurate information in a timely fashion.

- ***Lessons Learned Process***

The oddly-named tool "lessons learned process" describes meetings that help facilitate the gathering of the lessons learned. These then become an organizational process asset for use in future projects.

All lessons learned generally focus on one simple question: "Where were the variances on the project, and what would we do differently to avoid those variances if we had the project to do over again?" This question is evaluated across all components of the project.

- Outputs

    - *Organizational Process Assets Updates*

All of the information coming out of the Information Distribution process is potentially valuable to the organization for use in future projects.

-
    - *Requested Changes*

The changes described here typically represent changes to the types of information that are distributed or the way in which it was distributed. These requested changes should be processed back through the Integrated Change Control process.

Performance Reporting

WHAT IT IS

The process of Performance Reporting is another one that lives up to its name. This process reports to the stakeholders how the project is progressing against the plan. It is easy to confuse this process with its cousin, Information Distribution; however, they are different.

For instance, Performance Reporting belongs to the monitoring and controlling process group (Information Distribution, as you remember, is an executing process), and it is specifically focused on reporting against the performance baseline. In other words, performance reporting tracks how the project is doing against the plan, which is more specific than Information Distribution.

WHY IT IS IMPORTANT

Performance reporting is important because it updates stakeholders on how the project is doing against the plan, and this kind of communication is especially beneficial.

WHEN IT IS PERFORMED

Like its cousin, Information Distribution, Performance Reporting begins early on the project and takes on increasing importance as the project enters the construction phase where more resources and costs are expended.

HOW IT WORKS

- Inputs

  - *Work Performance Information*

This input comes from the process Direct and Manage Project Execution, and it covers practically all aspects of the project. Work performance information can include performance updates on scope, schedule, costs, quality, and the use of resources. This work performance information is brought in as the primary input into Performance Reporting, since it is compared with the various baselines to measure progress against the plan.

  - *Performance Measurements*

Performance measurements specifically refer to the cost variance (CV), schedule variance (SV), cost performance index (CPI), the cumulative CPI (CPI^C), and the schedule performance index (SPI). They are used to determine how the project is performing against the plan.

- *Forecasted Completion*

This input typically describes the estimate at completion (EAC) or the estimate to completion (ETC). It provides the best estimate as to how much the project will cost based on past performance.

- *Quality Control Measurements*

How a project is performing against the quality specifications is an important aspect of its overall performance. Measurements from quality control are brought into this process so the results may be compared with the plan.

- *Project Management Plan*

The project management plan is brought into this process, because it provides the planned baseline against which the actual performance is measured and reported.

- *Approved Change Requests*

Approved changes to the project which affect the baseline (whether scope, schedule, cost, or quality), should be factored into Performance Reporting. Note that this applies to approved changes and not to all changes. The reason for this is that approved changes become part of the baseline, while changes that were not approved do not. Performance reporting is done against the baseline, so changes that were not approved would show up negatively on the reporting.

- *Deliverables*

This one is a bit unusual here and probably won't show up on the exam. It is an input here, presumably so that performance may be measured against what was actually completed, but the more important input of work performance information (listed earlier in this process) would cover this more completely.

- Tools &Techniques

  - *Information Presentation Tools*

The use of the word "information" here primarily means graphical information. Project information, and specifically performance information, may be presented in a number of different ways, including text, graphics, animation, etc., and this type of presentation can make the information more readily understood. Tools that assist with this should be leveraged as part of Performance Reporting.

  - *Performance Information Gathering and Compilation*

Numerous sources may be used for gathering and compiling the information, including electronic, paper, and other sources. The takeaway from this tool is that information must be gathered and compiled before it is reported.

  - *Status Review Meetings*

Status meetings to review progress should happen at regular intervals.

  - *Time Reporting Systems*

A system should be used to track actual time spent on activities or work packages and on the project as a whole.

  - *Cost Reporting Systems*

A cost reporting system should be used to track actual costs spent on activities or work packages and on the project as a whole. These systems can be especially useful in breaking down or summarizing costs and comparing actual results with the plan.

- Outputs

  - *Performance Reports*

This is the essential output from Performance Reporting. The performance reports show how the project is progressing against the various baselines (scope, time, cost, quality). The performance reports are tailored to the audience. For instance, the project manager may produce only a one page executive summary for the sponsor, while a more detailed report may be prepared and distributed for the team.

  - *Forecasts*

Forecasts are future predictions combined with expert judgment based on past performance. Estimate at completion (EAC) and estimate to completion (ETC) are the most common financial forecasts, while the project's schedule is typically used to forecast a projected completion date.

  - *Requested Changes*

Changes to the project are a normal output of Performance Reporting. For example, if the project is progressing slower than anticipated or is costing more than planned, the project manager may elect to eliminate certain pieces of non-essential scope to get the project back on track. These changes should be factored back into the integrated change control system.

  - *Recommended Corrective Actions*

Corrective actions are anything done to bring future results in line with the plan. In the process of Performance Reporting, needed corrective actions will often become apparent.

- *Organizational Process Assets Updates*

Anything that may be reused on this project or on future projects should be considered an organizational process asset and should be updated and archived appropriately.

Manage Stakeholders

WHAT IT IS

As you spend time studying the PMBOK Guide, you will find that many knowledge areas (such as Communications Management) contain one or more planning processes, sometimes an executing process, and are often followed by a controlling process.

Controlling processes ensure that the executed results match up with the plan. In the case of Manage Stakeholders, you are ensuring that the stakeholders are heard and that their needs are addressed.

WHY IT IS IMPORTANT

This process helps to ensure that no issues raised by stakeholders mushroom into problems that could jeopardize the project. Instead, Manage Stakeholders works to identify and resolve stakeholder concerns in a timely manner.

WHEN IT IS PERFORMED

This process will be performed throughout the project, as soon as you begin communicating on the project (typically quite early).

HOW IT WORKS

- Inputs

  - *Communications Management Plan*

The project manager uses the communications management plan to set and manage stakeholder expectations in regards to project communications. This plan drives communication on the project, defining the who, what, when, where, and how of informational flow.

  - *Organizational Process Assets*

Any examples from previous projects of how stakeholder issues were resolved or records of issues being handled would be of benefit to bring into this process.

- Tools &Techniques

  - *Communications Methods*

The important take-away from this tool is that face-to-face meetings are the best ones to use where stakeholder issues are concerned.

A. Effective Meetings

While meetings are an important part of good communication, effective and productive meetings are rare in the real world. Meetings should be held for a clear purpose and should involve only the people necessary. Their main purpose is to make decisions and communicate decisions. Meetings with no clear purpose simply should not be held.

Following are ingredients to an effective meeting. Most of these should be intuitive, and test takers should focus on understanding the list rather than memorizing it:

- Clearly define the reason, issues, and processes for the meeting
- Establish clear objectives for the meeting
- Publish and follow a written agenda
- Have a structure for conducting the meeting
- Foster creative thinking
- Drive toward making decisions and not only toward discussion
- Listen and communicate collaboratively
- Control communication during the meeting. This does not mean that you micro-manage every detail of every discussion, but rather that you keep the discussion relevant and aligned with the topics
- Include all of the necessary people and only the necessary people. Meetings are important, but can be a tremendous waste of time.
- Document the meeting through written minutes

B. Kickoff Meeting

The kickoff meeting is an opportunity to bring the team together, along with key stakeholders, the sponsor, the customer, and senior management to discuss the project plan. Additionally, it is considered a good practice to share lessons learned from previous projects during the kickoff meeting.

- *Issue Logs*

A log of all issues in indispensable to the project manager for two reasons:

1. It helps keep track of all issues so that none are lost.

2. It gives stakeholders a way of seeing that their issue is being actively tracked and managed.

Note that an issue log is also a tool of Manage Project Team in Human Resources Management.

- **Outputs**

  - *Resolved Issues*

In an ideal world, all issues are resolved, and the identified resolutions are documented in the issue log.

- *Organizational Process Assets Updates*

Any information that could help future projects avoid issues your project had to resolve should be properly archived.

- *Approved Change Requests - See Chapter 2, Common Outputs*

- *Approved Corrective Actions - See Chapter 2, Common Inputs*

- *Project Management Plan Updates - See Chapter 2, Common Outputs*

Exercise

1. Calculate the number of communication channels that would exist between eight people.

With 8 individuals, there are 28 communication channels as proven by the formula 8 * (8-1) / 2 = 28

A graphical depiction of the 28 communication channels between the 8 people

Communications Management Questions

1. If there were 4 people on the project team and 9 more are added, how many additional channels of communication does this create?

   A. 6.

   B. 30.

   C. 36.

   D. 72.

2. The process to create a plan showing how all project communication will be conducted is known as:

   A. Communications Modeling.

   B. Communications Planning.

   C. Information Method.

   D. Communication Distribution Planning.

3. The responsibility of decoding the message rests with:

   A. The sender.

   B. The receiver.

   C. The communications management plan.

   D. The communications model.

4. Which of the following is FALSE regarding Information Distribution?

   A. Information Distribution is an executing process.

   B. Information Distribution ends when the product has been accepted.

   C. Information Distribution may involve unexpected requests from stakeholders.

   D. Information Distribution carries out the communications management plan.

5. Your latest review of the project status shows it to be more than three weeks behind schedule. You are required to communicate this to the customer. This message should be:

   A. Formal and written.

   B. Informal and written.

   C. Formal and verbal.

   D. Informal and verbal.

6. Which of the following statements is TRUE regarding issues?

   A. All issues must be resolved in order for the project to be closed.

   B. The issue log is a tool for stakeholders to manage project issues.

   C. Each issue should be assigned to a single owner.

   D. Issue management may be treated as a sub-project on larger, more complex projects.

7. The majority of a person's communication is:

   A. Verbal.

   B. Non-verbal.

   C. Documented.

   D. Unnecessary.

8. Which of the following communication techniques is the most effective for resolving conflict?

   A. Instant messaging.

   B. Conference calls.

   C. E-mail.

   D. Face-to-face communication.

9. Communications skills would be used most during which of the following processes?

   A. Information Distribution.

   B. Status Meetings.

   C. Performance Reporting.

   D. Communications Change Control.

10. The communications management plan typically contains all of the following EXCEPT:

    A. The expected stakeholder response to the communication.

    B. The stakeholder communication requirements.

    C. What technology will be used to communicate information.

    D. A glossary of terms.

11. You are about to attend a bi-weekly status meeting with your program manager when she calls and asks you to be certain to include earned value analysis in this and future meetings. Why is earned value analysis important to the communications process?

   A. It communicates the project's long-term success.

   B. It communicates how the project is doing against the plan.

   C. It communicates the date the project deviated from the plan.

   D. It communicates the value-to-cost ratio.

12. You receive a last minute status report from a senior member of the project team that you believe is incorrect. It shows tasks as complete that you are almost certain are no more than 60% complete, and it documents deliverables as having being turned over to the customer that you do not believe are even finished yet. You are walking into a communication meeting with key stakeholders. What is the BEST way to handle this problem?

   A. Ask the team member who wrote the report to sign the bottom of it.

   B. Ask the stakeholders to wait a few minutes while you try to verify the information.

   C. Summon the project team to the meeting and get to the bottom of the discrepancy.

   D. Do nothing with this status and provide an amended report at the next meeting with the stakeholders.

13. The MOST important skill for a project manager to have is:

A. Good administrative skills.

B. Good planning skills.

C. Good client-facing skills.

D. Good communication skills.

14. The best definition of noise is:

    A. Any unsupportable information that finds its way onto written or verbal project communications.

    B. Anything that interferes with transmission and understanding of a message.

    C. Any communication that takes place through unofficial project channels.

    D. A communications acronym for Normal Operational Informing of Select project Entities.

15. A project manager is holding a meeting with stakeholders related to the status of a large project for constructing a new runway at a major airport. The runway project has a CPI of 1.2 and an SPI of 1.25, and the manager is going to have to deliver the message to the stakeholders that a crucial quality test has failed. What kind of communication does this meeting represent?

    A. Formal verbal.

    B. Informal verbal.

    C. Paralingual.

    D. Non-verbal.

16. You have just taken over as the project manager for a new runway for a major airport. The project is already in progress, and there are over 200 identified stakeholders on the project. You want to know how to communicate with these stakeholders. Where should you be able to find this information?

   A. It depends on the type of project.

   B. The stakeholder management plan.

   C. The communications management plan.

   D. Communication requirements.

17. Mary is using forecasting to determine her project's estimate at complete. What would be the most likely place to include this information?

   A. The communications management plan.

   B. The project activity report.

   C. The performance reports.

   D. The stakeholder management report.

18. Marie is a project manager who is involved in a meeting with the customer. After the customer makes a statement, Marie carefully reformulates and restates the message back to them. What is Marie practicing in this case?

   A. Listening skills.

   B. Project communications management.

C. Professional courtesy.

D. Passive listening.

19. Lessons learned should contain:

A. The collective wisdom of the team.

B. Feedback from the customer as to what you could have done better.

C. Information to be used as an input into administrative closure.

D. Analysis of the variances that occurred from the project's baseline.

20. In which process would earned value analysis be used?

A. Communications Planning.

B. Information Distribution.

C. Performance Reporting.

D. Report Project Value.

Answers to Communications Questions

1. D. If you tried to take a shortcut here, chances are you missed this one and guessed 'C'. If there were 4 people, there would have been 6 communication channels. 9 more would create 13, which equals 78 communications channels. The question is asking how many additional channels were created, so the answer is 78-6 = 72

2. B. Communications Planning is the process for determining how the overall communication process will be carried out. It is the general plan for communications. None of the other three answers were terms used in this book, but the real giveaway was that only one answer 'B' was even the name of a process.

3. B. In the communications model, it is the sender that encodes, and the receiver decodes the message.

4. B. Did you get tricked by this one? Information Distribution doesn't always end when acceptance has occurred, so this is the answer that doesn't fit. Some stakeholders will need information distributed on the closure of the contracts and projects. 'A' is true, because Information Distribution is an executing process. 'C' is true because Information Distribution carries out predetermined communication, but also will be used to respond to unplanned requests from stakeholders. 'D' is true because Information Distribution is the process that executes the communications management plan.

5. A. Communication on schedule slippage, cost overruns, and other major project statuses should be formal and in writing. That doesn't mean you can't pick up the phone to soften the blow, but the formal and written aspects of the communication are what count here.

6. C. Each issue should be assigned to an owner and be assigned a target completion date. 'A' is incorrect since a project could be closed (successfully) and still have outstanding issues. Sometimes the issues are out of the project manager's control. 'B' is incorrect since the issue log is not for the stakeholders - it is for the project manager to use to manage issues. 'D' is incorrect because issues are managed within the context of a project. If you were even considering creating a separate project to manage issues, your project is probably beyond hope.

7. B. Most of a person's communication takes place non-verbally. It is body language that carries much of the message. 'A' is the opposite of the correct answer. 'C' is incorrect since most of the communication is non-verbal, but not written (documented). 'D' may well be true for some people, but it is not the right answer here.

8. D. Face-to-face communication is the most effective means of resolving conflict. This fits an overall theme that direct, clear, and personal communication is favored for project managers. If you guessed 'A', go find a place to hide in shame, or at least go sit in time out for a few minutes. 'B' and 'C' might seem like appropriate choices in some situations, but face-to-face is still more effective.

9. A. Your communications skills are used as a tool in Information Distribution. You should have eliminated 'B' and 'D' because they were not real processes. 'C' might have been tricky for you, but it is incorrect because the process of Performance Reporting mainly produces the performance reports, which factually state the status of the project, while Information Distribution, which can cover many more topics, requires more in the way of skill and communication ability.

10. A. The expected response you will receive is not part of the communications management plan. The communications management plan focuses on how you will communicate to stakeholders and not how they will communicate to you. 'B', 'C', and 'D' are all typically part of the communications management plan.

11. B. Earned value analysis is a communication tool, and it's all about how the project is doing against the plan.

12. B. This is a hard question, but be prepared for questions like this on the PMP! The reasoning behind it is this: a project manager should always communicate good information and should always report the truth. 'A' is wrong because it isn't about getting your team member to sign off. Accurate information is more important than accountability. 'C' is incorrect because it is not the team's job to go to these meeting. They should be doing the work on the project. 'D' is incorrect because waiting only postpones the situation and delays getting accurate information to the stakeholders. Choice 'B' is best in this case, because it is the only one that gets accurate information to the stakeholders as quickly as possible.

13. D. Good communication skills are the most important skills a project manager can have! Project managers spend more time communicating than anything else.

14. B. Noise is anything that interferes with the transmission and understanding of a message. If you guessed 'D', then you were indeed guessing.

15. B. Many people incorrectly guess 'A' for this question, but meetings are classified as informal verbal - even when the subject matter is important!

16. C. The stakeholder's communication needs are all contained in the communications management plan.

17. C. Performance Reporting uses the tool of forecasting and produces the performance reports. These performance reports often contain the estimate at complete and the estimate to complete, as well as the cost and schedule performance indexes. 'A' was the only other answer that contained a term that is used in this book, and it is not an appropriate match for this question.

18. A. Marie is practicing good listening skills to make sure she communicates well with her customer.

19. D. This is important! Lessons learned focus on variances from the plan and what would be done differently in the future in order to avoid those variances.

20. C. Earned value analysis is a tool of Performance Reporting, since earned value analysis factors in the difference between what was planned and the work that was actually accomplished. This information can then be distributed out to the appropriate stakeholders.

Risk Management

The processes of Project Risk Management with their primary outputs

Risk Management

Process	Primary outputs
Risk Mgt. Planning	Risk Management Plan
Risk Identification	Risk Register
Qualitative Analysis	Risk Register Updates
Quantitative Analysis	Risk Register Updates
Risk Response Planning	Risk Register Updates Contract Agreements
Risk Monitoring and Control	Risk Register Updates Requested Changes

Risk management is a very rich field, full of information and tools for statistical analysis. In the real world, actuaries anticipate risk and calculate the probability of risk events and their associated cost, and entire volumes are written on risk analysis and mitigation. In this section, you do not need to know every tool and technique associated with risk. Instead, focus your study on the high-level interactions within the different risk processes.

Risk is one of the rare sections of the PMBOK Guide where the names and basic functions of the processes have not changed from the previous editions of the PMBOK Guide. That is not to suggest that the material has not changed. There are significant differences in the details of the processes, but the good news is that this section has actually been significantly pared down, streamlined and simplified from the prevision edition of the PMBOK Guide!

Philosophy

By this time, you should have picked up on the fact that very little in PMI's methodology is reactive. The overriding philosophy is that the project manager is in control and proactively managing events, avoiding as many problems as possible.

The project manager must understand how to anticipate and identify areas of risk, how to quantify and qualify them, and how to plan for them.

Importance

Risk is one of the areas many people find difficult on the exam. The material may be new or unfamiliar, and the techniques may take some work in order to master.

Preparation

In order to pass this section of the exam, you need to understand the risk management plan and the terms related to risk. The 6 risk management processes contain 25 inputs, 23 tools and techniques, and 13 outputs.

All of these components to the processes are important, but the secret is that most of them are either common sense or they rarely show up on the exam. This chapter puts more emphasis on the essential elements that you need to know. Material is organized around these six processes, building upon the different components that go with each one.

There are six processes within project risk management. In the PMI framework, these processes touch two process groups: planning (Risk Management Planning, Risk Identification, Qualitative Risk Analysis, Quantitative Risk Analysis, Risk Response Planning), and monitoring and controlling (Risk Monitoring and Control).

Process Group	Risk Management Process
Initiating	(none)
Planning	Risk Management Planning, Risk Identification, Qualitative Risk Analysis, Quantitative Risk Analysis, Risk Response Planning

Executing	(none)
Controlling	Risk Monitoring and Control
Closing	(none)

The primary outputs associated with the six risk management processes are shown in the table below.

Process	Primary Outputs
Risk Management Planning	Risk management plan
Risk Identification	Risk register
Qualitative Risk Analysis	Risk register updates
Quantitative Risk Analysis	Risk register updates
Risk Response Planning	Risk register updates
Risk Monitoring and Control	Risk register updates Requested changes Recommended corrective actions Recommended preventive actions

Risk

PMI's usage of the word "risk" is different than many project managers and organizations may have encountered before. Risk has two characteristics that must be understood for the exam:

1. Risk is related to an uncertain event.

2. A risk may affect the project for good or for bad. Although risk usually has negative connotations, it may well have an upside.

Risk Management Planning

WHAT IT IS

Risk Management Planning is the process that is concerned with one thing: creating the risk management plan. Your understanding of that plan is the key to unlocking this process.

In Risk Management Planning, the remaining five risk management processes are planned. How they will be conducted is documented in the risk management plan, which is typically general and high-level in nature.

This means that Risk Management Planning usually doesn't concern itself with specific project risks. Instead, it focuses on how risk will be approached on the project.

WHY IT IS IMPORTANT

Think of this process as creating your roadmap for the five processes of Risk Identification, Qualitative Risk Analysis, Quantitative Risk Analysis, Risk Response Planning, and Risk Monitoring and Control. By creating a plan (the Risk Management Plan) for these five processes, you are being deliberate and proactive with risk on the project.

The more risk that is inherent on the project, and the more important the project is to the organization, the more resources you would typically apply to performing this process.

WHEN IT IS PERFORMED

This process is general and high-level in nature and therefore takes place early on the project, usually before many of the other planning processes are performed. The reason it usually takes place very early is that the results of this (and other risk processes) can significantly influence decisions made about scope, time, cost, quality and procurement.

HOW IT WORKS

- Inputs
    - *Enterprise Environmental Factors - See Chapter 2, Common Inputs*
    - *Organizational Process Assets - See Chapter 2, Common Inputs*
    - *Project Scope Statement - See Chapter 5, Scope Definition, Outputs*
    - *Project Management Plan - See Chapter 2, Common Inputs*
- Tools and Techniques
    - *Planning Meetings and Analysis*

As is stressed in this process, the risk management plan outlines the project's overall approach to risk, and this is not to be determined in a vacuum. As this tool suggests, the approach is plotted out through meeting with all appropriate stakeholders. This is followed up with a careful analysis to determine the appropriate level of risk and the approach warranted on the project.

- Outputs
    - *Risk Management Plan*

As stated in the introduction to this process, the risk management plan is the real purpose of this process. In fact, it is the process's sole output.

The risk management plan is a roadmap to the other five risk management processes. It defines what level of risk will be considered tolerable for the project, how risk will be managed, who will be responsible for risk activities, the amounts of time and cost that will be allotted to risk activities, and how risk findings will be communicated.

Another important part of the risk management plan is a description of how risks will be categorized. This will be a significant help in the subsequent risk processes. One tool for creating consistent risk categories is the risk breakdown structure (RBS). Because of the RBS's introduction and expanded treatment in this edition of the PMBOK Guide, you should expect to see it on the exam.

The RBS, like its cousin the WBS, is a graphical, hierarchical decomposition used to facilitate understanding and organization. In this case, however, we are breaking down the categories of risks and not the work. One important thing to note with the RBS is that we are not breaking down the actual risks (they won't be known until we perform the Risk Identification process). Instead, we are breaking down the categories of risks that we will evaluate.

Risk Identification

WHAT IT IS

Risk Identification is a planning process that evaluates the project to identify which risks could impact the project and to understand the nature of these risks.

There is a single output from this process, and that should help you understand it better. The output is the risk register, which is a list of all risks, their causes, and any possible responses to those risks that can be identified at this point in the project.

Be aware that Risk Identification, like many processes discussed in this book, is often performed multiple times on the project. This may be especially true for Risk Identification, since your understanding of risk, and the nature of the risks themselves, will change and evolve as the project progresses.

WHY IT IS IMPORTANT

Risk Identification builds the risk register, which is needed before the remaining four risk processes (Qualitative Risk Analysis, Quantitative Risk Analysis, Risk Response Planning, and Risk Monitoring and Control) may be performed. This list of risks will drive the planning processes.

WHEN IT IS PERFORMED

Although Risk Identification is typically performed early on the project, risks change over time, and new risks arise. It may be necessary to be perform this process multiple times throughout the project.

HOW IT WORKS

- Inputs
    - *Enterprise Environmental Factors - See Chapter 2, Common Inputs*
    - *Organizational Process Assets - See Chapter 2, Common Inputs*

- *Project Scope Statement - See Scope Definition - Outputs*
- *Risk Management Plan - See Risk Management Planning, Outputs*
- *Project Management Plan - See Chapter 2, Common Inputs*

- Tools
    - *Documentation Reviews*

A documentation review is a review of all project documentation that exists to date. The documentation is reviewed for completeness, correctness, and consistency. For instance, if the plan appears sketchy or quickly thrown together, that could identify a significant risk - especially if the project were of high importance.

- *Information Gathering Techniques*

There are numerous techniques for gathering information to create the risk register. The techniques most commonly discussed in the context of risk are: brainstorming, Delphi technique, expert interviews, root cause identification, and SWOT analysis.

SWOT analysis is particularly useful since it is a tool used to measure each risk's strengths (S), weaknesses (W), opportunities (O), and threats (T). Each risk is plotted, and the quadrant where the weakness (usually internal) and threats (usually external) are highest, and the quadrant where strengths (again, usually internal), and opportunities (usually external) are highest will represent the highest risks on the project.

SWOT analysis can give you another perspective on risk that will often help you identify your most significant project risks factors.

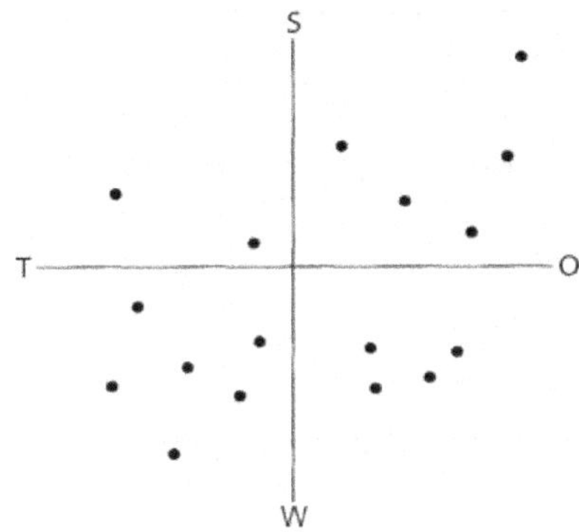

llustration of one type of SWOT analysis where each dot represents a specific risk

- ***Checklist Analysis***

Checklist analysis uses a Risk Breakdown Structure (RBS), either from this project or from a previous project, to check off items and ensure that all significant risks or categories are being evaluated. Although it may not be exhaustive, this tool provides structure to the Risk Identification process.

- ***Assumptions Analysis***

Assumptions should not only be documented; they should be analyzed and challenged if necessary.

- ***Diagramming Techniques***

Ishikawa diagrams, also called cause-and-effect diagrams and fishbone diagrams, are one way to show how potential causes can lead to risks.

Another diagramming method used to identify risks is influence diagrams. This category shows how one set of factors may influence another. For instance, late arrival of materials may not be a significant risk by itself, but it may influence other factors, such as triggering overtime work, or causing quality problems later on in the project due to inadequate time to properly perform the work.

Finally, flow charts are useful in identifying risks. Flow charts are graphical representations of complex process flows. They are especially useful, because they can break down something very complex into an understandable diagram.

- Outputs

  - ***Risk Register***

As stated earlier, the risk register provides a list of all identified risks on the project, what the possible reactions to this risk are, what the root causes are, and what categories the risks fall into. It is also common to update the RBS with the more specific information as the following example illustrates.

A more complete RBS, including identified risks

The risk register is an essential input into the remaining risk management processes and may be updated throughout the life of the project.

Risk ID	Risk	Responses	Root Cause	Categories
R001	Threat of being hacked	Firewall; intrusion detection software	Poorly designed security; Outdated technology	Security

Fragment of a Risk Register

Qualitative Risk Analysis

WHAT IT IS

The process of Qualitative Risk Analysis is usually done rapidly on the project in order to determine which risks are the highest priority on the project.

It takes each risk from the risk register and works to analyze its probability of occurring and impact to the project if it did occur. By using the probability and impact matrix (PIM), a prioritization and ranking can be created, which is updated on the risk register.

WHY IT IS IMPORTANT

This process helps you rank and prioritize the risks so that you can put the right emphasis on the right risks. It helps to ensure that time and resources are spent in the right risk areas.

WHEN IT IS PERFORMED

Qualitative risk analysis, like many other risk processes, is usually performed more than once on a project. The reason for this is twofold:

1. Qualitative risk analysis can usually be performed fairly quickly relative to other planning processes.

2. It is normal for risks and their underlying characteristics to change over the life of the project, making it important to revisit often.

HOW IT WORKS

- Inputs

  - *Risk Management Plan*

The risk management plan should define the overall approach to risk on the project, including how much risk is acceptable and who will be responsible for carrying out the analysis of the risks.

  - *Risk Register*

The risk register specifies each risk to be analyzed as part of this process.

  - *Organizational Process Assets - See Chapter 2, Common Inputs*

  - *Project Scope Statement - See Chapter 5, Scope Definition, Outputs*

- Tools

The next two tools, because of their similarities, will be discussed together. Although they are different, it is highly unlikely you would be called upon to differentiate between them for the exam.

- ***Risk Probability and Impact Assessment and Probability and Impact Matrix***

When evaluating risks to determine what the highest priorities should be, the two tools and techniques mentioned above can assist you.

The way they are used is that each risk in the risk register is evaluated for its likelihood of occurring and its potential impact on the project. Each of these two values is given a ranking (such as low, medium, high, or 1 through 10) and are multiplied together to get a risk score. This resulting score is used to set the priorities.

- ***Risk Data Quality Assessment***

The data used should be objectively evaluated to determine whether or not it is accurate and of acceptable quality. For instance, if you were evaluating weather risk for a construction project, you would need to evaluate the quality of the weather data you were using.

- ***Risk Categorization***

Categorizing the detailed risks can help you build a better big-picture of the risks. This may help you understand which parts of the project have the highest degree of uncertainty.

- ***Risk Urgency Assessment***

Urgent risks are those that cannot wait. As you are evaluating the risks, it is important to determine which risks are the most urgent, requiring immediate attention. For instance, if you determined that a building's structural support may be inadequate, that would require immediate attention, while other risks, such as future weather threats, although equally important, might be less urgent.

- **Outputs**

- *Risk Register Updates*

The risk register, developed earlier in the Risk Identification process, contains a list of all the risks. Now this register is further completed, including the priority of the risks, the urgency of the risks, the categorization of the risks, and any trends that were noticed while performing this process.

Quantitative Risk Analysis

WHAT IT IS

This process is easy to confuse with the previously covered Qualitative Risk Analysis, and in reality, the processes have quite a lot in common; however, Quantitative Risk Analysis seeks to assign a projected value to (quantify) the risks that have been ranked by Qualitative Risk Analysis. This likely value is typically specified in terms of cost or time.

WHY IT IS IMPORTANT

Quantitative Risk Analysis updates the risk register, and this information will be used by the subsequent two processes (Risk Response Planning and Risk Monitoring and Control). Without performing this process, the information about the identified risk is less complete and less useful.

WHEN IT IS PERFORMED

Quantitative Risk Analysis relies on the prioritized list of risk from the Qualitative Risk Analysis process. Therefore, it is usually performed right after Qualitative Risk Analysis; however, in some cases they may be performed at the same time.

HOW IT WORKS

- Inputs
  - *Organizational Process Assets – See Chapter 2, Common Inputs*
  - *Project Scope Statement - See Scope Definition, Outputs*
  - *Risk Management Plan - See Risk Management Planning, Outputs*
  - *Risk Register - See Risk Identification, Outputs*
  - *Project Management Plan - See Chapter 2, Common Inputs*
- Tools
  - *Data Gathering and Representation Techniques*

A. Interviewing

Interviewing uses a structured interview to ask experts about the likelihood and impact of identified risks. After interviewing several experts, for instance, the project manager might create pessimistic, optimistic, and realistic values associated with each risk.

B. Expert Judgment

People with expertise in the areas of risk that you are evaluating are one of the richest sources of data gathering. Asking experts to review your data and your methodology can be very useful.

  - *Quantitative Risk Analysis and Modeling Techniques*

A. Sensitivity Analysis

Sensitivity analysis is used to analyze your project and determine how sensitive it is to risk. In other words, you are analyzing whether the occurrence of a particular negative risk event would ruin the project, or merely be an inconvenience.

B. Tornado Diagrams

Tornado diagrams, named for the shape of their bars, are one way to analyze project sensitivity to cost or other factors.

A tornado diagram, used to depict risks

The tornado diagram above depicts the effects of a 10% change in labor costs on the project. If labor costs increase by 10% and all other costs hold steady, development will be affected the most. Specifically, if the costs rise by 10%, then the development costs will rise by approximately $13,000. If the labor costs fall by 10%, development costs will fall by approximately $7,000. This shows how sensitive each analyzed area of the project is to risk (in this case, the risk of a cost increase).

A tornado diagram ranks the bars from greatest to least on the project so that the chart takes on a tornado-like shape.

C. Expected Monetary Value Analysis

Expected monetary value analysis takes uncertain events and calculates a most likely monetary value (i.e. dollar amount). It is typically calculated by using decision trees, covered next.

D. Decision Tree Analysis

Decision trees are used to show probability and arrive at a dollar amount associated with each risk.

The numbers for this decision tree work out as follows:

	Initial Cost	Risk Cost	Probability	Total

Commercial Package	$2,250,000	$5,000,000	10%	$2,750,000
Custom Software	$1,325,000	$5,000,000	30%	$2,825,000

The totals above were calculated by first multiplying the risk cost by the probability and then adding it to the initial cost. e.g. $5,000,000 * 10% = $500,000. $2,250,000 + $500,000 = $2,750,000.

In the preceding example, the analysis shows that the custom software has a higher overall cost due to the higher risk. Decision trees can get very complex as they grow larger, but this method is a good overall tool to quantify the outcome of various options.

- *Modeling and Simulation*

There are almost as many types of simulation as there are projects; however, one technique for simulating the schedule is Monte Carlo analysis.

This is a favorite topic on the exam. Monte Carlo analysis, also discussed in Chapter 6 - Time Management, is a tool that takes details and assembles a big picture. Performed by computer, Monte Carlo analysis throws large numbers of scenarios at the schedule to see the impact of certain risk events.

This technique will show you what is not always evident by simply looking at the schedule. It will often identify tasks that may not appear inherently high risk, but in the event they are delayed, the whole project may be adversely affected.

- Outputs

- *Risk Register Updates*

The risk register is updated with the probabilities associated with each identified risk and the probability of meeting the project's cost and time projections. Additionally, the priorities of the risks should be updated, and any trends that have been observed should be noted.

Do not forget that risk may be beneficial! As an example, when you purchase a lottery ticket, you are running the risk that you will win money, and that risk can be quantified. If you are constructing a building, you would plan for a certain amount of bad weather to impact the construction schedule; however, you run the risk that the weather will be worse than anticipated as well as the risk that the weather will be better than anticipated.

This usage of the word risk is counter-intuitive to many people but it is correct within the domain of project management, because the risk is in the uncertainty, not just in the outcome.

- *Prioritized List of Quantified Risks*

Lists of risk should be prioritized by the numbers in this process. For instance, based on the expected value (probability x $ impact), the following 4 risks are ranked in the table. The rankings would show you how to manage these risks.

Risk	Probability	Impact	Expected Value
Risk B	25%	$1,750,000.00	$437,500,00
Risk C	50%	$600,000.00	$300,000.00
Risk D	80%	$200,000.00	$160,000.00
Risk A	13%	$152,200.00	$19,786.00

Risk Response Planning

WHAT IT IS

Earlier, in the process of Risk Management Planning, we created a general approach to risk (the risk management plan). Then, in Risk Identification, we created a list of risks and started our risk register. Then we qualitatively and quantitatively analyzed the risks, and now we are ready to create a detailed plan for managing the risk. That is precisely what Risk Response Planning does; it creates a plan for how each risk will be handled. Remember that risk can be a positive or negative event (e.g. there is a risk that the project will finish late, but there is also a risk that it will finish early). Therefore, careful consideration must be given to each risk, whether the impact of that risk is positive or negative.

WHY IT IS IMPORTANT

Up to this point, all we have done is identify and analyze the risks, but now that the analysis is complete, we need to create a specific plan. resulting plan is actionable, meaning that it assigns specific tasks and responsibilities to specific team members.

WHEN IT IS PERFORMED

One of the helpful things about the risk management processes is that the processes are typically conducted in the order that they are presented in this book. This means that you begin with Risk Management Planning, continue with Risk Identification, proceed with Qualitative Risk Analysis, followed by Quantitative Risk Analysis, and then by this process, Risk Response Planning.

HOW IT WORKS

- Inputs
    - *Risk Management Plan - See Risk Management Planning, Outputs*

- *Risk Register - See Risk Identification, Outputs*

- Tools and Techniques

You will almost certainly see exam questions related to the risk responses. Be sure you can identify different responses based on behaviors described in the test questions.

- *Strategies for Negative Risks or Threats*

A. Avoid

Avoidance is a very appropriate tool for working with undesirable risk in many circumstances. For instance, a software project may choose to avoid the risk associated with using a particular piece of cutting edge technology in favor of using a slower but more reliable technology.

B. Transfer

To transfer a risk to another party is to make it their responsibility. Contractual agreements and insurance are common ways to transfer risks.

C. Mitigate

Mitigating a risk simply means to make it less. For instance, if you were concerned about the risk of weather damage to a construction project, you might choose to construct the building outside of the rainy season.

- *Strategies for Positive Risks or Opportunities*

As is stressed throughout this chapter, risks can be positive or negative, and where positive risks are concerned, the project manager wants to take steps to make them more likely. The following are specific strategies taken to capitalize on the positive risks.

A. Exploit

The definition for risk is uncertainty. Where the strategy of exploitation is concerned, you are trying to remove any uncertainty. For instance, if a positive risk of finishing the project early is identified, then adding enough people to ensure that the project is completed early would be an example of exploiting the risk.

B. Share

In order to share a positive risk, the project seeks to improve their chances of the risk occurring by working with another party. For instance, if a defense contractor identifies a positive risk of getting a large order, they may determine that sharing that risk by partnering with another defense firm, or even a competitor, would be an acceptable strategy.

C. Enhance

Enhancing a positive risk first requires that you understand the underlying cause(s) of the risk. By working to influence the underlying risk triggers, you can increase the likelihood of the risk occurring.

- ***Strategies for Both Threats and Opportunities***

A. Acceptance

Acceptance is often a perfectly reasonable strategy for dealing with risk, whether positive or negative. When accepting a risk you are simply acknowledging that the best strategy may not be to avoid, transfer, mitigate, share, or enhance it. Instead, the best strategy may be simply to accept it and continue with the project. Many people miss questions on the exam related to this, because they don't have the mindset that acceptance may be the best strategy if the cost or impact of the other strategies is too great.

- *Contingent Response Strategy*

A contingent response strategy is one where the project team may make one decision related to risk, but make that decision contingent upon certain conditions. For example, a project team may decide to mitigate a technology risk by hiring an outside firm with expertise in that technology, but that decision might be contingent upon the outside firm meeting intermediate milestones related to that risk.

- Outputs

  - *Risk Register Updates*

Now that a specific plan has been created for each risk in the risk register, the register should be updated to reflect this new information.

  - *Project Management Plan Updates - See Chapter 2 Common Inputs*

  - *Risk-Related Contractual Agreements*

One of the tools for this process is to transfer the risk to another party. This tool can often result in contractual agreements. For instance, a company that has little experience in mission-critical software development may contract out that part of the project, but risk-related contractual agreements would typically accompany this decision. In this example, it might be appropriate to specify cost, time, and performance targets for the subcontracting software company to meet in order to receive full payment.

Risk Monitoring and Control

WHAT IT IS

At this point, you should have detected a pattern with the knowledge areas. All of them have a controlling process that looks back over the plans and any execution that has taken place and compares them. In these controlling processes, you are asking questions such as, "Did we plan properly?" "Did the results come out the way we anticipated?" "If the results did not match the plan, should we take corrective action by modifying the plan or changing the way we are executing?" "Are there lessons learned that we need to feed into future projects?"

WHY IT IS IMPORTANT

Plans have to be reassessed and reevaluated. Where risk is concerned, we've done quite a bit of planning, identifying, analyzing, and predicting, but the process of Risk Monitoring and Control takes a look back to evaluate how all of that planning is lining up with reality.

WHEN IT IS PERFORMED

Risk Monitoring and Control is a process that is performed almost continually throughout the project. That is not to say that you do these activities without stopping or that someone is assigned full time, but rather that monitoring and controlling the risk is an ongoing concern.

HOW IT WORKS

- Inputs
    - *Risk Management Plan - See Risk Management Planning, Outputs*
    - *Risk Register - See Risk Identification, Outputs*
    - *Approved Change Requests - See Chapter 2, Common Inputs*
    - *Work Performance Information*

Work performance information is used as an input here since controlling processes compare the plan to the results. The plan is brought in as an input above, and the work performance information provides information on the results. For instance, the status of a deliverable provides helpful information related to schedule risk, cost risk, or other areas of concern.

- *Performance Reports*

The performance reports do not focus so much on what has been done as they do on how it was done. For instance, whereas the work performance information provides information on the status of deliverables, the performance reports focus on cost, time, and quality performance. Where the performance reports are concerned, the actual results are compared against the baselines to show how the project is performing against the plan.

- Tools

  - *Risk Reassessment*

As you perform a project, your information about risks, and even the very nature of the risks, change. You should reassess this information as often as necessary in order to make sure that the risk needs of the project are current and accurate. Note that a constant reassessment may not be required by all projects. Some projects may not need to reassess their risk information at all, while others may need more frequent updates.

  - *Risk Audits*

The important thing to know about risk audits is that they are focused on overall risk management. In other words, they are more about the top-down process than they are about individual risks.

Periodic risk audits evaluate how the risk management plan and the risk response plan are working as the project progresses and also whether the risks which were identified and prioritized are actually occurring.

- *Variance and Trend Analysis*

Variance analysis focuses on the difference between what was planned and what was executed. Trend analysis shows how performance is trending. The reason trend analysis is important is that a one-time snapshot of cost may not cause concern, but a trend showing worsening cost performance may indicate that things are steadily worsening and may indicate that a problem is imminent.

- *Technical Performance Measurement*

Performance can take on many flavors. In this case, technical performance measurement focuses on functionality, looking at how the project has met its goals for delivering the scope over time.

- *Reserve Analysis*

The project's reserve (also called contingency) can apply to schedule, time or cost. Periodically, the project's reserve, whether cost or time, should be evaluated to ensure that it is sufficient to address the amount of risk the project expects to encounter.

- *Status Meetings*

This particular technique is not necessarily suggesting that you have specially called status meetings related to risk. Instead, it is suggesting that you create a project culture where bringing up items related to risk is always acceptable and risk is discussed regularly.

- Outputs

- *Risk Register Updates*

New risk information, whether it is changes to your risk estimates or actual numbers (such as costs related to weather damage) should be regularly updated in the risk register.

- *Requested Changes*

When risk events occur, change requests to the project are a normal outcome. In addition, even when the events do not occur, the project may be changed as a result of new risk-related information gathered during this process.

- *Recommended Corrective Actions*

Corrective action is one of the most important concepts for the exam. It is anything done to bring future results in line with the plan, and in this case, corrective action to the project could include any actions taken to ensure that the future project results and the plan were in alignment. That could mean changing the plan, or it could mean changing the way the plan is executed or even changing things about the environment. The most important thing to keep in mind is that whatever changes are affected, they are intended to help your risk plans and the future line up. These recommendations are typically funneled in through the integrated change control system.

- *Recommended Preventive Actions*

Whereas corrective action (above) is about making sure a problem does not reoccur, preventive action is about making sure a problem never occurs.

- *Organizational Process Assets Updates*

All of the risk plans and information related to them, including analysis and corrective actions, can be archived for use on future projects. Over time, these plans become an asset to assist in future project planning and development.

- *Project Management Plan Updates*

Any changes that are made to the project should be updated in the project management plan.

Risk Management Questions

1. You are managing the construction of a data center, but the location is in an area highly prone to earthquakes. In order to deal with this risk, you have chosen a type of building and foundation that is particularly earthquake resistant. This is an example of:

   A. Risk transfer.

   B. Risk avoidance.

   C. Risk mitigation.

   D. Risk acceptance.

2. You are evaluating the risk by trying to produce a risk score for each risk. This is an example of which tool?

   A. Monte Carlo analysis.

   B. Probability impact matrix.

   C. RACI Chart.

D. Cause and effect diagrams.

3. As part of your project, you have identified a significant risk of cost overrun on a software component that is integral to the product. Which represents the BEST strategy in dealing with this risk?

   A. Outsource the software development.

   B. Insure the cost.

   C. Double the estimate.

   D. Eliminate the need for this component.

4. Planning meetings and analysis are used in which process?

   A. As part of Risk Monitoring and Control.

   B. As part of Quantitative Risk Analysis.

   C. As part of Risk Response Planning.

   D. As part of Risk Management Planning.

5. Refer to the diagram below. What is the expected value of Result A?

   A. $200,000.

   B. $100,000.

   C. $50,000.

D. $25,000.

6. Refer to the diagram from the previous question. What risk management tool is employed in this diagram?

   A. Earned value management.

   B. Sensitivity analysis.

   C. Decision tree analysis.

   D. Flowcharting.

7. Marie is meeting with her project team to evaluate each identified risk and try to assign an estimated dollar amount or time impact estimate to it. Which process is her team performing?

   A. Quantitative Risk Analysis.

   B. Qualitative Risk Analysis.

C. Risk Response Planning.

D. Risk Monitoring and Control.

8. If a project manager is recommending that immediate corrective action be taken, which process is he involved in?

   A. Qualitative Risk Analysis.

   B. Risk Management Planning.

   C. Risk Identification.

   D. Risk Monitoring and Control.

9. What is the BEST source of information about potential risk on your project?

   A. Computer risk analysis.

   B. Interviews with team members from other projects.

   C. Historical records from similar projects.

   D. Your own experience in this industry.

10. You have just finished a thorough Monte Carlo analysis for your project. Which of the following would the analysis MOST likely identify?

    A. Divergent paths causing risk.

    B. Points of schedule risk.

C. Points of schedule conflict that lead to risk.

D. Gaps in the project path that could create risk.

11. Your company is beginning a new building project and has assigned you the role of project manager. During the first few meetings with stakeholders you become aware of several risks that are of concern to the project sponsor. The topic of risk management, however, has yet to be addressed. What is the first thing you should do to address the project risks?

A. Develop a risk management plan.

B. Identify project risks.

C. Plan responses to project risks.

D. Determine how risks will be controlled.

12. Risk Management Planning is:

A. A process of identifying potential risks for a project.

B. Deciding how risk management activities will be structured and performed.

C. Assessing the impact and likelihood of project risks.

D. Numerical analysis of the probability- of project risks.

13. Which of the following is an output of Risk Identification?

A. Risk register.

B. Probabilistic analysis.

C. Risk-related contractual agreements.

D. Recommended cash reserves.

14. Which of the following would NOT be a strategy for dealing with negative risk?

A. Avoid.

B. Transfer.

C. Share.

D. Mitigate.

15. A risk probability and impact assessment is used in:

A. Risk Identification.

B. Qualitative Risk Analysis.

C. Quantitative Risk Analysis.

D. Risk Response Planning.

16. The BEST definition of risk management is:

A. The process of identifying, analyzing, and responding to risk.

B. The process of reducing risk to the minimum level possible for the project.

C. The process of proactively ensuring that all project risk is documented and controlled.

D. Creation of the risk response plan.

17. Which of the following is NOT a tool or technique for gaining expert opinion as it relates to risk?

A. Brainstorming.

B. Delphi technique.

C. Monte Carlo analysis.

D. Expert interviews.

18. Which of the following would NOT be contained in the risk management plan?

A. A risk breakdown structure.

B. A description of the overall approach to risk on the project.

C. Risk roles and responsibilities.

D. A list of identified risks.

19. Which of the following statements is TRUE regarding risk?

A. All risk events must have a planned workaround.

B. All risk events are uncertain.

C. All risk events are negative.

D. All risk events should be covered by a contingency budget amount.

20. Which of the following is NOT a valid way to reduce risk?

A. Select a contract type that reduces risk.

B. Insure against the risk.

C. Create a workaround for the risk.

D. Plan to mitigate the risk.

Answers to Risk Management Questions

1. C. The best answer here is risk mitigation since you are taking steps to lessen the risk. 'A' is incorrect because you are not transferring the risk to anyone else. 'B' is incorrect because you would need to relocate in order to completely avoid the risk of earthquake. 'D' is incorrect because you are not merely accepting the risk- you are taking steps to make it less severe.

2. B. The probability impact matrix (PIM) derives a risk score by multiplying the probability of the risk by its impact (both of these numbers are estimated). This resulting risk score may be used to help prioritize the risk register.

3. A. Outsourcing the software development could allow you to cap the cost. 'B' is not a good choice because costs for development such as this cannot be insured in a cost-effective manner. 'C' is not correct because doubling the estimate does not deal with the root of the problem. It only arbitrarily changes the estimate. 'D' is incorrect because you cannot simply eliminate every high risk component in the real world.

4. D. Planning meetings and analysis are a tool of the process of Risk Management Planning.

5. D. The way this problem is solved is by multiplying out the probabilities times the value. In this case, it the 50% probability of Risk 1 X the 25% probability of Risk 2 X the $200,000 value of result A. .5 X .25 * 200,000 = $25,000.

6. C. This is an example of decision tree analysis.

7. A. You should have seen the fact that Marie's team was quantifying the risks by seeking to assign a dollar or time estimate to them, and Quantitative Risk Analysis is the process that does this.

8. D. Recommended corrective action is the result of monitoring and controlling processes, and the only monitoring and controlling process in the choices was Risk Monitoring and Control.

9. C. Historical records from similar projects would provide you with the best source of information on potential risks. 'A', 'B', and 'D' are all good inputs or tools, but they would not be as pertinent or helpful as the records from other similar projects. Historical information gets brought into your planning processes as an organizational process asset.

10. B. One of the things Monte Carlo analysis would show you is where schedule risk exists on the project. 'A' is incorrect, because it is typically convergent and not divergent tasks that create schedule risk. 'C' is incorrect because it is not looking for schedule conflicts- those would be corrected in Schedule Development. 'D' is incorrect because gaps in the project path do not, by themselves, cause risk.

11. A. You should develop the risk management plan. A risk management plan will outline how all risk planning activities and decisions will be approached. Methods of identification, qualification, quantification, response planning, and control will all follow the development of the risk management plan.

12. B. Risk management planning is not planning for actual risks (which include choices 'A', 'C', and 'D'); it is the PROCESS of deciding how all risk planning activities and decisions will be approached. It is the plan for how to plan.

13. A. The risk register is the only output of Risk Identification, and it is updated in Qualitative Risk Analysis, Quantitative Risk Analysis, Risk Response Planning, and Risk Monitoring and Control.

14. C. There are three identified strategies for dealing with negative risks. They are: mitigate, transfer, and avoid. The reason that 'C' was incorrect is that share is a strategy for managing a positive risk or opportunity.

15. B. The hardest part about risk is keeping the various processes straight. Since the outputs are similar (for the most part), focus more of our study on the processes themselves and their tools. In this question, the tool of risk probability and impact assessment is a tool of Qualitative Risk Analysis.

16. A. The process of identifying, analyzing, and responding to risk is the definition of risk management.

17. C. Monte Carlo analysis, which is a computer-based analysis, might be useful for revealing schedule risk, but it would not be useful for gaining expert opinion. Choices 'A', 'B', and 'D' are all tools used as part of project risk management.

18. D. The risk management plan does not contain the identified risks. It is more general and high-level than that. The identified risks will be listed in the risk register, produced after the risk management plan.

19. B. Risk events are, by definition, uncertainties. These could either be positive or negative. 'A' is incorrect because some risks are simply accepted and have no workaround. 'C' is incorrect because a risk may be positive or negative. 'D' is incorrect because not all risks are budgeted. Some are transferred to other parties or are too small or unlikely to consider.

20. C. Read these questions carefully! The workaround is what you do if the risk occurs, but it does not reduce the risk as the question specified. 'A', 'B', and 'D' all focus on reducing risk by transferring or mitigating it.

Procurement Management

The processes of Project Procurement Management with their primary outputs

Procurement Management

- **Plan Purchases and Acquisitions** — Procurement Mgt Plan; Contract S.O.W
- **Plan Contracting** — Procurement Docs; Evaluation Criteria
- **Request Seller Responses** — Qualified Sellers List
- **Select Sellers** — Contract; Contract Mgt. Plan
- **Contract Administration** — Risk Register Updates; Contract Agreements
- **Schedule Control** — Risk Register Updates; Requested Changes

Procurement management is the set of processes performed to obtain goods, services, or scope from outside the organization. Procurement management can be a very challenging knowledge area on the exam. One reason it can be so difficult is that relatively few people have formal procurement training in their background.

Philosophy

PMI's procurement management approach is steeped in formal government procurement practices. In fact, many of the processes, tools, techniques, and outputs found here are near duplicates of those used by many government and military institutions in the United States.

The overarching philosophy of procurement management is that it should be formal. Many people's practical experience may differ from this rigid approach, but it is necessary to understand it and to be able to apply PMI's philosophy on the exam.

Importance

Several questions on the exam will be drawn from procurement management. If formal procurement is new to you, this material is especially important. There are not any formulas to learn or complex techniques to apply; however, there is a significant amount of material presented in this chapter.

Preparation

As mentioned earlier, it would be wise to take special care in this section if you do not have a background in formal procurement activities. Keep in mind as you approach this material that the PMBOK was not written to be memorized.

It was written to be practiced and applied. For this chapter in particular, understanding is more important than memorizing. In this chapter, key terms, concepts, and the processes and their components are provided.

Procurement Processes

There are six processes in procurement management, tying with risk management for the most processes of any knowledge area on the exam.

These processes are displayed in the figure at the beginning of the chapter and summarized in the tables below.

Process Group	Procurement Management Process
Initiating	(none)
Planning	Plan Purchases and Acquisitions, Plan Contracting
Executing	Request Seller Responses, Select Sellers
Controlling	Contract Administration
Closing	Contract Closure

Process	Key Outputs
Plan Purchases and Acquisitions	Procurement Mgt. Plan, Contract S.O.W.
Plan Contracting	Procurement Documents, Eval. Criteria
Request Seller Responses	Proposals
Select Sellers	Selected Sellers, Contract, Contract Mgt. Plan
Contract Administration	Contract Documentation, Updates
Contract Closure	Closed Contracts

Procurement Roles

In procurement management, there are two primary roles defined, and the project manager could play either of these roles. In fact, it is not uncommon for project managers to play both roles on the same project. The roles are:

BUYER

The organization or party purchasing (procuring) the goods or services from the seller.

SELLER

The organization or party providing or delivering the goods or services to the buyer.

Plan Purchases and Acquisitions

WHAT IT IS

This process involves looking at the project and determining which components or services of the project will be made or performed internally and which will be "procured from an external source." After that decision is made, the project manager must determine the appropriate type of contracts to be used on the project.

WHY IT IS IMPORTANT

Currently best practices in the field of project management favor buying externally vs. building internally, all other things being equal; however, there are numerous factors that should go into your decision on whether to "make or buy." Carefully planning what to procure, and how to go about the processes of procurement, will ensure that right things are procured in the right way.

WHEN IT IS PERFORMED

Because a project may have multiple subcontractors, potentially in every phase of the project, any of the procurement processes could be performed repeatedly and at any time throughout the project.

HOW IT WORKS

- Inputs
  - *Organizational Process Assets - See Chapter 2, Common Inputs*
  - *Project Management Plan - See Chapter 2, Common Inputs*
  - *Enterprise Environmental Factors*

There may be factors at work in an organization that have a strong influence on procurement. For instance, an organization may have a strong culture of "build internally vs. buy" or they could have a strong culture of buying from a few trusted sellers. All of this should be factored in when making procurement decisions.

- *Project Scope Statement*

The project scope statement defines the scope of the project. This information will be useful to review when considering what components of the scope should be procured (i.e. performed by groups outside the organization).

- *Work Breakdown Structure*

The WBS is an organization of all of the deliverables on the project. Oftentimes when reviewing the WBS, an entire segment that may be "broken off' and procured becomes apparent.

- *WBS Dictionary*

The WBS dictionary always accompanies the WBS. It provides expanded information on each node of the WBS.

- Tools

  - *Make-or-Buy Analysis*

Make-or-buy analysis can be difficult to sum up succinctly. The analysis looks at all of the factors that could sway the decision toward making internally or buying externally, including risk factors, cost, releasing proprietary information, and a host of other decision points.

When using this tool, the project must often look past itself. For instance, writing a software component may not make as much sense as procuring it externally where only the project is concerned; however, if the performing organization has an interest in developing the capability to build that kind of software, it may make sense for the project to make vs. buy.

- *Expert Judgment*

Where procurement planning is concerned, expert judgment can be of tremendous value. There will often be scope considerations, technical considerations, legal, cost, and schedule considerations, and bringing in people with expertise in this area can help the project team make the best decisions.

- *Contract Type*

When procuring goods or services, the type of contract that governs the deal can make a significant difference in who bears the risk. There are three categories of contracts you must know for the test. They are listed below with information on each one:

A. Fixed Price Contracts (also known as Lump Sum Contracts)

Fixed price contracts are the easiest ones to understand. There is generally a single fee, although payment terms may be specified so that the cost is not necessarily a lump sum payable at the end. This type of contract is very popular when the scope of work is thoroughly defined and completely known. Two types of fixed price contracts are:

- **Fixed Price Incentive Fee**

The price is fixed, with an incentive fee for meeting a target specified in the contract, (such as finishing the work ahead of schedule).

- **Fixed Price Economic Price Adjustment**

This type of contract is popular in cases where fluctuations in the exchange rate or interest rate may impact the project. In this case, an economic stipulation may be included to protect the seller or the buyer. The economic stipulation may be based on the interest rate, the consumer price index, cost of living adjustments, currency exchange rates, or other indices.

B. Cost Reimbursable Contracts

There are two common types of cost reimbursable contracts:

- **Cost Plus Fixed Fee**

The seller passes the cost back to the buyer and receives an additional fixed fee upon completion of the project.

- **Cost Plus Incentive Fee**

The seller passes the cost back to the buyer and gets an incentive fee for meeting a target (usually tied back to keeping costs low) specified in the contract.

C. Time and Materials Contracts

In a time and materials contract, the seller charges for time plus the cost of any materials needed to complete the work.

Type of Contract	Who Bears the Risk	Explanation
Fixed Price	Seller	Since the price is fixed, cost overruns may not be passed on to

		the buyer and must be borne by the seller.
Cost Plus Fixed Fee	Buyer	Since all costs must be reimbursed to the seller, the buyer bears the risk of cost overruns.
Cost Plus Incentive Fee	Buyer and Seller	The buyer bears most of the risk here, but the incentive fee for the seller motivates that seller to keep costs down.
Time and Materials	Buyer	The buyer pays the seller for all time and materials the seller applies to the project. The buyer bears the most risk of cost overruns.

- Outputs

  - *Procurement Management Plan*

The procurement management plan is an important output of the Plan Purchases and Acquisitions process. It defines how all of the other procurement management processes will be carried out. This includes defining what will be procured on the project, how a seller will be selected, what types of contracts will be used, how risk will be managed, and how sellers will be managed, including how their performance will be measured.

  - *Contract Statement of Work*

Many people find this document confusing when first approaching it, but it is highly important for the exam! If you recall the project scope statement, produced as part of the Scope Definition process, then you will know that the project's scope has been defined at this point.

The statement of work is not merely a replication of what we did in Scope Definition. Instead, the contract statement of work is important because it helps explain a section of the scope to potential sellers in enough detail so that they can decide whether they want to (or are qualified to) pursue the work in question.

One key point to understand about the contract statement of work is that it is not the entire project's statement of work. Instead, it only defines the scope that the potential seller would need to understand to complete their part of the sub-project.

- *Make-or-Buy Decisions*

During this process, the project team make-or-buy analysis as a tool, and now it is time to act upon the data gathered. The decisions should be made, documented, and enough information to justify why these decisions were made should be included.

- *Requested Changes*

Sometimes changes need to be made to the project in order to facilitate procurement. For instance, a software application may be redesigned in order to allow a subcontractor to use their own database and keep their work and data separate from that of the core team. These requested changes should be channeled back through Integrated Change Control to evaluate their impact on the rest of the project.

Plan Contracting

WHAT IT IS

Now that the initial (and somewhat high-level) procurement planning has taken place, it is time to create the detailed documents, such as the request for proposal (RFP), that will allow the sub-project to be bid out and managed.

WHY IT IS IMPORTANT

The Plan Contracting process is important because it provides the definition and details needed to conduct the next few procurement processes. These documents allow potential sellers to know exactly what is expected of them in terms of the work to be done, how it is to be done, cost, time, and quality expectations, etc.

WHEN IT IS PERFORMED

This process, like all of the procurement management processes, may be performed more than once on a project. For instance, if you planned to contract out five components of the project, you would likely perform this process five times.

HOW IT WORKS

- Inputs
  - *Procurement Management Plan - See Plan Purchases and Acquisitions, Outputs*
  - *Contract Statement of Work - See Plan Purchases and Acquisitions, Outputs*
  - *Make-or-Buy Decisions - See Plan Purchases and Acquisitions,Outputs*
  - *Project Management Plan - See Chapter 2, Common Inputs*
- Tools
  - *Standard Forms*

Keep in mind that the purpose of this process is to create the procurement documents such as the RFP (see outputs below for more information on the documents produced). When an RFP is created, it is generally unheard of to start with a blank sheet of paper. Instead, it is more common to begin with a standard template or form that provides standard language or at the very least a checklist of items to consider.

- *Expert Judgment*

In the area of creating procurement documents, very few project managers have all of the expertise needed to create an RFP or any of the other documents needed to support solicitation.

- **Outputs**

  - *Procurement Documents*

The procurement documents should be produced by the buyer. They may be called several things, including proposals, bids, and similar names. These documents may be general or detailed, depending on the need.

For instance, if you were a manufacturing company, and your request for proposal (RFP) was related to computer networking, you may choose to keep the proposal fairly general so that your potential sellers could provide specifics that were in their areas of expertise. If, however, you were an automaker and wanted to procure the braking assembly as part of a new automobile, the RFP for that would likely be highly specific and detailed.

If the item being procured is not within the buyer's area of expertise, the buyer may want to ensure that the procurement documents allow for enough flexibility to get creative responses from the sellers. These procurement documents are invitations and requests for potential sellers to evaluate the work and prepare a proposal.

- *Evaluation Criteria*

Before taking the PMP Exam, make sure you understand that evaluation criteria are created by the buyer at the same time as the procurement documents. In other words, at the same time you assemble the procurement documentation, you decide how you will select the winning bid. It is not always mandatory that the evaluation criteria be objective, or even that you reveal it to the potential sellers. For instance, you might wish to select a winning bid based on their ability to work well with your team (note that this could be illegal on some public-sector projects, and rules regarding evaluation criteria will differ among organizations and industries).

- *Contract Statement of Work updates*

The easiest way to think about this is that the contract statement of work is the general document, and the procurement documents provide the specifics. As you are documenting the details out in the RFP (or similar procurement document), it is normal that you may want to make some changes to the more general contract statement of work.

Request Seller Responses

This is a somewhat unusual process in that most of the work here is being performed by the potential sellers. Request Seller Responses is the process concerned with distributing the procurement documents out to the potential sellers and having them respond with how they would perform the work and their qualifications to perform the work.

- Inputs
  - *Organizational Process Assets - See Chapter 2, Common Inputs*
  - *Procurement Management Plan - See Plan Purchases and Acquisitions, Outputs*

- *Procurement Documents - See Plan Contracting, Outputs*

- Tools

  - *Bidder Conferences*

The important point regarding bidder conferences is that all interested potential sellers are given the procurement documents (typically in advance of the conference) and are allowed to ask questions. Using this type of meeting, all bidders are kept on a level playing field, and none has more or less access to people or information than the others.

  - *Advertising*

Oftentimes bids must be advertised, either in order to ensure that an adequate number of qualified sellers know about the bid or to meet legal requirements. For instance, some governments require that certain bids be advertised to the public.

These ads may be distributed through public means, such as periodicals or the Internet, or they may be advertised only to the qualified sellers list.

  - *Develop Qualified Sellers List*

If the performing organization requires that sellers be on a list of qualified sellers, then the list should be used. If the buyer's organization has no such list, then one may be developed for the purpose of the project.

- Outputs

  - *Qualified Sellers List*

As mentioned above, a qualified sellers list may be developed for this project if one does not already exist.

- *Procurement Document Package*

This is the formal package of documents that includes the packaging of procurement documents created earlier (e.g. Request for Proposal, Invitation for Bid, or Request for Quotation). These are distributed to the potential sellers.

- *Proposals*

The proposal is the answer to the procurement document package described above. It is prepared by the seller and includes whatever information was requested by the buyer. Typically, the seller's qualifications are included, along with how they would deliver the goods or services, the price, their projected performance, and any helpful technical details that would assist the buyer in deciding to use the seller. These proposals are often followed up by interviews with the buyer which allow the buyer and seller to explore the potential seller's proposal in greater detail.

Select Sellers

WHAT IT IS

Most of the processes in the PMBOK Guide do exactly what they sound like. In this case, the process of Select Sellers does just that. It picks a seller who will be awarded a contract.

WHY IT IS IMPORTANT

This one gets the ball rolling with procurement. So far you have planned what you want to procure, written an RFP, issued it to potential sellers and held your bidder's conferences, and you have received proposals from your potential sellers. Now you are going to select one of those proposals and issue a contract. You know, however, that real life isn't that simple, and this can be a very tricky area on the PMP Exam as well.

WHEN IT IS PERFORMED

To state the obvious, this process is performed when you are ready to select a seller. This means that it will occur after Plan Purchases and Acquisitions, Plan Contracting, and Request Seller Proposals.

Like the other procurement processes, Select Sellers is performed as needed. It may be performed multiple times if there were multiple contracts, or it would not be performed at all if the project is not procuring anything.

HOW IT WORKS

- Inputs
  - *Organizational Process Assets - See Chapter 2, Common Inputs*
  - *Procurement Management Plan - See Plan Purchases and Acquisitions, Outputs*
  - *Evaluation Criteria - See Plan Contracting, Outputs*
  - *Procurement Document Package - See Request Seller Responses, Outputs*
  - *Proposals - See Request Seller Responses, Outputs*
  - *Qualified Sellers List - See Request Seller Responses, Outputs*
  - *Project Management Plan - See Chapter 2, Common Inputs*
- Tools
  - *Weighting System*

When the buyer receives proposals from the seller, there needs to be some way to rank and quantify the information. The reason for this is to be as objective as possible.

For instance, if you specified in your procurement documents that you wanted networking experience, then you need some way to rank your potential seller responses based on their networking experiences. You may do this by assigning a 1 through 10 ranking or some other scale.

Following this procedure for all key information in the procurement process will help make seller selection easier.

- *Independent Estimates*

The buyer may go and commission independent estimates to validate the proposals and bids received from the potential sellers. For example, the buyer may go out and ask an independent group to prepare an estimate so that they have an objective source of information. These independent estimates are often referred to as "should-cost" estimates.

- *Screening System*

Screening systems set the bar at a certain level so that they can filter out certain non-qualified sellers. For instance, a buyer may require that all sellers be ISO-9002 certified in order to submit a bid. A screening system is put into place to allow only the bids of ISO-9002 certified companies through.

- *Contract Negotiations*

Negotiations between the buyer and the potential seller are performed with the goal of reaching mutual agreement on the contractual terms and conditions. Where the exam is concerned, the goal of negotiations is to reach a win- win scenario between the buyer and seller. When the buyer exerts too much pressure on the seller or negotiates an untenable deal, this can backfire on both of them in the long run.

Keep in mind that the project manager may or may not be the person who conducts these negotiations, as this can be a highly specialized area. On large and complex contracts, professional negotiators are the norm.

- *Seller Rating Systems*

An organization may keep a grading system on the past performance of sellers. For example, a company may keep records on how well a seller previously did, and this can be used as a rating system that may be used as part of future seller selection.

- *Expert Judgment*

In the area of selecting a seller, the buyer can use expert judgment to evaluate the bids and proposals.

- *Proposal Evaluation Techniques*

There is no one way to evaluate proposals. In reality, there are as many techniques as there are buyers. This tool is essentially a combination of the other tools in the Select Seller process, used to evaluate the seller's bid or proposal.

- Outputs
  - *Selected Sellers*

The RFP has been generated, the sellers have submitted their proposals, the negotiations have taken place, and now a seller is selected to provide goods or services on the project.

- *Contract*

The contract is a formal document governing the relationship between the buyer and seller. Contracts are legal documents with highly specialized and technical language and should be written and changed only by people specializing in that field (e.g. the contracting officer, procurement office, lawyer, legal counsel).

The project manager should not attempt to write, negotiate, or change the contract on his own. The contract describes the work to be performed and perhaps the way in which that work will be performed (e.g. location, work conditions). It may specify who will do the work, when and how the seller will be paid, and delivery terms.

In reality, there is very little that the contract cannot specify in one way or another, so long as the terms and conditions are legal and they are mutually agreed upon by buyer and seller. Several other legal factors may come into play depending on the country that governs the contract, legal consideration, and other technical legal matters that are out of scope of this discussion.

An important component to include with the contract is how disputes (also called claims) will be resolved. This includes the process of dispute resolution, the parties who will be involved, and where the dispute resolution will take place.

- *Contract Management Plan*

This component of the project management plan describes how the contract will be managed and how changes to the contract will be managed. There may be more than one contract management plan on a project if there is more than one contract on the project.

- *Resource Availability*

As resources and their availability are modified as a result of the contract, they should be updated. For instance, if a contract specifies that a team will work on site or that a piece of equipment will be available for a range of dates, these terms need to be documented, and appropriate resource calendars should be updated.

- *Procurement Management Plan Updates - See Plan Purchases and Acquisitions, Outputs*

- *Requested Changes - See Chapter 2, Common Outputs*

Contract Administration

WHAT IT IS

In a nutshell, Contract Administration is the process where the buyer and seller review the contract and the work results to ensure that the results match the contract. This typically includes a review of:

- Were the goods or services delivered?
- Were the goods or services delivered on time?
- Were the right amounts invoiced or paid for the goods or services?
- Were any additional conditions of the contract met?

The process of Contract Administration is a process performed by the buyer and the seller, and because of the ramifications of any issues here, the project managers from both the buyer and seller should use whatever resources are necessary to ensure that all of these ramifications are fully understood.

WHY IT IS IMPORTANT

When looking at this from a project management perspective, the contract may be viewed as a plan (albeit a very specialized and binding type of plan). The process of Contract Administration ensures that the results of the project match this plan and that all conditions of the contract are met.

WHEN IT IS PERFORMED

Contract Administration, like the other procurement management processes, may be performed throughout the project if goods or services are being procured. It typically occurs for a given contract at predefined intervals, but may also be performed as requested or needed.

HOW IT WORKS

- Inputs
  - *Contract - See Select Sellers, Outputs*
  - *Contract Management Plan - See Select Sellers, Outputs*
  - *Selected Sellers - See Select Sellers, Outputs*
  - *Performance Reports - See Performance Reporting, Outputs*
  - *Approved Change Requests - See Chapter 2, Common Inputs*
  - *Work Performance Information - See Direct and Manage Project Execution, Outputs*
- Tools
  - *Contract Change Control System*

The contract change control system is a component of the integrated change control system, and you will likely have a separate one for each contract. It was defined in the Select Sellers process. This system describes the procedures for how the contract may be changed, and it would typically include the people and the steps that need to be taken.

- *Buyer-Conducted Performance Review*

This is a periodic review, initiated by the buyer but including the seller, where the seller's progress is measured against the contract and any other applicable plans. The seller is shown any areas where they are compliant as well as any areas where performance is a problem.

- *Inspections and Audits*

This tool focuses on the product itself and its conformance to specifications. One important difference between this and the previous tool is that although inspections and audits do not seek to measure the seller's performance (i.e. how quickly or cost-effectively they are delivering the results), the buyer may use them to help the seller find problems in the way they are delivering the work results.

- *Performance Reporting*

Performance reporting is an excellent tool to help measure the seller's performance against the plan. This may include such items as earned value, the cost performance and schedule performance indices, and trend analysis. Oftentimes there are contract conditions tied to performance (e.g. seller must deliver at least 50% of the quantity of motors within 30 days of executing this contract).

- *Payment System*

A payment system helps ensure that invoices and payments match up and that the right amount is being invoiced for the right deliverables at the right time. Additionally, payment systems will help avoid duplicate payments (or invoices if you are the seller). This tool can be of particular importance when payments are specified in the contract.

- *Claims Administration*

Claims are basically disagreements. They may be about scope, the impact of a change, or the interpretation of some piece of the contract. The essential (legally binding) elements of the process for claims administration are defined in the contract itself, but there may be additional components in the contract management plan. The most important thing about claims administration is to understand that they must be managed and ultimately resolved and that the process for doing so should be defined in advance of the claim.

- *Records Management System*

The project manager uses the records management system to keep track of all communication that would be relevant to the contract.

- *Information Technology*

Contract administration can be administratively burdensome, especially where larger contracts are concerned, and information technology can be used as a tool to help manage the large stream of information.

- Outputs

  - *Contract Documentation*

In short, contract documentation includes everything relevant to the contract, and it is kept by the buyer and the seller (although they would certainly keep separate documentation as well). Items such as supporting detail on deliverables and performance should be kept, as well as financial records. Also, any changes from the original contract should be included in the contract documentation.

- *Organizational Process Assets Updates*

There are many items that come out of contract administration that may be assets, but one of the most common ones to be updated as part of this process is the information on the seller's performance during the project. This could assist future projects in deciding whether to use this seller in the future.

- *Requested Changes - See Chapter 2, Common Outputs*

- *Recommended Corrective Action - See Chapter 2, Common Outputs*

- *Project Management Plan Updates - See Chapter 2, Common Outputs*

Contract Closure

WHAT IT IS

Contract Closure is the process where the contract is completed by the buyer and seller. Ideally, the contract is terminated amicably with the seller delivering all of the contracted items and the buyer making all payments on time, but it isn't always done that way. For instance, the buyer may consider a seller to be in default and cancel the contract, or a buyer and seller may mutually agree to terminate the contract for any reason.

WHY IT IS IMPORTANT

Every contract must be closed. Contracts inadvertently left open potentially put both the buyer and seller at risk. This process is tightly linked to the Close Project process in integration management as it contributes to the closure of the entire project.

WHEN IT IS PERFORMED

This process is performed at the end of each contract, whether or not that end is successful. When the contract is completed or terminated for any reason, this process is performed.

HOW IT WORKS

- Inputs

  - *Procurement Management Plan - See Plan Purchases and Acquisitions, Outputs*

  - *Contract Management Plan - See Select Sellers, Outputs*

  - *Contract Documentation - See Contract Administration, Outputs*

  - *Contract Closure Procedure*

The procedures for closing the contract are typically specified in the contract and the contract management plan.

- Tools

  - *Procurement Audits*

The point of a procurement audit is to capture lessons learned from a contracting perspective.

  - *Records Management System*

Contract documentation should always be archived for future reference. Many organizations have records management systems to facilitate the storing, archiving, and retrieval of records like this.

- **Outputs**

  - ***Closed Contracts***

The closed contracts are the essential outputs from Contract Closure.

  - ***Organizational Process Assets Updates***

In addition to any other assets that may need updating, the lessons learned captured in this process (during the procurement audit) are formally documented here.

Procurement Management Questions

1. The contract type that represents the highest risk to the seller is:

   A. Fixed price plus incentive.

   B. Cost reimbursable.

   C. Fixed price.

   D. Cost reimbursable plus incentive.

2. You have been tasked with managing the seller responses to a request for proposal issued by your company. The seller responses were numerous, and now you have been asked to rank the proposals from highest to lowest in terms of their response. What are you going to use as a means to rank the sellers?

A. Evaluation criteria.

B. Request for quotation.

C. Seller response guidelines.

D. Seller selection matrix.

3. Make or buy analysis is a tool used in which process?

   A. Plan Purchases and Acquisitions.

   B. Request Seller Responses.

   C. Plan Contracting.

   D. Procurement Analysis.

4. Which of the following represents the right sequence of processes?

   A. Plan Contracting, Plan Purchases and Acquisitions, Request Seller Responses, Select Sellers, Contract Administration.

   B. Plan Contracting, Plan Purchases and Acquisitions, Select Sellers, Request Seller Responses, Contract Administration.

   C. Plan Purchases and Acquisitions, Plan Contracting, Contract Administration, Request Seller Responses, Select Sellers.

   D. Plan Purchases and Acquisitions, Plan Contracting, Request Seller Responses, Select Sellers, Contract Administration.

5. You are managing a large software project when the need for a new series of database tables is discovered. The need was previously unplanned, and your organization's staff is 100% utilized, so you decide to go outside the company and procure this piece of work. When you meet with prospective sellers, you realize that the scope of work is not completely defined, but everyone agrees that the project is relatively small, and your need is urgent. Which type of contract makes the MOST sense?

   A. Fixed price.

   B. Time and materials.

   C. Open ended.

   D. Cost plus incentive fee.

6. Your project plan calls for you to go through procurement in order to buy a specialty motor for an industrial robot. Because of patent issues, this motor is only available from one supplier that is across the country. After investigation, you believe that you could procure the motor from this company for a price that is within your budget. What is your BEST course of action?

   A. Revisit the design and alter the specification to allow for a comparable motor.

   B. Procure the motor from this source even though they are the sole source.

   C. See if the component may be produced in another country, avoiding your country's patent issues.

   D. Take the product out of the procurement management process.

7. You have a supplier that is supplying parts to you under contract. The terms and

conditions give you the right to change some aspects of the contract at any time, and you need to significantly lower the quantities due to a change in project scope. How should you notify the supplier?

A. Take them to lunch and explain the situation gently to preserve the relationship.

B. Have your attorney call their attorney.

C. Communicate with the supplier via e-mail.

D. Send them a formal, written notice that the contract has been changed.

8. Your project has been terminated immediately due to a cancellation by the customer. What action should you take FIRST?

A. Call a meeting with the customer.

B. Enter contract closure.

C. Ask your team leads for a final status report.

D. Verify this change against the procurement management plan.

9. You are evaluating proposals from prospective sellers. What process are you involved in?

A. Plan Contracting.

B. Request Seller Responses.

C. Select Sellers.

D. Contract Administration.

10. Your project scope calls for a piece of software that will control a valve in a pressurized pipeline. Your company has some experience with this type of software, but resources are tight, and it is not part of your company's core competency. You are considering involving other sellers but want to decide whether it is a better decision to produce this within your company or source it externally. What activity are you performing?

A. Source selection.

B. Make or buy analysis.

C. Rational project procurement.

D. Source evaluation.

11. Your organization is holding a bidders conference to discuss the project with prospective sellers, and a trusted seller you have worked with many times in the past has asked if they can meet with the project manager the day before the conference to cover some questions they do not wish to ask in front of other sellers. Should your organization meet with the seller?

A. Yes, the more that prospective sellers know about the project, the better.

B. Yes, they are your primary seller, and past history should be factored in.

C. No, prospective suppliers should be kept on equal footing.

D. No, that would represent an illegal activity.

12. The most important thing to focus on in contract negotiations is:

A. To negotiate the best price possible for your project.

B. To maintain the integrity of the scope.

C. To negotiate a deal that both parties are comfortable with.

D. To make sure legal counsel or the contract administrator has approved your negotiating points.

13. If a project manager was performing Contract Administration, which of the following duties might he be performing?

A. Approving seller invoices.

B. Negotiating the contract.

C. Closing the contract.

D. Weighing seller responses.

14. Who generally bears the risk in a time and materials contract?

A. The buyer.

B. The seller.

C. The buyer early in the project and the seller later on.

D. It depends on the materials used.

15. Your company is outsourcing a project in an area where it has little experience. The procurement documents should be:

A. Completely rigid to ensure no deviation from sellers.

B. Flexible enough to encourage creativity in seller responses.

C. Informal.

D. Reviewed by senior management.

16. The contract statement of work for an item being procured should provide:

A. Enough detail for the prospective seller to complete the project.

B. Enough detail to describe the product, but no so much as to divulge trade secrets or sensitive information.

C. Enough detail to perform make or buy analysis.

D. Enough detail for the prospective seller to know if they are qualified to perform the work.

17. You are the project manager for a seller who has been selected to construct an industrial kitchen for a large food services company. Before the contract negotiations, the buyer confides in you that design is not finalized, and they want you to begin work with incomplete specifications. What type of contract should you ask for in negotiations?

A. Fixed price.

B. Cost plus incentive fee.

C. Time and materials.

D. Cost plus fixed fee.

18. The product or result of the project is created during which process group?

    A. Project lifecycle.

    B. Contract administration.

    C. Project executing.

    D. Work package processing.

19. Your customer has asked to meet with you and inspect the work you have completed to date on the project to ensure that it meets the contractual agreements. What is your customer manager engaged in?

    A. Contract Closure

    B. Seller efficiency audit.

    C. Seller administration.

    D. Buyer-conducted performance review.

20. You have completed a project and delivered the full scope of the contract. The buyer agrees that you have technically satisfied the terms of the contract but is not satisfied with the end results. In this case, the contract is:

    A. Contested.

    B. Complete.

C. Poorly written.

D. Lacking terms and conditions.

Answers to Procurement Management Questions

1. C. This question was easier than it may have first appeared to be. Fixed price is the highest risk to the seller since the seller must bear the risk of any cost overruns. Choice 'B' would provide the highest risk to the buyer.

2. A. The evaluation criteria will provide the guidelines by which you can evaluate and rank responses. The RFQ would not necessarily help you pick the seller, as it is their proposal that is of interest. 'C' and 'D' were terms that were made up for this question.

3. A. Make or buy analysis is a tool used during the Plan Purchases and Acquisitions process, where you are deciding which deliverables should be procured and what should be created internally. 'D' is not a real process, and 'B' and 'C' are incorrect, because by the time you plan contracting or request responses from sellers, you need to already know what you are going to make and what you are going to buy.

4. D. Plan Purchases and Acquisitions, Plan Contracting, Request Seller Responses, Select Sellers, Contract Administration. A partial explanation that may help you understand why this sequence is correct and the others are not is that the outputs earlier processes are used as inputs into one or more subsequent processes. In this case, Plan Purchases and Acquisitions produces the Procurement Management Plan, which becomes an input into Plan Contracting. Plan Contracting produces the procurement documents, which are an input into Request Seller Responses. Request Seller Responses produces the procurement document package, which is used as an input into Select Sellers. Select Sellers produces the contract, which is an input into Contract Administration. This should help you understand why the processes must take place (at least initially) in this order.

5. B. Time and Materials. Choice 'A' is incorrect because the scope is not defined enough to establish a fair fixed price. Choice 'C' is a made up type of contract. Choice 'D' would not make sense in this case since the seller's costs are not abundantly clear, and this type of contract would create too much risk.

6. B. This one may trick some who think that it is wrong to use a sole source. In many cases it is the only choice. 'A' would not be good since the design has nothing to do with this. 'C' is not necessary in this case, since the issue is not a legal issue. Choice 'D' would be completely invalid since the item is still being procured outside of your organization.

7. D. Choices 'A' and 'B' are verbal. Contract changes should always be made in writing! 'C' is written, but e-mail is not the proper forum for making important contractual changes.

8. B. There are plenty of questions on the PMP Exam that defy common sense. This one asks for your FIRST action, and the most appropriate action is to enter Contract Closure. Choices 'A' and 'C' may be appropriate at some point, but if a project is terminated, Contract Closure needs to be performed.

9. C. If you are evaluating seller responses, you are performing the process of Select Sellers. The seller proposals are brought into this process, and they are screened, weighted, rated, and evaluated against the criteria.

10. B. Make or buy analysis is the process where an organization decides whether it should produce the product internally or outsource it. This is done as part of the Plan Purchases and Acquisitions process.

11. C. If you are involved in formal procurement, you should make every effort to keep sellers on equal footing. If one seller is provided with an advantage, it negates much of the value of the procurement process.

12. C. The most important point is to create a deal that everyone feels good about. 'A' sounds like a good choice, but it is incorrect, since the best possible price might not be fair to your seller, and that could create a bad scenario for the project in the future. 'B' is important, but that is not the primary focus of negotiations. 'D' may or may not be necessary, depending on the situation.

13. A. One of the activities in Contract Administration is to pay seller invoices or generate invoices if you are the seller. Make sure to learn the primary process inputs, tools, techniques, and outputs for the processes.

14. A. In a time and materials contract, the buyer has to pay the seller for all time and materials, and often times it involves an incomplete scope definition. Therefore, the buyer is the one most at risk.

15. B. In this scenario, you want sellers to respond with their own ideas. Procurement documents should be rigid enough to get responses to the same scope of work, and flexible enough to allow sellers to interject their own good ideas and creativity. Many people incorrectly choose 'A' because they assume that a rigid approach is almost always correct, but in this example, you do not have sufficient experience to rigidly manage the process, and you want your sellers to give you some guidance in their proposals. 'C' was clearly incorrect, as procurement is something that should be done formally. 'D' is incorrect because senior management has many functions in an organization, but they would not be expected to review procurement documents.

16. D. A contract statement of work should be as complete and concise as possible. At a minimum, it should contain enough information for the seller to determine if they are qualified to do the work.

17. C. The major clue here is that the scope of work is not completely defined and they want you to begin work anyway. In that case, the project is at a higher risk, and a time and materials contract shifts much of that risk back to the buyer.

18. C. This question ties back to Chapter 2- Foundational Terms and Concepts. The actual work packages are completed (or executed) during the executing process group.

19. D. Buyer-conducted performance reviews are a tool of Contract Administration where the buyer arranges a meeting with the seller to review the seller's performance against the plan. This question presents a near-textbook case of this.

20. B. If the scope of the contract is complete and no other terms were breached, then the contract is complete. 'C' is not a good guess here, since you don't have enough information to state that the problem was that the contract was poorly written.

Professional Responsibility

In recent versions of the PMP Exam, 15% of your overall grade came from the category of Professional Responsibility, but that is no longer the case. That number has been trimmed down. Still, you should pay special attention to this chapter. Once you have mastered this, you should have little difficulty with the professional responsibility questions on the exam.

Philosophy

This chapter is all about philosophy, and each of the sections below builds upon the philosophical base that drives the questions and answers. The philosophy behind professional responsibility is that the project manager should be a leader, should deal with issues in a direct manner, should act ethically and legally, and should be open and up front. For each of these questions, don't ask, "What would I do in my organization?" but rather, "What should I do?" The PMP is expected to be professional, and that means following the processes outlined in the PMBOK Guide.

Hard choices are a favorite tactic for the questions in this chapter. You may be presented with a small ethical violation that will be painful to resolve. Always look for the answer that resolves it quickly, openly, and fairly. You may be given a situation that lets you ignore a problem instead of confronting it. Always look for the choice that will let you deal with the problem directly.

Importance

Nine percent of the questions on the exam relate directly to this chapter, and many more questions will relate indirectly, but none of this information is referenced in the PMBOK

Guide. By studying this chapter, you will learn PMI's approach to professional responsibility, and you will be able to answer these questions, and perhaps more importantly, eliminate many of the incorrect answers from questions in other sections. Nine percent represents a large allocation of questions and can make a tremendous contribution toward a passing or failing score on the exam, so invest the time to study and understand this chapter.

Preparation

After studying, most students find these questions among the easiest on the exam. Because it is impractical to try and cover every possible ethical scenario, the preparation in this chapter is focused on a general understanding of the project manager's professional responsibility.

PMI Code of Conduct

Carefully review the code of conduct printed below. PMP applicants sign this statement when applying to take the exam. Most of the questions in this section will pertain to this code, either directly or indirectly.

RESPONSIBILITIES TO THE PROFESSION

- **Compliance with all organizational rules and policies**

1. Responsibility to provide accurate and truthful representations concerning all information directly or indirectly related to all aspects of the PMI Certification Program, including but not limited to the following: examination applications, test item banks, examinations, answer sheets, candidate information and PMI Continuing Certification Requirements Program reporting forms.

2. Upon a reasonable and clear factual basis, responsibility to report possible violations of the PMP Code of Professional Conduct by individuals in the field of project management.

3. Responsibility to cooperate with PMI concerning ethics violations and the collection of related information.

4. Responsibility to disclose to clients, customers, owners or contractors, significant circumstances that could be construed as a conflict of interest or an appearance of impropriety.

- **Candidate/ Certificant Professional Practice**

1. Responsibility to provide accurate, truthful advertising and representations concerning qualifications, experience and performance of services.

2. Responsibility to comply with laws, regulations and ethical standards governing professional practice in the state/province and/or country when providing project management services.

- **Advancement of the Profession**

1. Responsibility to recognize and respect intellectual property developed or owned by others, and to otherwise act in an accurate, truthful and complete manner, including all activities related to professional work and research.

2. Responsibility to support and disseminate the PMP Code of Professional Conduct to other PMI certificants.

RESPONSIBILITIES TO CUSTOMERS AND THE PUBLIC

- **Qualifications, experience and performance of professional services**

1. Responsibility to provide accurate and truthful representations to the public in advertising, public statements and in the preparation of estimates concerning costs, services and expected results.

2. Responsibility to maintain and satisfy the scope and objectives of professional services, unless otherwise directed by the customer.

3. Responsibility to maintain and respect the confidentiality of sensitive information obtained in the course of professional activities or otherwise where a clear obligation exists.

- **Conflict of interest situations and other prohibited professional conduct**

1. Responsibility to ensure that a conflict of interest does not compromise legitimate interests of a client or customer, or influence/interfere with professional judgments.

2. Responsibility to refrain from offering or accepting inappropriate payments, gifts, or other forms of compensation for personal gain, unless in conformity with applicable laws or customs of the country where project management services are being provided.

Categories

There are five categories on the exam for professional responsibility. They are listed below with the rough distribution you can expect on the exam:

Subcategory	Approximate # of Questions
Ensure Integrity and Professionalism	5
Contribute to Knowledge Base	2
Enhance Individual Competence	3

Balance Stakeholder Interests	3
Interact with Team and Stakeholders	5

ENSURE INTEGRITY AND PROFESSIONALISM

Some of the questions in this section may surprise the PMP applicant who has not thoroughly prepared. Many of the questions may create very difficult situations that would be easier to ignore or dodge in real life; however, PMI requires that Project Management Professionals deal with these situations in a direct and open manner.

If there were one simple phrase that sums up this section of the test, it would be "Do the right thing," even if it is painful or would be tempting to avoid. If a test question offers you an easy way out, beware! If the exam presents you with an option that represents a shortcut, do not take it.

The key subtopics for this section are:

- Laws

You should always follow the laws and customs of the state, municipality, or country where you are working.

International law can be a fairly tricky and sometimes ambiguous area, but the questions on the exam are generally straightforward in this regard. The general rule for the exam is this: if you are asked to do something in another country that is not customarily done in your culture, you should first evaluate and investigate it, determine if it is unethical or illegal, and then act accordingly. This may be difficult for those who are not accustomed to international dealings. For instance, if you are asked to make a payment to a city council in another country in order to get a work permit, evaluate whether or not the payment is a bribe. If it is a bribe, do not make the payment. If it is not a bribe, and it is customary, or even the law, then make the payment.

A short way of looking at it is that if it is illegal or unethical in any way, then it's wrong. Otherwise, the custom in the country where the work is being performed may prevail.

There are infinite possibilities as to what may be asked here. Questions of bribery, discrimination, and illegal activity are among the favorites, but by following the thought process previously described, these questions should present no problem. Only make sure that you are not thinking so concretely that you think anything is wrong if it is different from the customary practices in your home country.

- Policies

Your organization's policies must be followed. If you have an interest that conflicts with a policy, the policy is to be considered first. If your organization has a policy that all travel must be booked through the company's travel agency but you find that you can get a cheaper rate through your mother's travel agency, you should adhere to the company policy and use the corporate travel agency.

INTEGRITY

Integrity may be defined as sticking to high moral principles. For the test, the important concept is that you should do what you said you would do, deal with problems openly and honestly, and do not put personal gain ahead of the project.

Company and professional politics may play a prominent role in your life or company, and many project managers learn to be quite adept at them, but they are relegated to a low position on the exam.

Keep in mind that carrying out the right choice for questions in this section of the PMP exam would often involve standing up to the company's president, refusing an order from your boss, telling the customer the whole truth, and many other things that might have unpleasant consequences.

If a choice appears sneaky, underhanded, or dishonest in any way, it is probably not the correct answer. If the behavior is not direct, open, and straightforward, it is probably not the right behavior - even if it would ultimately appear to help the project! This will help you eliminate the wrong choices on many questions.

PROFESSIONALISM

Questions about professionalism on the exam are testing your knowledge of how a Project Management Professional acts as a professional in the workplace. According to the code of professional conduct, a project manager is to follow the process and act with respect toward others.

- **Processes**

One key to these questions is to take the 44 processes outlined in this book and the PMBOK seriously. The PMI process framework is not just a formality or a theoretical best-case scenario. It is a serious set of processes, inputs, tools and techniques, and outputs that will reduce risk and improve time, cost, and efficiency.

That said, the process is not painless. For the exam, however, the process must be followed. If your customer asks you to cut corners on the process in order to save money, you should not agree.

- Respect

The project manager shows respect to others. This extends not only to individuals, but to cultures. PMI is a strong advocate of multiculturalism, and questions on the test will often reflect this bias. Just as PMI does not advocate forcing in other areas, the project manager is not to force or impose their culture or personal beliefs upon others.

Multicultural and "politically correct" answers are usually good choices for questions related to professionalism.

Another area of respect that often appears on the test is the respect for confidentiality. This covers confidentiality of client information, trade secrets, project information, and any personal information that may be disclosed during the course of the project.

CONTRIBUTE TO KNOWLEDGE BASE

PMI expects Project Management Professionals to stay engaged and further the profession. The PMP certification was not designed for people to earn and never use. This is reflected on the exam with questions about activities that don't necessarily relate directly to a project.

Any time a project manager has an opportunity to further the project management training or learning of someone else, such as sharing lessons learned, mentoring, teaching, or leading in best practices, there is a very good chance that is the correct choice.

PMPs are encouraged to publish, teach, write, and disseminate the methodology and process as much as they can. Answers that offer a variant of these activities as the choice are often correct.

ENHANCE INDIVIDUAL COMPETENCE

PMI considers the PMP to be quite a milestone in one's career, but it is by no means the end. Project Management Professionals are expected to continue to study, learn, and grow professionally. As you have no doubt seen, any one of the knowledge areas could consume an entire career, so do not assume that you have learned all there is to know.

Even one topic, such as quality, or risk, could provide more than a lifetime's worth of material to study and master.

Another favorite key here is that you know your own areas of weakness and continue to develop them. Do you know what your professional weaknesses are? Are you strong at planning but weak in communication? Are you good at planning tasks but poor at leading people? Everyone has strong suits as well as areas that need to be developed. It is important that the Project Management Professional knows what his or her professional growth needs are and pays attention to them.

Additionally, project managers can contribute more to the body of knowledge if they are familiar with their industries. PMPs should study their industries, learn them well, and thus enhance their ability to apply the project management processes to their work.

You should expect to see questions that put you in a situation of taking a hard look at your abilities, or learning where it is that you are weak. In these scenarios, the project manager should strive for continued improvement, growth, and increased proficiency.

BALANCE STAKEHOLDER INTERESTS

This one is certainly one of the most difficult areas to master in real life. Stakeholders may not care about each other, and thus their interests may collide or conflict. The project manager must accurately identify the stakeholders, then understand them, and then seek to balance their interests. This can be nearly impossible at times! For the exam, keep these principles in mind:

- Be fair to everyone and respect the differences of the group.
- Resolve stakeholder conflict in favor of the customer.
- Be open and honest about the resolution. Don't hide things from one stakeholder in order to please another.
- Do the ethical thing in all decisions.

INTERACT WITH TEAM AND STAKEHOLDERS

This section can present difficulties for those who go in to the test unprepared or with the wrong mindset, especially in the area of work ethics. It is tempting for some people to approach questions with a sense of "fairness" about how hard someone should work, but work ethics vary from country to country, and project managers should take that into account.

That does not mean that laziness or negligence should be tolerated, but it does mean that different cultures place different values upon work, and it is not the project manager's job to force them to the level of his home country.

As stated previously, cultural differences on the team should be respected (notice how many times the concept of "respect" is mentioned in this chapter), and multiculturalism is something PMI promotes heavily. Your dealings with your team and the many stakeholders on a project should be professional and mindful of their customs.

As in other areas, communication should be open and regular so that people are aware of what is going on with the project.

SOCIAL-ECONOMIC-ENVIRONMENTAL SUSTAINABILITY

An additional topic included here is that organizations are accountable for social, environmental, and economic impacts to their project. Project managers should factor in the interests of the community, the environment, and society when making decisions.

On the exam, you may encounter a question that poses a situation where the project would benefit but society would suffer. The project manager should avoid all such situations and scenarios. If the situation becomes untenable or unresolvable, the project manager should disclose the situation and as a last resort, resign the project.

Professional Responsibility Questions

1. You have reviewed the schedule and have discovered that the project is going to be later than originally communicated. Your boss has asked you not to tell this to your customer even though he agrees that there is no way to shorten the schedule. You have an upcoming status meeting with the customer later on that same day. What should you do?

    A. Cancel the status meeting with the customer.

    B. Call the customer and explain your dilemma to them in confidence.

C. Explain to your boss that it is unethical to knowingly report an incorrect status to anyone on the project.

D. Ask your boss to put his request to you in writing.

2. You are in a PMP study group when a coworker, who is also a friend, tells you that she has acquired the questions from the actual exam. What would be the MOST appropriate course of action?

   A. Study from the questions, compare them to the questions on the PMP exam and determine if any further action is warranted.

   B. Ask your friend to surrender the questions and turn them over to your boss.

   C. Decline the offer and change to a different study group.

   D. Contact PMI and explain the situation to them.

3. You have started managing a project in a country that observes nearly double the amount of holidays that your home country takes. Additionally, each team member from this country receives three more weeks of vacation each year than you do. What is your MOST appropriate course of action?

   A. Authorize the same amount of vacation time for your team in order to preserve team equality.

   B. Use the other country's vacation and holidays as schedule constraints.

   C. Allow the extra vacation for that country's team, but request that they produce the same or better results as your team.

D. Try to get this part of the project relocated to a different country with a stronger work ethic.

4. Your customer instructs you that they would like to bypass creating a work breakdown structure in favor of creating a more detailed activity definition list. You have expressed disagreement with the customer on this approach, but they remain insistent. What is the BEST way to handle this situation?

   A. Document your disagreement with the customer and do it their way.

   B. Follow the customer's wishes, but create the activity list so that it is as close as possible to a work breakdown structure.

   C. Have your project sponsor explain the necessity of the work breakdown structure to the customer.

   D. Have PMI call your customer.

5. You are working in a foreign country, trying to procure a piece of expensive industrial machinery as you evaluate the bids, one of the potential sellers calls you and mentions that he could probably get 25% of the price knocked off his bid, making his by far the most attractive price. In return, he asks you to pay him one fifth of that amount as a show of your appreciation. What should you do?

   A. Do what is necessary to secure the lowest bid for your customer.

   B. Ignore the offer and evaluate the bids as if this had not happened.

   C. Eliminate this seller's bid, and notify the organization.

D. Ask the seller if he could reduce the price by even more and assure him he will be well rewarded for his efforts.

6. A project manager has confided in you that he is struggling with the whole concept of project management. He is constantly being handed down unreasonable demands by the customer and is trying to balance that with unreasonable and often opposing demands from senior management and stakeholders. He is worried that the two projects he is managing may fail and asks you for help. What is the BEST thing you can do in this case?

A. Provide your friend with a copy of the PMBOK.

B. Encourage your friend to get project management training.

C. Tell your friend that he should not be managing projects.

D. Have a private talk with your friend's boss about the problem.

7. You have just taken over a project that is executing the work packages. As you review the status of the project, you become aware that the previous project manager reported milestones as being hit that have not been reached. Additionally, the CPI is 0.88 and the SPI is 0.81. Management is unaware of all of these facts. What should you do?

A. Crash the schedule to try to get things back in line.

B. Explain the revised status to management.

C. Refuse to take on this project assignment.

D. Ask the key stakeholders for direction.

8. Two of your coworkers ask if they can confide in you. They tell you that they have been using the chemical lab that is a part of your project's construction to manufacture illegal drugs. The drugs have made them quite a lot of money, and they want to know if you would like to participate in this venture with them. What should you do in this situation?

   A. Take the opportunity to mentor these resources.

   B. Investigate the situation and determine whether or not it is really illegal.

   C. Report the situation to the authorities.

   D. Insist that they seek help, but do not report them since this was shared in confidence.

9. Two stakeholders have begun fighting over functionality on your project. Who should resolve this conflict?

   A. Senior management.

   B. The sponsor.

   C. The customer.

   D. The project manager.

10. You discover that a new pipeline project you are managing poses a previously undiscovered threat to the environment. What is the MOST appropriate course of action?

   A. Seek guidance from government officials.

B. Seek guidance from the media.

C. Seek guidance from an attorney.

D. Resign the project.

11. Your project team has begun execution of the work packages, when you discover that two of the team members do not have all of the skills needed to finish the project. The team members will need to undergo some training in order to complete their tasks. What is the BEST thing you can do?

A. Help them get the necessary training.

B. Contact human resources to find out why they did not have necessary training.

C. Let them learn by doing the work.

D. Ask the customer to invest in training the resources.

12. A colleague confesses that he has been posing as a PMP. He says that the prestige is wonderful, but pursuing the certification has never interested him. What should you do in this situation?

A. Contact the police.

B. Contact PMI and alert them to the situation.

C. Ask your colleague to pursue the real PMP Certification.

D. Report this to your colleague's human resources department.

13. You are working on a project to construct an office building in another country when you discover that you must obtain a second building permit from the local authority. When you inquire about this, you find out that the local permit will take 6 to 8 weeks to be issued; however, an official mentions that you can speed this up by paying a $500 "rush fee". The fee is within the project's budget, and the need is urgent. What is the BEST thing to do?

   A. Pay the $500.

   B. Do not pay the $500 and wait the 6 to 8 weeks.

   C. Ask the official if he would accept less than $500.

   D. Ask someone on your team to pay the fee for you.

14. You are employing a subcontractor to complete a critical piece of work when the project manager for that subcontractor asks you out on a date, making it clear that this is a personal engagement and will not affect business. What would be the BEST course of action?

   A. Accept the invitation if you are so inclined.

   B. Politely refuse the invitation.

   C. Ask your manager for permission to go on the date.

   D. Consult your human resource department for guidance.

15. You have been assigned to a project for a new software product that you do not believe will succeed. You have made these feelings clear to your company, but they wish you to work on the project anyway. What would be the BEST course of action in this case?

    A. Excuse yourself from the project.

    B. Escalate the situation to the customer.

    C. Manage the project.

    D. Resign from the company.

16. As project manager, you have met with the stakeholders and certain high profile stakeholders have provided you with a series of change requests that fundamentally alter the project and raise the risk to an unacceptably high level. The stakeholders have heard your explanation of this, but they remain insistent that the new functionality is needed. Which option below would be MOST appropriate?

    A. Explain to the stakeholders that their request cannot be fulfilled.

    B. Stall the stakeholders until the request becomes irrelevant.

    C. Remove these stakeholders from your list of stakeholders and exclude them from further project correspondence.

    D. Call a meeting with the project team to help resolve the problem.

17. As project manager, you have been asked to serve on a product selection committee. When you join, you find out that one of the products that the company is evaluating is manufactured by the company where your wife works as a sales director. What is the BEST way to handle this situation?

   A. Disclose the situation to the project selection committee.

   B. Quietly excuse yourself from the committee.

   C. Remain completely impartial throughout the product selection.

   D. Vote against the product in the interest of fairness.

18. You and your project team work for a large and well-known consulting company and are compensated based on how many hours you bill. In the make or buy analysis you performed; you find that you can save the project nearly $300,000 if you procure a software module rather than have your consulting team write it. What is the MOST appropriate way to resolve this conflict?

   A. Resolve the conflict in favor of you and your team if you are within budget.

   B. Resolve the conflict in favor of your employer.

   C. Let the team vote on how to resolve the conflict.

   D. Resolve the conflict in favor of the customer.

19. The project you are currently managing is highly similar to one you worked on for a different customer six months ago. On the previous project, you performed a detailed feasibility study that would save two calendar months if you could use it on this project. You signed a non-disclosure agreement with the previous customer. What is the BEST course of action in this situation?

    A. Disregard the previous feasibility study and perform a new feasibility study for your current customer.

    B. Ask your previous customer for permission to use the feasibility study.

    C. Use the feasibility study from the first project, since such non-disclosure agreements are not legally enforceable.

    D. Ask your project office for guidance.

20. A conflict of interest should be:

    A. Resolved quietly.

    B. Disclosed openly.

    C. Monitored carefully.

    D. Ignored deliberately.

Answers to Professional Responsibility Questions

1. C. The code of professional conduct reads: "Responsibility to provide accurate and truthful representations to the public in advertising, public statements and in the preparation of estimates concerning costs, services and expected results." You cannot ethically lie to a customer, management, your boss, or your team, even if ordered to do so.

2. D. The code of professional conduct reads: "Upon a reasonable and clear factual basis, responsibility to report possible violations of the PMP Code of Professional Conduct by individuals in the field of project management. "

3. B. Professional responsibility- questions tend to have a strong multicultural slant. This is because respect for other cultures is highly valued. Any answer other than 'B' does not reflect this value.

4. C. Here is a great example of where the customer is not always right. Did you remember that PMI considers it unethical to deviate from the stated process? Skipping something as foundational as the work breakdown structure would not further the project, your reputation, or the customer's interests. This may make sense, but why should the sponsor explain this to the customer? Because the sponsor can act as a liaison to the customer in difficult circumstances like this.

5. C. This seller wants you to bribe him with a 25% kickback. You cannot ethically do this, even though it may save your customer or your project money. In this case, the best thing to do is avoid the situation and disclose it.

6. B. The best answer here is to encourage your friend to get training in project management. Answer 'A' would not help because the PMBOK would not help your friend be successful. 'C' is not a good answer at all. Just because someone struggles does not mean that they should not be doing the job of managing projects. 'D' is a bad choice because you should confront the problem directly (with your friend).

7. B. Based on these numbers, you know that the schedule and costs are slipping. Any time a status changes you should alert stakeholders (in this case, management). Before any other action, you should report the status. 'A' is incorrect because crashing would only make the cost situation worse. 'C' is incorrect because there is no reason to refuse the project simply because it is not tracking well. A good project manager is exactly what is needed here. 'D' is incorrect because you should be evaluating and giving direction to the project instead of asking the stakeholders to do this.

8. C. You have a responsibility to report any illegal activity to the authorities.

9. D. It is the project manager's responsibility to balance stakeholder's interests. The project manager should attempt to resolve this conflict.

10. A. Because you have a socio-economic-environmental responsibility, you should alert the government officials and seek guidance. They have a vested interest in helping resolve this quickly, and this would deal with the root of the problem. Answers 'B', 'C', and 'D' do not deal directly with the problem.

11. A. You have a responsibility to mentor and help your team get training. Remember that the cost of training is typically born by the performing organization's budget and not by your project's budget, and not by the customer.

12. B. The PMP Code of Professional Conduct states that "Upon a reasonable and clear factual basis, responsibility to report possible violations of the PMP Code of Professional Conduct by individuals in the field of project management."

13. A. This does not constitute a bribe. You should pay the fee. If the official wanted you to slip the money under the table, that would be different, but in this case the fee has a legitimate purpose.

14. B. This would be a conflict of interest, and it is the project manager's duty to avoid all conflicts of interest. Consulting the human resources department or your manager would not avoid a conflict of interest.

15. C. This is a difficult scenario, but you cannot always pick and choose projects that appeal to you. The best thing is to manage the project and follow the project management processes. If the project will not be successful, it should become apparent sooner rather than later.

16. A. It is the project manager's job to balance competing stakeholder interests. It may be tempting to try to accommodate everyone's requests, but some simply must be refused. In this case, you must explain to the stakeholders that their requests cannot be fulfilled.

17. A. The code of conduct states that conflicts of interest should be clearly disclosed. None of the other choices really satisfies that rule. 'B' might be a tempting choice, but it is not the best choice, since you should be upfront and open about such situations.

18. D. As a general rule, conflicts should be resolved in favor of the customer. In this case, that is particularly true.

19. B. You should ask permission from the previous customer. 'A' is incorrect because you are not working in the best interest of your current customer if you do this. It may become necessary to do the work over again, but you should try to negotiate that with your previous customer first rather than rushing to "reinvent the wheel." 'C' is incorrect because even if the non-disclosure agreement was not legally enforceable, it is still agreement between you and your previous customer, and you are ethically bound to comply with it. The PMP code of conduct gives specific guidance against improper use of proprietary information. 'D' is incorrect because more guidance is not what is needed here, as the issues are clear. Your project office's approval to use the previous material would not make it acceptable.

20. B. The PMP Code of conduct states "Responsibility to disclose to clients, customer, owners, or contractors, significant circumstances that could be considered as a conflict of interest or an appearance of impropriety."

How To Pass The PMP

Passing the PMP on your first try has nothing to do with good luck. It is all about preparation and strategy. While the other chapters in this book are all about the preparation, this chapter focuses on the test strategy itself. The following are techniques on how you can be sure to avoid careless mistakes during your exam.

Reading the Questions

A critical step to passing the PMP is to read and understand each question. Questions on the exam may be long and have many twists and turns. They are often full of irrelevant information thrown in intentionally to distract you from the relevant facts. Those who pass the PMP know to read the questions carefully. Many times, the only relevant information is contained at the very the end. Consider the following example:

Q: Mark has a project where task A is dependent on the start and has a duration of 3. Task B is dependent on start and has a duration of 5. Task C is dependent on A and has a duration of 4. Task D is dependent on B and has a duration of 6, and the finish is dependent on tasks C and D. Mark is using his project network diagram to help create a schedule. The schedule for the project is usually created during which process?

A. Cost Estimating.

B. Cost Budgeting.

C. Schedule Control.

D. Schedule Development.

Questions like the one above are not uncommon on the PMP Exam. If you take the time to draw out a complex project network diagram, you will have wasted valuable time, when the question was only asking you to pick the process (the answer is D).

On lengthy questions, the best practice is to quickly skip down to the last sentence for a clue as to what the question is asking. Then read the entire question thoroughly. Most of them have a very short final sentence that will summarize the actual question. Make sure, however, to read the entire question at least once! Don't simply rely on the last sentence.

Just as important as carefully reading the questions is reading each of the four answers. You should never stop reading the answers as soon as you find one you like. Instead, always read all four answers before making your selection.

A Guessing Strategy

By simply reading the material in this book, you will immediately know how to answer many of the questions on the exam. For many others, you will have an instinctive guess. If you have studied the other chapters, you should trust that instinct. It is not there by chance. Your instinct was created by exposing yourself to this material in different ways. Your mind will begin to gravitate toward the right answer even if you are not explicitly aware of it.

Guessing on the PMP does not have to be left purely to chance. If you do not know the answer immediately, begin by eliminating wrong answers, or ones you suspect are wrong. Let's take a fairly difficult question as an example:

Q: Organizational Process Assets are used as an input to all of the following processes EXCEPT:

A. Activity Definition.

B. Create Project Charter.

C. Direct and Manage Project Execution.

D. Quantitative Risk Analysis.

Unless you have memorized all of the inputs to all of the processes (the PMBOK Guide lists over 592 inputs, tools, and outputs), you are going to have to guess at this one. However, if you throw up your hands and pick one, you only have a 25% chance of getting it right. Instead, you should think about what is being asked. Organizational process assets are used as an input to processes all over the PMBOK Guide, so that doesn't offer help, but when you stop to consider that it is used primarily in planning processes, suddenly the picture becomes a little clearer.

Now you can see that 'A' and 'B' are probably not the right answer. Both of these would be a good fit for historical information, which is an organizational process asset. Now you have a 50% chance of guessing between 'C' and 'D'. Look at them more carefully and ask yourself where would historical information most likely be used an input?

Quantitative risk analysis is a good guess, since you might use past results (an organizational process asset) to help you analyze and quantify risk. So now, you are left with choice 'C', Direct and Manage Project Execution, as the one that looks least likely to have an organizational process asset as an input. It may be a guess, but it is a very educated one.

The method here is simply to think about each answer and eliminate ones that are obviously wrong. Even if you only knock off one wrong answer, you have significantly increased your odds of choosing the right one. You will find that most times you can knock off at least two, evening your chances of answering the question correctly.

Spotting Tricks and Traps

The PMP does have trick questions. They are designed specifically to catch people who are coming in with little formal process experience, those who have thumbed through the PMBOK a few times and are now going to take the exam. These people try to rely on their work experience, which often does not line up with PMI's prescribed method for doing things. As a result, they typically don't even come close to passing the exam.

At times, however, these trick questions can also fool a seasoned pro! Listed below are some techniques you can use so that you will not fall into these traps.

FOLLOW THE PROCESS

This is always the right answer. There will be questions on your exam that give you "common sense" scenarios that will give you a seemingly innocent way to skip the formal process and save time, or perhaps avoid some conflict by not following procedure. This is almost certainly a trap. The right answer is to follow PMI's process! Do not give in to pressure from irate customers, stakeholders, or even your boss to do otherwise.

DON'T TAKE THE EASY WAY OUT

There will often be choices that allow you to postpone a difficult decision, dodge a thorny issue, or ignore a problem. This is almost never the right thing to do for questions on the exam.

ACT DIRECTLY AND SAY WHAT YOU MEAN

In PMI's world, project managers communicate directly. They do not dance around the issue, gossip, or imply things, and they do not communicate through a third party. If they have bad news to tell the customer, they go to the customer and tell them the facts - and the sooner, the better. If they have a problem with a team member, they confront the person, usually directly, although at times it may be appropriate to get the team member's functional manager involved.

STUDY THE ROLES

By the time you take the exam, you should be confident about the roles of stakeholders, sponsors, customers, team members, functional managers, the project office, and most importantly, the project manager (plus the other roles that are discussed in Chapter 2 - Foundational Terms and Concepts). Expect several "who should perform this activity" type questions. If you have absolutely no clue, guessing the "project manager" is a good idea.

Additionally, understand the difference between the different types of organizations (projectized, matrix, and functional). Most of your questions will pertain to matrix organizations, so focus your study on that one.

PROJECT MANAGER'S ROLE

Expanding on the previous point, project managers are the ones who make decisions and carry them out. They have the final decision on most points, can spend budget, can change schedules, and can approve or refuse scope. For the test, assume that the project manager is large and in charge!

Another attribute of project manager is that they are proactive in their approach to managing tasks and information. They do not wait for changes to occur. Instead, they are actively influencing the factors that contribute to change. Instead of waiting for information to come to them, they are actively communicating and making certain they have accurate and up-to-date information.

Don't Get Stuck

You should expect to find a few questions on the exam that you do not know how to answer. You will look at it and see 4 correct answers, making it impossible to pick just 1. In this case, do not agonize. Even using every good technique, you will still have to make an educated guess at some questions. Some test takers can get quite upset at this, and it can undermine their confidence. If a question stumps you, simply mark it for review and move on. Never spend 15 minutes staring at a single question unless you have already answered all the others. One question is only worth one a fraction of a percent on the exam, so if you do not know the answer, do not obsess over it.

You may even discover that a block of questions seems especially difficult to you. This experience can be discouraging and may cause your confidence to waver. Don't be alarmed if you happen upon several difficult questions in a row. Keep marking them for review and keep moving, until you come to more familiar ground. You may find that questions later in the test will offer you hints or jog your memory, helping you with those you initially found difficult.

Exam Time Management

You will have a few minutes at the beginning to go through a tutorial. You probably won't find much value in the actual tutorial; however, you absolutely should take it. After going through it, you will be given a chance to wait before taking the test. Use that time to write down essential formulas and processes on your scratch paper.

Scratch Paper

You will be given five sheets of blank paper when you walk into the exam. You may not carry your own paper into the test. When you sit down to begin the exam, you should write down a few key things. Regardless of how well you know this material, at a minimum, write the following on your first sheet of scratch paper.

1. Earned Value formulas (EV, PV, AC, CPI, CPI^C, SPI, CV, SV, BAC, EAC, ETC) from Chapter 7. You will probably need to refer to these several times during the exam, and it will save time and improve your accuracy if you have written them out.

2. The time management formulas for the three-point estimates and standard deviation.

3. The communication channels formulas described in Chapter 10.

Even if you are tempted to skip this step, don't! When you come to a lengthy and confusing question that requires you to calculate several different values, you will be glad that you already have your formulas written down for review. This will free your mind to concentrate on the specific question rather than on recalling a formula.

Managing Your Review

When you make a review pass through the exam, you will come across questions that you missed the first time but that are apparent when you look at them again. This is normal, and you should not hesitate to change any answers that you can see you missed. Many people change as many as 10% of the answers on their review. If you catch yourself changing more than that, be careful! You may be second guessing yourself and actually do more harm than good.

When you go through your review pass on the exam, do not take the whole test over again. Instead, employ three rapid fire steps:

1. Did you read the question correctly the first time?

2. Did your selected answer match what was being asked?

3. Perform a complete check of your math where applicable.

Difficulty

Everyone wants to know what the hardest topic on the exam is. That is a difficult question to answer for two reasons:

1. The PMBOK Guide and the test are divided differently. There are nine knowledge areas in the PMBOK Guide, each given their own chapter. But all of this material also fits into one of five processes (plus professional responsibility). To make it more confusing, the PMBOK Guide is arranged by knowledge area, but the test is arranged by process group. This can heighten the confusion when evaluating your test results!

2. The other reason that difficulty is hard to predict is that everyone's experience will differ. No two tests are alike just as no two people have identical backgrounds. If you have heavy finance and accounting experience, time and cost may be easy subjects for you. If you have a strong legal background or have worked for the government then questions about procurement may be easy for you. If you have human resources or psychology training, human resources management may be easiest.

Based on a typical profile of someone who had earned an undergraduate degree in management and had non-industry specific experience managing projects, the material would roughly rank as shown below:

<table>
<tr><td>Process Framework</td><td>1</td><td rowspan="4">Harder</td></tr>
<tr><td>Foundational Concepts</td><td>2</td></tr>
<tr><td>Integration</td><td>3</td></tr>
<tr><td>Cost</td><td>4</td></tr>
</table>

Time	5	
Risk Procurement Quality Scope	7 8 9 10	Medium Difficulty
Human Resources Communications Professional Responsibility	11 12 13	Easier

This scale is relative, since very little of the material on the exam is considered "easy" to everyone. Ultimately, your professional experience and study can significantly change the order of this list for you.

Managing Anxiety

Finally, if test-taking has always been a fear-inducing activity for you, there is one simple strategy that may help you manage the physical symptoms of anxiety so that your thinking and memory are not impaired: Take a deep breath. This may sound like obvious advice, but it is based on sound research. Studies in the field of stress management have shown that feelings of anxiety (the "fight or flight" response") are linked with elevated levels of adrenaline and certain brain chemicals.

One way to bring your brain chemistry back into balance is to draw a deep breath, hold it for about six seconds, and slowly release it. Repeat this breathing pattern whenever you begin to feel panicky over particular questions. It will help to slow your heart rate and clear your mind for greater concentration on the task at hand.

Another thing to remember is that many people who take the exam do not pass - especially on their first attempt. While no one wants to fail the test, you can turn around immediately and apply to take it again (at a reduced rate). You can take the exam up to three times in a year starting with the time you receive your letter of eligibility from PMI.

If you do have to take the exam again, use it as a learning experience. You will have a much higher chance for success on your next attempt, and you will have your score sheet that gives you a breakdown of where you need to study. As the inspired Jerome Kern once penned to music, "Take a deep breath; pick yourself up; dust yourself off; start all over again."

Final Exam

Instructions

This simulated PMP Exam may be used in several ways. If you take it as a final, you will get a very good idea how you would do if you walked right in to take the PMP Exam. In that way, it can be a very good readiness indicator.

Perhaps the best way to use this exam is to take it again and again, reviewing the answers that go with each question. The answers and explanations will give your insight into the formation of each question and the thought process you should follow to answer it.

Prior to taking the PMP, the best strategy is to take this exam repeatedly, reviewing the answers, until you can make a score of 90% or better.

If you are taking this as a final exam, you have 4 hours (240 minutes) to complete the following 200 questions, including any breaks you may take.

Each question has only one best answer. Mark the one best answer on your answer sheet by filling in the circle next to A, B, C, or D.

1. During testing, multiple defects were identified in a product. The project manager overseeing this product's development can best use which tool to help prioritize the problems?

   A. Pareto diagram.

   B. Control chart.

C. Variance analysis.

D. Order of magnitude estimate.

2. You are the manager of an aircraft design project. A significant portion of this aircraft will be designed by a subcontracting firm. How will this affect your communications management plan?

   A. More formal verbal communication will be required.

   B. Performance reports will be more detailed.

   C. More formal written communication will be required.

   D. Official communication channels will significantly increase.

3. What officially creates the project?

   A. The project initiation document.

   B. The kickoff meeting.

   C. The project charter.

   D. The statement of work.

4. Refer to the table at the right. What is the critical path?

Task	Dependency	Duration
Start	None	0
A	Start	3
B	A	2
C	B	2
D	Start	4
E	D	1
F	Start	5
G	F	7
H	B,E	3
I	C,G,H	4
Finish	I	0

   A. Start-A-B-C-I-Finish

   B. Start-A-B-H-I-Finish

C. Start-D-E-H-I-Finish

D. Start-F-G-I-Finish

5. The Delphi technique is a way to:

   A. Analyze performance.

   B. Gather expert opinion.

   C. Resolve conflict.

   D. Estimate durations.

6. The work authorization system makes sure that:

   A. All the work and only the work gets performed.

   B. Work gets performed in right order and at the right time.

   C. Work is done completely and correctly.

   D. Functional managers are allowed complete control over who is assigned and when.

7. Your team is hard at work on their assigned project tasks when one team member discovers a risk that was not identified during risk planning. What is the FIRST thing to do?

   A. Halt work on the project.

   B. Update the risk management plan.

C. Look for ways to mitigate the risk.

D. Assess the risk.

8. The activity duration estimates should be developed by:

   A. The person or team doing the work.

   B. The project manager.

   C. Senior management.

   D. The customer.

9. The project plan should be all of the following EXCEPT:

   A. A formal document.

   B. Distributed to stakeholders in accordance with the communications management plan.

   C. Approved by all project stakeholders.

   D. Used to manage project execution.

10. You have been asked to take charge of project planning for a new project, but you have very little experience in managing projects. What will be the best source of help for you?

   A. Your education.

   B. Your on-the-job training.

C. Historical information.

D. Your functional manager.

11. The majority of the project budget is expended on:

A. Project plan development.

B. Project plan execution.

C. Integrated change control.

D. Project communication.

12. Corrective action is:

A. Fixing past anomalies.

B. Anything done to bring the project's future performance in line with the project management plan.

C. The responsibility of the change control board.

D. An output of project plan execution.

13. Outputs of Direct and Manage Project Execution include:

A. Deliverables and performance reports.

B. Deliverables and corrective action.

C. Deliverables and work performance information.

D. Performance reports and requested changes.

14. Your original plan was to construct a building with six stories, with each story costing $150,000. This was to be completed in four months; however, the project has not gone as planned. Two months into the project, earned value is $400,000. What is the budgeted at completion?

    A. 450,000

    B. 600,000

    C. 800,000

    D. 900,000

15. Project integration is primarily the responsibility of:

    A. The project team.

    B. The project manager.

    C. Senior management.

    D. The project sponsor.

16. One of your team members has discovered a way to add an extra deliverable to the project that will have minimal impact on the project schedule and cost. The project cost performance index is 1.3 and the schedule performance index is 1.5. The functionality was not included in the scope. How should you proceed?

    A. Conform to the project scope and do not add the deliverable.

B. Deliver the extra work to the customer since it will not increase their costs.

C. Reject the deliverable because you are behind schedule.

D. Ask senior management for a decision.

17. If a project manager is unsure who has the authority to approve changes in project scope, she should consult:

A. The customer.

B. The scope statement.

C. The sponsor.

D. The scope management plan.

18. An end user has just requested a minor change to the project that will not impact the project schedule. How should you, the project manager, respond?

A. Authorize the change quickly to ensure that the schedule can truly remain unaffected.

B. Deny the change to help prevent scope creep.

C. Evaluate the impact of the change on the other project constraints.

D. Submit the change request to the customer for approval.

19. Which of the following tools is NOT used in initiating a project?

A. Project selection methods.

B. Project management methodology

C. Expert judgment.

D. Earned value analysis.

20. You overhear a casual conversation between two team members in which one confides to the other some problems he is having in completing his part of the project work. You realize that the work being discussed is on the project's critical path and that the information you overheard could mean a significant delay for your project. What should you do?

    A. Let the team member know that you heard his conversation and discuss the work problems with him immediately.

    B. Begin analyzing ways to compress the project schedule in anticipation of the potential delay.

    C. Ask a third team member to get involved immediately and encourage the two other team members to come to you with the delay.

    D. Ask human resources for help in resolving the problem.

21. In which group of processes should the project manager be assigned his or her role in the project?

    A. Initiating.

    B. Planning.

    C. Executing.

D. Controlling.

22. A project charter should always include:

    A. Historical information.

    B. The business need underlying the project.

    C. A detailed budget.

    D. The scope management plan.

23. Your project team has just received the sponsor's approval for the scope statement. What is the NEXT step that needs to be taken?

    A. Develop the product description.

    B. Create a work breakdown structure.

    C. Hold the kickoff meeting.

    D. Create the network diagram.

24. Which of the following is NOT an input into Scope Definition?

    A. Accepted deliverables.

    B. Organizational process assets.

    C. Project charter.

    D. Preliminary project scope statement.

25. The key function of the project manager's job in project integration is:

    A. Minimizing conflict to promote team unity.

    B. Making key decisions about resource allocation.

    C. Communicating with people of various backgrounds.

    D. Problem-solving and decision-making between project subsystems.

26. In which of the following documents could the sponsor find work package descriptions?

    A. The work breakdown structure dictionary.

    B. The project charter.

    C. The scope management plan.

    D. The project scope statement.

27. The process in which project deliverables are reviewed and accepted is called:

    A. Scope planning.

    B. Scope verification.

    C. Initiation.

    D. Scope change control.

28. A statement of work is:

A. A type of contract.

B. A description of the project's product.

C. Necessary for every project.

D. A description of the part of a product to be obtained from an outside vendor.

29. A commercial real-estate developer is planning to build a new office complex. He contracts with a construction firm to build one of the buildings for the actual cost of providing the materials and services plus a fixed fee for profit. What type of contract does this scenario represent?

    A. Independent vendor.

    B. Fixed price.

    C. Cost-reimbursable.

    D. Time and materials.

30. You and your spouse both work for large companies in different industries. However, one day you learn that your company will be soliciting bids for a project and your spouse's company intends to bid. Your spouse will not be involved in the bidding process or any of the work it might produce if won. What should you do?

    A. Request to be transferred off the project.

    B. Inform management of the situation.

    C. Say nothing and go on with the bidding process.

D. Say nothing but set up a system of checks and balances to ensure that your team selects the contractor impartially.

31. Which of the following is NOT a purpose that Create WBS serves?

    A. To increase the accuracy of estimates.

    B. To help facilitate roles and responsibilities.

    C. To document the relationship between the product and the business need.

    D. To define a baseline for project performance.

32. You are assigned to replace a project manager on a large software project for a telecommunications company in the middle of executing the work. Portions of the software are being supplied by subcontractors working at your company's offices. You would like to know what performance metrics are going to be tracked for these contract workers. Where could you find such information?

    A. The project charter.

    B. The procurement management plan.

    C. The work breakdown structure.

    D. The organizational chart.

33. Which of the following is NOT a type of contract?

    A. Cost-revisable.

    B. Fixed-price.

C. Cost-reimbursable.

D. Time and materials.

34. Your team has identified a component that they need for a project. There is some concern that they have never constructed a component like this one, but there are similar components available from sellers. Which of the following procurement activities would be MOST appropriate to perform?

A. Solicitation.

B. Make-or-buy analysis.

C. Benefit/cost analysis.

D. Source selection.

35. The activity list serves as an input to:

A. Create WBS.

B. Activity Definition.

C. Activity Duration Estimating.

D. Resource Planning.

36. The person or group that formally accepts the project's product is:

A. The quality team.

B. The customer.

C. The project team.

D. Senior management.

37. The schedule activity list:

A. Serves as an extension of the work breakdown structure.

B. Is synonymous with the work breakdown structure.

C. Is used to create the project scope statement.

D. Is included in the project charter.

38. Float refers to:

A. A method for decreasing risk on a project.

B. How long an activity can be delayed without affecting the critical path.

C. A time lapse between a project communication and the response that follows.

D. The difference between the budgeted cost and actual cost.

39. If a project scope requires goods or services that must be obtained outside the project organization, what management process will be used in obtaining them?

A. Project contract management.

B. Project solicitation management.

C. Project procurement management.

D. Project source management.

40. You are producing a training video for your company's human resource department. After the project is underway, a member of senior management requests that you use a copyrighted piece of music as background in the video. This video will not be sold or viewed outside of your company. As the project manager, you should:

A. Use the song as requested by management.

B. Investigate obtaining permission from the music publisher to use the song.

C. Submit the request to the change control board.

D. Produce the video without the song as specified in the project plan.

41. What is the most important function the project manager serves?

A. Staffing.

B. Motivating.

C. Team building.

D. Communicating.

42. If a task has been estimated at O = 4 days, P = 9 days, and M = 7, what is the standard deviation?

A. 5/6 of a day.

B. 6.83 days.

C. 1/3 of a day.

D. 1/2 of a day.

43. Refer to the table at the right. If task H were increased from 3 to 7, what impact would this have on the project?

Task	Dependency	Duration
Start	None	0
A	Start	3
B	A	2
C	B	2
D	Start	4
E	D	1
F	Start	5
G	F	7
H	B,E	3
I	C,G,H	4
Finish	I	0

A. The project would finish later.

B. The project would finish earlier.

C. The schedule risk would decrease.

D. The critical path would change, but the finish date would not change.

44. All of the following are needed for creating the project budget except:

A. Activity cost estimates.

B. Work breakdown structure.

C. Cost management plan.

D. Deliverables.

45. Your company's CIO has requested a meeting with you and two other project managers for a status update on your various projects. What is the BEST document you can bring with you to this meeting:

A. The milestone chart for this project.

B. The network diagram for this project.

C. Copies of the most recent status reports from the team members.

D. The project charter.

46. Resources are estimated against which project entity:

A. The work packages.

B. The schedule activities.

C. The elements of scope.

D. The level set by the project office.

47. Analogous estimating uses:

A. Estimates of individual activities rolled up into a project total.

B. Actual costs from a previous project as a basis for estimates.

C. Computerized estimating tools.

D. Parametric modeling techniques.

48. If the optimistic estimate for an activity is 15 days and the pessimistic estimate is 25 days, what is the realistic estimate?

A. 19 days.

B. 20 days.

C. 21 days.

D. Unknown.

49. What does the standard deviation tell about a data set?

A. How diverse the population is.

B. The mean of the population as it relates to the median.

C. The specification limits of the population.

D. The range of data points within the population.

50. Quality management theory is characterized by which of the following statements:

A. Inspection is the most important element for ensuring quality.

B. Planning for quality must be emphasized.

C. Contingency planning is a critical element of quality assurance.

D. Quality planning quantifies efforts to exceed customer expectations.

51. Which of the following is NOT emphasized in project quality management?

A. Customer satisfaction.

B. Team responsibility.

C. Phases within processes.

D. Prevention over inspection.

52. Scatter diagrams, flowcharts, run diagrams, control charts, and inspection are all techniques of what quality process?

    A. Perform Quality Assurance.

    B. Perform Quality Control.

    C. Perform Quality Execution.

    D. Perform Quality Definition.

53. Which of the following is NOT an input to quality planning?

    A. Enterprise environmental factors.

    B. Organizational process assets.

    C. Quality baseline.

    D. Project management plan.

54. The procurement management plan provides:

    A. Templates for contracts to be used.

    B. A formal description of how risks will be balanced within contracts.

    C. A description of procurement options.

    D. The types of contracts to be used for items being procured.

55. Ultimately, responsibility for quality management lies with the:

A. Project team.

B. Quality team.

C. Project manager.

D. Functional manager.

56. All of the following are tools used in Perform Quality Control EXCEPT:

A. Benchmarking.

B. Pareto charts.

C. Histograms.

D. Statistical sampling.

57. Control charts are:

A. Used in product review.

B. Used to chart a project's expected value.

C. Used to determine if a process is in control.

D. Used to define a statistical sample.

58. Which of the following statements regarding stakeholders is TRUE?

A. They have some measurable financial interest in the project.

B. Their needs should be either qualified or quantified.

C. Key stakeholders participate in the creation of the stakeholder management plan.

D. They may either be positively or negatively affected by the outcome of the project.

59. A probability and impact matrix is useful for:

A. Risk Identification.

B. Qualitative Risk Analysis.

C. Risk Response Planning.

D. Risk Monitoring and Control.

60. Your company must choose between two different projects. Project X has a net present value of $100,000. Project Y has a net present value of $75,000. What is the opportunity cost of choosing Project X?

A. $100,000

B. $75,000

C. $25,000

D. $50,000

61. A risk register is created during:

A. Risk Management Planning.

B. Risk Monitoring and Control.

C. Risk Assessment.

D. Risk Identification.

62. A workaround is:

A. A technique for conflict management.

B. An adjustment to the project budget.

C. A response to an unplanned risk event.

D. A non-critical path on the network diagram.

63. You are managing a team developing a software product. You have contracted out a portion of the development. Midway through the project you learn that the contracting company is entering Chapter I I. A manager from the subcontracting company assures you that the state of the company will not affect your project. What should you do FIRST?

A. Perform additional risk response planning to control the risk this situation poses.

B. Stop all pending and future payments to the subcontractor until the threat is fully assessed.

C. Contact your legal department to research your options.

D. Meet with senior management to apprise them of the situation.

64. In a functional organization:

A. Power primarily lies with the project manager.

B. Power primarily lies with the functional manager.

C. Power is blended between functional and project managers.

D. Power primarily lies with the project office.

65. The process of identifying, documenting, and assigning roles, responsibilities, and reporting relationships for a project is called:

A. Project Interfacing.

B. Organizational Breakdown.

C. Staff Management Planning.

D. Human Resource Planning.

66. A Responsibility Assignment Matrix (RAM) does NOT indicate:

A. Who does what on the project.

B. Job roles for team members.

C. Job roles and responsibilities for groups.

D. Project reporting relationships.

67. A manager who follows Theory X believes:

A. Employees can be trusted to direct their own efforts.

B. Project success requires that project objectives must align with company objectives.

C. Workers must be closely supervised because they dislike work.

D. Effective quality management requires the use of performance measurements.

68. Which of the following is NOT a constructive team role?

A. Withdrawer.

B. Information seeker.

C. Clarifier.

D. Gate keeper.

69. If a project team is experiencing conflict over a technical decision that is negatively affecting project performance, the BEST source of power the project manager could exert to bring about cooperation would be:

A. Legitimate.

B. Penalty.

C. Referent.

D. Expert.

70. You have asked a team member to estimate the duration for a specific activity, and she has reported back to you with three estimates. The best-case scenario is that the activity could be completed in 18 days; however, her most likely estimate for the task is 30 days. She has also indicated that there is the possibility the task could take as long

as 60 days. What is the three-point estimate for this task?

A. 4 days.

B. 7 days.

C. 33 days.

D. 49 days.

71. Which of the following types of conflict resolution provides only a temporary solution to the problem?

A. Withdrawal.

B. Compromising.

C. Forcing.

D. Problem-solving.

72. The communications management plan:

A. Should include the performance reports.

B. Should always be highly detailed.

C. Should include the project's major milestones.

D. Should detail what methods should be used to gather and store information.

73. When communication links are undefined or broken:

A. The communications management plan should be rewritten.

B. Conflict will increase.

C. The project manager's power will decrease.

D. Project work will stop.

74. For two days you have been asking a member of your team for a status report on one of the key deliverables of your project. You finally get the report thirty minutes before you are to meet with your manager for a project update. A quick review of the status report reveals some information that you know is incorrect. What action should you take?

A. Fix what you can in the report before the meeting starts and try to steer discussion away from the areas you don't have time to fix.

B. Bring your team member with you to the meeting and confront her with the inaccuracies in her report.

C. Reschedule the meeting for a later date and have the team member rewrite the report.

D. Cancel the meeting, fix the report yourself, and circulate the new report to senior management in lieu of the original meeting.

75. There were 10 people on your project, and 5 more people were added last week. How many additional paths of communication were created?

A. 10.

B. 45.

C. 60.

D. 105.

76. A project manager is having difficulty getting resources from a functional manager. Which of the following would be the MOST appropriate to help resolve this problem?

A. Senior management.

B. The customer.

C. Key stakeholders.

D. The sponsor.

77. Communicating via e-mail is considered:

A. Formal written communication.

B. Informal written communication.

C. Formal electronic communication.

D. Informal non-verbal communication.

78. Smoothing, forcing, and withdrawing are all forms of:

A. Organizational power.

B. Communication.

C. Conflict resolution.

D. Schedule compression.

79. You have a team member who is habitually late to meetings with the customer. The customer has expressed dissatisfaction with the situation and has asked you to resolve it. Your BEST course of action is:

A. Issue a formal written reprimand to the team member

B. Meet with the team member to discuss the problem and ask for solutions.

C. Meet with the team member and the customer to promote further understanding.

D. E-mail the team member to bring the situation to his attention.

80. You are midway through managing a project with a sponsor-approved budget of $850,000. Using earned value management, you have determined that the project will run $125,000 under budget. You have also determined that if the project is delivered that far under budget you will not make your bonus since you are compensated for the hours you bill. What is your BEST course of action?

A. Add extra features to the project scope that take advantage of the available budget, and increase customer satisfaction.

B. Meet with the project sponsor to inform him of your findings.

C. Maintain current project activities, and bill for the original amount.

D. Ask the sponsor to approve additional features, given the available budget.

81. The most important factor in project integration is:

    A. A clearly defined scope.

    B. Timely corrective action.

    C. Team buy-in on the project plan.

    D. Effective communication.

82. You are a project manager for a software development firm. You are in the final stages of negotiation with a third-party vendor whose product your company is considering implementing. You discover by chance that one of your employees has scheduled a product demonstration with herself, the vendor, and your boss, but you have not been notified about the meeting. What do you do?

    A. Show up at the meeting unannounced and discuss the situation with the employee later in private.

    B. Report this employee's actions to your boss using the company's formal reporting procedure.

    C. Discuss the employee's actions with her before the meeting.

    D. Report the employee's actions to her functional manager.

83. Your team has encountered recent unanticipated problems. After extensive earned value analysis, you determine that the project has a schedule performance index of .54 and a cost performance index of 1.3. Additionally, your customer has just requested a significant change. What should you do?

A. Alert management about the schedule delays.

B. Alert management about the cost overruns.

C. Alert management about the scope change.

D. Reject the requested change.

84. You are in the middle of bidding on a large and complex project that will produce a great deal of revenue for your company should you win the bid. The buyer has specified several conditions that accelerate some of the key dates and milestones. You have done extensive planning with your team and have determined that there is no way that your organization can perform the required scope of work under these new deadlines. Your boss, however, is not convinced that you are right and is also concerned that if you don't agree to the new dates you will lose the contract. What should you do?

A. Appeal to the buyer for additional time to estimate.

B. Ask your boss to make the commitment on behalf of the team.

C. Adhere to the estimates your team has made.

D. Agree to the dates the customer has requested.

85. Your company is soliciting bids from advertising firms for the marketing of the product your project will produce. A sales representative from one of the bidding firms calls to invite you to attend a sporting event with him at his expense so that he can ask you some questions about the product to help him put together his bid. How should you respond?

A. Attend the event and discuss the project with the representative.

B. Accept his offer but tell him in advance that you cannot answer questions about the project.

C. Attend the event and answer all questions, but then write a report on the questions and answers and submit the FAQ to the other bidders.

D. Reject the offer and invite the sales representative to pose his questions at the bidders' conference.

86. Analogous estimating is also called:

A. Vendor bid analysis.

B. Bottom-up estimating.

C. Scalable model estimating.

D. Top-down estimating.

87. You are beginning construction of a bridge in another country when you discover that this country requires that one of its licensed engineers sign off on the plans before you break ground. Your senior engineer on the project is licensed in your own country and is probably more qualified than anyone to sign off on this, and their engineer is not available to review the plans for another three weeks. The customer has stressed that this project must not be delayed. What do you do?

A. Have your engineer sign off on the plans, forward them to the other engineer and begin construction.

B. Have your engineer sign off on the plans since he is licensed in your country and begin construction.

C. Wait to begin construction until the country's engineer signs off on the plans.

D. Forward the plans to the country's engineer for his signature and start construction.

88. Refer to the table at the bottom. What is the length of the critical path?

Task	Duration
Start-A	1
A-B	3
B-C	4
A-E	3
E-F	1
C-D	6
E-C	0
B-F	7
F-G	7
G-Finish	5
D-Finish	2

A. 16

B. 21

C. 22

D. 23

89. You are managing the installation of a new oil well pump in a very productive well. Due to very efficient management and significant personal effort, the project is completed several days ahead of schedule. The customer is ecstatic and offers you a $2,500 "appreciation fee." What should you do?

A. Accept the fee, but notify management.

B. Do not accept the fee and notify management.

C. Put the gift into the project's reserve fund.

D. Donate the gift to charity.

90. Decomposition is a technique used in:

A. Activity Definition

B. Activity Duration Estimating

C. Activity Sequencing

D. Schedule Development

91. A fixed-price contract offers the seller:

A. A higher risk than the buyer.

B. A risk level equal to that of the buyer.

C. A lower risk than the buyer.

D. Reimbursement of actual costs.

92. The individual on a project with limited authority who handles some communication and ensures that tasks are completed on time is:

A. The project manager.

B. The project leader.

C. The project coordinator.

D. The sponsor.

93. At what point in project planning would you decide to change the project scope in order to avoid certain high-risk activities?

A. Risk Identification.

B. Risk Qualification.

C. Risk Monitoring.

D. Risk Response Planning.

94. Which of the following techniques is used in Scope Planning?

A. Project selection methods.

B. WBS templates.

C. Expert judgment.

D. Inspection.

95. Which of the following statements is NOT TRUE of integration management?

A. Integration is primarily concerned with making sure various elements of the project are coordinated.

B. Integration is a discrete process.

C. The project management information system is used to support all aspects of the project.

D. The project manager must make tradeoffs between competing project objectives.

96. Maslow's Hierarchy of Needs theory concludes that:

   A. Higher needs cannot be realized until the lower needs are satisfied.

   B. Hygiene factors are those that provide physical safety and emotional security.

   C. Psychological needs for growth and fulfillment are ineffective motivators.

   D. The greater the financial reward, the more motivated the workers will be.

97. A project is all of the following EXCEPT:

   A. Progressively elaborated.

   B. Has never been done by this company before.

   C. Interrelated activities.

   D. Strategic to the company.

98. Your manager has asked to review the quality management plan with you to ensure that it is being followed appropriately. In which process is your boss involved?

   A. Perform Quality Control.

   B. Perform Quality Management.

   C. Quality Planning.

   D. Perform Quality Assurance.

99. Variance analysis is a tool used to:

A. Measure variance between actual work and the baseline.

B. Measure variance between planned value and schedule variance.

C. Measure variance between earned value and actual cost.

D. Measure variance between earned value and cost variance.

100. In a strong matrix organization:

A. More power is given to the functional manager.

B. More power is given to the project manager.

C. More power is given to the project expeditor.

D. More power is given to the project coordinator.

101. Team performance assessments are made during which process?

A. Human Resource Planning.

B. Acquire Project Team.

C. Develop Project Team.

D. Manage Project Team.

102. The Project Scope Statement is typically:

A. A definitive list of all the work and only the work to be done on the project.

B. Issued by senior management.

C. Progressively elaborated.

D. Defined before the functional specifications.

103. Which of the following is a negative team role?

A. Initiator.

B. Information seeker.

C. Devil's advocate.

D. Gate keeper.

104. During Direct and Manage Project Team, your project team should be:

A. Focused on making sure earned value is equal to planned value.

B. Communicating work results to the stakeholders.

C. Ensuring that all project changes are reflected in the project plan.

D. Executing the work packages.

105. The system that supports all aspects of the project management processes from initiating through closing is:

A. Information technology.

B. The information distribution system.

C. The project management information system.

D. The work authorization system.

106. In a typical project, most of the resources are utilized and expended during:

A. Initiation processes.

B. Planning processes.

C. Executing processes.

D. Monitoring and controlling processes.

107. A project manager is taking the product of his project to his customer for verification that it meets the scope. The customer and the project manager are working together to carefully compare the product to the project's scope to ensure that the work was done to specification. The tool that the customer and the project manager are using is:

A. Inspection.

B. Perform Quality Assurance.

C. Perform Quality Control.

D. User acceptance testing.

108. Which of the following is NOT a project quality management process?

A. Perform Quality Assurance.

B. Quality Planning.

C. Quality Improvement.

D. Perform Quality Control.

109. Schedule constraints would likely include all of the following EXCEPT:

A. Imposed dates.

B. Key events.

C. Major milestones.

D. Leads and lags.

110. Which of the following choices fits the definition for benchmarking?

A. Comparing planned results to actual results.

B. Comparing actual or planned results to those of other projects.

C. Statistical sampling of results and comparing them to the plan.

D. Comparing planned value with earned value.

111. The activity list should include:

A. The schedule activities on the project that are on the critical path.

B. A subset of all schedule activities.

C. All of the schedule activities defined on the project.

D. A superset of the schedule activities for this and related projects.

112. Close Project should be performed:

   A. At the end of each phase or the end of the project.

   B. Before formal acceptance of the project's product.

   C. As a safeguard against risk.

   D. By someone other than the project manager.

113. Release criteria for team members is defined as part of:

   A. The organizational management plan.

   B. The human resources management plan.

   C. The staffing management plan.

   D. The work authorization system.

114. Which of the following is NOT an input into Activity Definition?

   A. The work breakdown structure.

   B. Organizational process assets.

   C. Project scope statement.

   D. The activity list.

115. A project to lay 10 miles of a petroleum pipeline was scheduled to be completed today, exactly 20 weeks from the start of the project. You receive a report that the

project has an overall schedule performance index of 0.8. Based on this information, when would you expect the project to be completed?

A. 2 weeks early.

B. In 2 more weeks.

C. In 5 more weeks.

D. In 10 more weeks.

116. The term “slack” is also known as:

A. Lag.

B. Lead.

C. Float.

D. Free float.

117. Which of the following BEST describes the project plan?

A. A formal, approved document used to guide project execution, control, and closure.

B. The aggregate of all work performed during planning.

C. The work breakdown structure, schedule management plan, budget, cost management plan, and quality management plan.

D. The document that outlines all of the work and only the work that must be performed on a project.

118. Which of the following is NOT a primary goal of integrated change control?

   A. Influencing factors that cause change.

   B. Determining that a change has occurred.

   C. Managing change as it occurs.

   D. Denying change whenever possible.

119. You are providing project management services in a foreign country. In an attempt to improve employment, this country has enacted a law limiting the number of hours foreigners may work per week. You are behind schedule on the project and need to work overtime. What is the BEST way to handle this situation?

   A. Work the overtime only if your own country's laws do not prohibit this.

   B. Work the overtime if it is not constituted overtime by your country's definition.

   C. Do not work the overtime since it is prohibited by law.

   D. Speak with legal representation to find out if the law is enforceable.

120. "The features and attributes that characterize a product" describes which of the following?

   A. The product scope.

   B. The project scope.

C. The work breakdown structure.

D. The critical success factors.

121. You have assumed responsibility for a project that has completed planning and is executing the work packages of the project. In one of your first status meetings, a member of the project team begins to question the validity of the duration estimates for a series of related tasks assigned to her. What action should you take FIRST?

A. Remind the team member that planning has been completed and ask her to do her best to adhere to the estimates.

B. Temporarily suspend execution and ask the team member for updated estimates.

C. Review the supporting detail for the estimates contained in the project plan to understand how the estimates were originally derived.

D. Ask another team member with expertise in this area to perform a peer review on the estimates to validate or invalidate the concern.

122. Approved budget increases should be:

A. Added to the schedule management plan.

B. Added to the project's cost baseline.

C. Added to the project's reserve fund and used only if needed.

D. Added to the lessons learned.

123. An organization where the project manager is in charge of the projects and has

primary responsibility for the resources is:

A. Functional.

B. Projectized.

C. Matrix.

D. Hierarchical.

124. Based on the table at the right, which path listed below represents the LEAST schedule risk?

Task	Dependency	Duration
Start	None	0
A	Start	3
B	A	2
C	B	2
D	Start	4
E	D	1
F	Start	5
G	F	7
H	B, E	3
I	C,G,H	4
Finish	I	0

A. Start-A-B-C-I-Finish.

B. Start-A-B-H-I-Finish

C. Start-D-E-H-I-Finish

D. Start-F-G-I-Finish

125. As part of your project, your customer needs a software module to handle credit card payments. Senior management informs you that they already own the rights to such a module and would like for you to use this piece of software on the project; however, upon investigation, you determine that your company's software is not a good match for this customer's needs. After informing your senior management, they maintain their request that you use their software. How should you resolve this conflict?

A. Look to build or procure another solution for the customer.

B. Act in favor of the performing organization.

C. Use your company's software module as requested as long as it does not jeopardize the project.

D. Involve an objective outside party to help resolve the dispute.

126. Leads and lags are evaluated and adjusted in which process?

A. Performance Reporting.

B. Cost Estimating.

C. Schedule Control.

D. Schedule Development.

127. You have received a report showing that your overall schedule performance index is 1.5. How should you interpret this information?

A. You are earning value into your project at 1.5 times the rate you had planned.

B. You are spending $1.50 for every dollar you planned to spend at this point in the schedule.

C. You are earning $1.50 of value back into your project for every $1.00 you spend.

D. You are earning $0.67 of value back into your project for every $1.00 you planned to earn.

128. The project organization chart should include which of the following:

A. The performing organization's organizational structure.

B. A representation of all identified stakeholders.

C. The reporting structure for the project.

D. The organizational structure for all entities related to the project.

129. Cost estimating should be performed:

A. Before the work breakdown structure is created and before the budget is developed.

B. Before the work breakdown structure is created and after the budget is developed.

C. After the work breakdown structure is created and before the budget is developed.

D. After the work breakdown structure is created and after the budget is developed.

130. A list of the risks that could affect the project is developed as part of which process?

A. Risk Management Planning.

B. Risk Identification.

C. Risk Response Planning.

D. Risk Monitoring and Control.

131. Which process focuses on producing a list of the activities needed to produce the deliverables and sub-deliverables described in the work breakdown structure?

A. Activity Definition.

B. Activity Sequencing.

C. Activity List.

D. Activity Duration Estimating.

132. Whose job is it to resolve competing objectives and goals between parties on the project?

A. The stakeholders.

B. The project manager.

C. Senior management.

D. The sponsor.

133. As project manager, you have made the decision to outsource a part of your project to an outside organization with whom you have never previously worked. You are ready to begin negotiating the contract. What should be your goal in the contract negotiations?

A. Having a lawyer or the legal department review each clause in the proposed contract prior to sharing it with any outside entity.

B. Negotiating the best possible price for your customer or organization.

C. Arriving at mutually agreeable terms for the contract between your organization and the subcontracting organization.

D. Shifting as much of the project risk to the subcontracting organization as possible.

134. You have been working under contract on a very large automotive project that has spanned several years. As part of this project, you have been privileged to practically all of the project information. As you are transitioning off of the project, you would like to use the extensive work breakdown structure as an input into future projects. You were never asked to sign a non-disclosure agreement with the current organization. What is the appropriate thing to do in this case?

A. Use the work breakdown structure as you are not bound by any agreement not to.

B. Use the work breakdown structure, but do not share it with others.

C. Do not use the work breakdown structure without the organization's permission.

D. Do not use the work breakdown structure because it would still be illegal.

135. Which of the following is NOT an output of Cost Control?

A. Recommended corrective action.

B. Forecasted completion.

C. Required changes.

D. Project budget.

136. If the schedule variance = $0.00, what must also be true?

A. Earned value must be equal to planned value.

B. The cost performance index must be equal to 1.

C. The schedule performance index must be greater than 1.

D. The estimate at complete must be equal to budgeted at complete.

137. Schedule activities are a further decomposition of which of the following:

A. The statement of scope.

B. The work packages.

C. The project network diagram.

D. The functional specification.

138. Which statement is TRUE regarding the staffing management plan?

A. It is used as an input into Human Resource Planning.

B. It is a component of the project plan.

C. It should name every human and material resource who will be working on the project.

D. It should contain an organization chart for the performing organization.

139. Which of the following statements is TRUE regarding risk management?

A. Negative risks should be quantified, while positive risks should be qualified.

B. Identified risks should be added to the risk register.

C. All known risks should be listed in the risk management plan.

D. The risks that cannot be mitigated must be avoided.

140. You have a friend in another organization who has shared with you that he is having difficulty understanding the value of doing a scope statement. Your friend's boss is not familiar with formal project management processes and does not want to waste time on the project performing unnecessary activities. What is your MOST appropriate response?

A. Do not get involved since this is not within your organization.

B. Pay a visit to your friend's project office and educate them on the value of a scope statement.

C. Mentor your friend on the value of project management processes.

D. Encourage your friend to change organizations.

141. You are managing a project to construct 25 miles of highway at an estimated cost of $1.2 million per mile. You have projected that you should be able to complete the project in 5 weeks. What is your planned value for the end of the 3rd week of the project?

A. $12,000,000

B. $18,000,000

C. $24,000,000

D. $30,000,000

142. Which term below describes the amount of time a schedule activity may be delayed before it affects the early start date of any subsequent schedule activity?

A. Float.

B. Slack.

C. Free float.

D. Lead.

143. When are the resource requirements estimated?

A. As soon as the scope has been adequately defined.

B. After the schedule has been defined, but before the budget has been created.

C. After the work packages have been defined, but before the activities have been defined.

D. After the activities have been defined and before the schedule has been developed.

144. You have an unfavorable project status to report to your customer at a weekly meeting; however, you are reasonably certain that you can correct the situation by next week's meeting. The customer will not be pleased to hear the current status and based on past history, will likely overreact. How should you handle this situation?

A. Report the current status to the customer.

B. Report your anticipated project status for next week to the customer.

C. Omit the information from your meeting and cover it next week when the news improves.

D. Ask you project office for guidance.

145. What type of process is Select Sellers?

A. Planning.

B. Executing.

C. Monitoring and controlling.

D. Closing.

146. Who is responsible for providing funding for the project?

A. The qualified financial institution.

B. Senior management.

C. The sponsor.

D. The project manager.

147. The project has been successfully completed when:

A. All of the work has been completed to specification within time and budget.

B. The customer is happy.

C. The sponsor signs off on the project.

D. Earned value equals planned value.

148. What is the PRIMARY objective of the project manager?

A. To follow PMI's processes.

B. To deliver maximum value for the organization.

C. To deliver the agreed upon scope of the project within the time and budget.

D. To delight the customer.

149. Which of the following would NOT be an organizational process asset?

A. Project plan templates.

B. Methodology guides.

C. Previous activity lists from projects.

D. Strong communication skills.

150. Defective deliverables are repaired as part of which process?

A. Direct and Manage Project Execution.

B. Perform Quality Assurance.

C. Monitor and Control Project Work.

D. Integrated Change Control.

151. What is indicated by an activity's late finish date?

A. The latest the activity can finish without delaying a subsequent activity.

B. The latest the activity can finish without delaying the project.

C. The latest probable date that the activity will finish.

D. The worst-case or pessimistic estimate for an activity.

152. A speech given at a trade show is an example of which kind of communication?

A. Informal written.

B. Formal written.

C. Informal verbal.

D. Formal verbal.

153. Which of the following is FALSE regarding the contract change control system?

A. It is primarily used during the Contract Administration process.

B. It is part of the integrated change control system.

C. It should be defined in the contract.

D. It should include contract dispute resolution procedures.

154. Herzberg's theory of motivation states that:

A. Hygiene factors must be present for motivational factors to work.

B. Motivation to work on the project must be related back to the individual's need.

C. An individual's higher needs will not emerge until the lower needs are met.

D. Individuals are motivated by a desire to reach proficiency.

155. If there are multiple critical paths on the project, which of the following must also be true?

A. Only one path will ultimately emerge as the true critical path.

B. The schedule risk will be higher with multiple critical paths than with one.

C. The schedule should be crashed in order to resolve the conflict.

D. The schedule should be fast-tracked in order to resolve the conflict.

156. Which projected payback period below is the MOST desirable?

A. 24 months.

B. 52 weeks.

C. 3 years.

D. 1000 days.

157. Which of the following would be an output of Activity Sequencing?

A. Mandatory dependencies.

B. An activity on arrow diagram.

C. Discretionary dependencies.

D. External dependencies.

158. While executing the project plan, you discover that a component was missed during planning. The project schedule is not in danger, but the component is not absolutely critical for go-live. What should you do?

A. Treat the component as a new project.

B. Reject the component as it would introduce unacceptable risk.

C. Appeal to the project sponsor for guidance.

D. Return to planning processes for the new component.

159. The goal of duration compression is to:

A. Reduce time by reducing risk on the project.

B. Reduce cost on the project.

C. Reduce the scope by eliminating non-critical functionality from the project.

D. Reduce the schedule without changing the scope.

160. You are managing two projects for two different customers. While meeting with one customer, you discover a sensitive piece of information that could help your other customer, saving them a significant percentage of their project budget. What should you do?

A. Act in accordance with any legal documents you have signed.

B. Disclose your conflict of interest and keep the information confidential.

C. Share the information with the other customer if it increases project value.

D. Excuse yourself from both projects if possible.

161. What is the difference between a standard and a regulation?

A. A standard is issued by ANSI, and a regulation is issued by the government.

B. A standard is an input into quality planning, while a regulation is an input into initiation.

C. A standard usually should be followed, and a regulation must be followed.

D. There is no appreciable difference between a standard and a regulation.

162. Evaluation criteria are used to:

A. Select a qualified seller.

B. Measure conformance to quality.

C. Determine if a project should be undertaken.

D. Evaluate performance on the project.

163. Richard is a project manager who is looking at the risks on his project and developing options to enhance the opportunities and reduce the threats to the project's objectives. Which process is Richard performing?

A. Risk Management Planning.

B. Risk Identification.

C. Qualitative Risk Analysis.

D. Risk Response Planning.

164. If you are soliciting bids for a project, which of the following would be an appropriate output from this process?

A. Proposals from potential sellers.

B. Change requests.

C. Contracts.

D. Qualified seller lists.

165. The work results of the project:

A. Are always products.

B. Are products, services, or results.

C. Are only considered work results if quality standards have been met.

D. Are an output of the work authorization system.

166. You have been directed by your customer, your sponsor, and senior management to manage a project that you believe will have a very negative impact on the economy and society. You have shared your concerns, but all parties continue to insist that you

proceed. What should you do?

A. Manage the project because all parties agree.

B. Refuse to manage the project.

C. Manage the project, but document your objections.

D. Contact PMI.

167. The quality policy is important in quality management because:

A. It defines the performing organization's formal position on quality.

B. It helps benchmark the project against other similar projects.

C. It provides specific quality standards that may be used to measure the output of the project.

D. It details the constraints and assumptions the project must take in to consideration.

168. During Cost Control, performance reviews are PRIMARILY used to:

A. Discuss the project and give the team a chance to voice any concerns.

B. Evaluate new budget change requests to determine if they would have an adverse effects on the project's performance.

C. Review the status of cost information against the plan.

D. Meet with the customer to evaluate and enhance satisfaction.

169. A project manager has a problem with a team member's attendance, but every time the project manager schedules a meeting to discuss the problem, the team member comes up with a reason they cannot attend at the last minute. Which conflict management technique is the team member exhibiting?

A. Passive-aggressive

B. Covering up

C. Diverting

D. Withdrawing

170. In which of the following organizations is the project manager MOST likely to be part-time?

A. Weak matrix

B. Strong matrix

C. Functional

D. Projectized

171. Which output would be used to show project roles and responsibilities?

A. A resource histogram.

B. An organization chart.

C. A RACI chart.

D. A staffing management plan.

172. Kim is managing a multi-million dollar construction project so that is scheduled to take nearly two years to complete. During one of the planning processes, she discovers a significant threat to her project's budget and schedule due to the fact that she is planning to build during the hurricane season in a high-risk area. After carefully evaluating her options, she decides to build earlier in the season when there is less of a risk of severe hurricane damage. This is an example of:

A. Risk avoidance.

B. Risk transference.

C. Risk acceptance.

D. Risk mitigation.

173. You are managing the development of a software product to be created under procurement. The team will span three countries and five time zones, and because of the size of the project, you are very concerned about cost. Which of the following types of contract would BEST help keep cost down?

A. Time and materials.

B. Cost plus fixed fee.

C. Cost plus incentive fee.

D. Variable conditions.

174. If there were 16 people on the project, and that number increases to 25, which of

the following must also be true?

A. The development of the communications management plan will be more difficult.

B. Stakeholder analysis will be more difficult.

C. Information distribution will be more difficult.

D. Controlling communication will be more difficult.

175. You are a PMP, and PMI has contacted you regarding an investigation of your best friend who is also a PMP at your company. How should you proceed?

A. Cooperate fully with PMI.

B. Protest that this is conflict of interest.

C. Protect your friend.

D. Find out why PMI thinks your friend should be investigated.

176. A project is scheduled to last for 6 months and cost $300,000. At the end of the 1st month, the project is 20% complete. What is the Earned Value?

A. $50,010.

B. $60,000.

C. $100,020.

D. $120,000.

177. Which of the following roles typically has the LEAST power?

A. Project coordinator.

B. Project expeditor.

C. Project manager.

D. Project director.

178. Configuration management is:

A. A technique used in Develop Project Management Plan.

B. Used to ensure that the product scope is complete and correct.

C. Formally defined in initiation.

D. A procedure to identify and document the functional and physical characteristics of an item or system.

179. Which two processes are tightly linked when performing a project and are often performed at the same time?

A. Scope Planning and Cost Planning.

B. Direct and Manage Project Execution and Acquire Project Team.

C. Perform Quality Assurance and Manage Project Team.

D. Perform Quality Control and Scope Verification.

180. A company in the middle of a new product development merges with another company in the middle of the project. The project is terminated because the new company already offers a similar product. What is the FIRST thing the project manager should do?

A. Make sure your lessons learned are communicated to the manager of the existing product.

B. Obtain a written project termination.

C. Perform a comparative product analysis.

D. Perform the Close Project process.

181. A benefit-cost ratio of 1.5 tells you that:

A. The payback period will be one and one half years.

B. The project cannot pay for itself.

C. The project will yield revenue that is 1.5 times its cost.

D. The project will cost 1.5 times the revenue it produces.

182. The MOST important input into Activity Resource Estimating is:

A. The activity list.

B. The cost estimates.

C. The project charter.

D. The constraints and assumptions.

183. Life-cycle costing involves:

A. Determining what physical resources will be needed to complete a project.

B. Considering the overall costs of a project during and after its completion.

C. Assigning a value to each activity on the project.

D. Calculating the internal rate of return.

184. A construction company is building a new office complex and has contracted with a computer hardware firm to install computer networking cables throughout the buildings. The contract states that the construction company will buy the materials and pay an hourly rate to the computer firm to cover their time on the project. Who is assuming the primary cost risk in this contract?

A. The companies share the risk equally.

B. There is not enough information given to answer the question.

C. The seller.

D. The buyer.

185. Linda is managing a multi-national project that utilizes several sub- project teams. Before the last team meeting, the sponsor asked her to bring a Control Account Plan for him to review. Which of the following statements is MOST correct regarding the Control Account Plan?

A. A control account plan is a node on the WBS that will be used during Risk Monitoring and Control.

B. A control account plan is a node on the WBS used for planning when the work packages cannot currently be defined.

C. A control account plan is a plan of how subcontractor costs will be charged, managed, and controlled.

D. A control account plan should be defined and costs should be controlled at the activity level.

186. Stakeholder analysis is performed as part of:

A. Schedule Development.

B. Human Resource Planning.

C. Scope Definition.

D. Risk Management Planning.

187. Alex is a project manager who wants to motivate his team by offering them a week of paid vacation if the project is delivered on time. When presenting this to the team, he spends extra time convincing the team that this goal is very achievable. What theory BEST explains Alex's behavior?

A. Expectancy theory.

B. Contingency theory.

C. Achievement theory.

D. Stimulus/Response theory.

188. All of the following are needed for creating the project budget except:

A. Cost estimates.

B. Schedule.

C. Resource calendar.

D. Organizational policies.

189. The fact that a software program must be written before it can be tested is an example of a:

A. Mandatory dependency.

B. Discretionary dependency.

C. External dependency.

D. Milestone dependency.

190. What is the difference between a Gantt chart and a milestone chart?

A. A Gantt chart is a project plan and a milestone chart is not.

B. A milestone chart is a project plan and a Gantt chart is not.

C. A milestone chart shows interdependencies between activities.

D. A milestone chart shows only major events, while a Gantt chart shows more information.

191. Your team has created a project scope statement and a work breakdown structure. What is the NEXT step that needs to be taken?

   A. Create a network diagram.

   B. Develop the schedule.

   C. Determine the critical path.

   D. Create the activity list.

192. You have assumed the position of project manager for a project that is nearly 50% complete. Because this project has reportedly had many quality problems, you want to review what trends and variances have occurred over time. Which tool would be most helpful to you?

   A. Cause and effect diagrams.

   B. Run charts.

   C. Inspection.

   D. Statistical sampling.

193. A cable company is installing new fiber optic cables in a community. You are the project manager and you will be using subcontractors to provide some of the installations. You have completed the procurement management plan and the statement of work, and you have gathered and prepared the documents that will be distributed to the sellers. What process should happen next?

A. Scope Planning.

B. Select Sellers.

C. Plan Purchases and Acquisitions.

D. Request Seller Responses.

194. Evaluating overall project performance on a regular basis to ensure that it meets expectations takes place in which process group of project management?

A. Planning.

B. Executing.

C. Monitoring and controlling.

D. Closing.

195. The MAIN use of a project network diagram is to:

A. Create the project plan.

B. Show activity percentages complete.

C. Show activity sequences and dependencies.

D. Create paths through the network.

196. Who issues the quality policy?

A. The performing organization.

B. The stakeholders.

C. The project manager.

D. A standards body.

197. You are managing a project when an unplanned risk event occurs. You meet with experts and determine a workaround to keep the project on track. In which process are you engaged?

A. Qualitative Risk Analysis.

B. Quantitative Risk Analysis.

C. Risk Response Planning.

D. Risk Monitoring and Control.

198. Which definition below best describes cost aggregation?

A. A risk management technique to view costs and their risks by an aggregated group instead of by an individual work package.

B. Adding the costs associated with each work package back up to the parent nodes on the WBS to ultimately get a total project

C. Multiplying the costs with each work package by a multiplier to calculate the needed reserve.

D. Aggregating the costs associated with work packages back to a single point in time to eliminate the time value of money.

199. A project is scheduled to last 4 months and cost $300,000. At the end of the 1st month, the project is 20% complete. What is the schedule performance index?

A. 0.80

B. 1.02

C. 1.18

D. 2.15

200. Bringing the entire project team and the customer on site to work together is an example of:

A. Communication control.

B. Collocation.

C. Active participation.

D. Collective team distribution.

Final Exam Answers

1. A. Pareto charts are column charts that rank defects based on the number of occurrences from highest to lowest. Because this tool is based on frequency, it prioritizes the most common causes. 'B' is used to determine whether or not a process is in control. 'C' is used to measure the difference between what was planned and what was done, and 'D' is a type of estimate used in cost management.

2. C. Because this is being done under contract, you will need to use more formal, written communication. Many people incorrectly guess 'D' on this one, but official channels of

communication could just as easily decrease since you are using another company and will probably have a single point of contact as opposed to your own team of many people doing the work. 'B' is not correct, since performance reports should be detailed regardless of who is doing the work.

3. C. The charter is the document that officially creates the project, names the project manager, and gives him authority on the project.

4. D. This problem should be solved in 3 steps. First, draw out the network diagram based on the table. Your representation should resemble the one below:

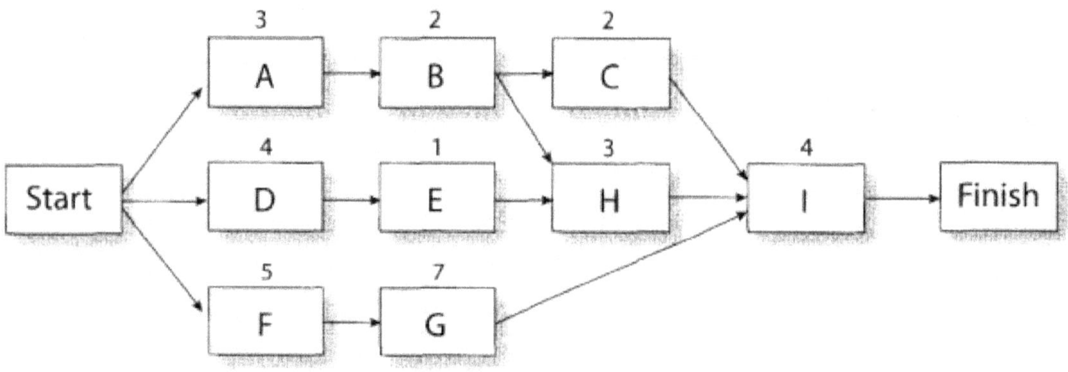

The next step is to list out all of the possible paths through the network. In this example, they are:

Start-A-B-C-I-Finish

Start-A-B-H-I-Finish

Start-D-E-H-I-Finish

Start-F-G-I-Finish

The last step is to add up the values associated with each path. Using the paths above, they are:

Start-A-B-C-I-Finish = 11 units

Start-A-B-H-I-Finish = 12 units

Start-D-E-H-I-Finish = 12 units

Start-F-G-I-Finish = 16 units

The critical path is the longest one, in this case, Start-F-G-I-Finish.

5. B. The Delphi technique is a way to solicit expert opinion by hiding the identities of group members from each other. This prevents the group from forming a single opinion or from letting one person dominate the group.

6. B. The work authorization system (WAS) is a system used during project integration management to ensure that work gets done at the right time and in the right sequence.

7. D. Typically you should assess, investigate, understand, and evaluate before acting. There may be exceptions to this rule, but in general you probably want to select the answer that lets you get all of the information. If there is a series of answers and you are asked what to do FIRST (as in this example), selecting the answer that allows you to be fully informed is usually best. 'A' is incorrect because halting work would probably send risks skyrocketing. 'B' is a good thing to do, but not until you have fully evaluated the risk. 'C' is something you may or may not choose to do, but you should not take action until you have fully assessed the risk.

8. A. This is frequently missed because people do not fully understand the role of the

team in planning. The team should help estimate and should support those estimates. If possible, the person who will be doing the work should have a say in the estimate. 'B' is incorrect because the project manager cannot possibly estimate all of the activities on the project, nor should he even try. 'C' and 'D' are incorrect since management and the customer are probably not aware of all of the low-level details needed, and their estimates would not be accurate.

9. C. The stakeholders do not have to approve the project plan. They may be very interested in the final product, but the actual plan cannot undergo approval by all stakeholders. Some stakeholders may never approve of the plan since they may be against the project! 'A' is incorrect because the project plan must be a formal. 'B' is incorrect because the communications management plan will specify to whom you should communicate the plan. 'D' is incorrect since the purpose of the project plan is to guide execution and control.

10. C. Your best source of help would be the information you can use from past projects. 'A' is incorrect since education may or may not be applicable to this project, and although it is good, education is principally theoretical, while historical information contains practical, hard data. 'B' is incorrect, since much practical experience only reinforces bad practices. 'D' is incorrect since functional managers have expertise in domains but not necessarily in the management of projects.

11. B. The majority of a project's budget is expended in the execution of the work packages. You may have guessed 'A' due to the number of processes that occur as part of planning, but most projects do not involve as many people or resources in planning as they do in execution. Choice 'D' is incorrect because, while a project manager spends 90% of his time communicating, that does not take most of the team's time or the project budget.

12. B. This is a very important definition. Corrective action is making adjustments to avoid

future variances. 'A' is more in line with the definition of rework. 'C' is incorrect since there may or may not be a change control board, and their job would be to approve or reject change requests that have been forwarded by the project manager. 'D' is incorrect because corrective action does not come out of execution, but out of various control processes.

13. C. This is the only one where both parts of the answer fit the definition. 'A' is incorrect since performance reports do not come out of the execution process. 'B' is incorrect since corrective action typically comes out of monitoring and controlling processes. 'D' is incorrect because performance reports do not come our of the Direct and Manage Project Execution process.

14. D. This type of question often appears on the PMP. It is easier than it first appears. Did you think the question was asking for the estimate at completion (EAC)? It is the originally budgeted amount, or budgeted at completion (BAC) that the question wants. That is calculated simply by taking 6 stories and multiplying it by $150,000/story. This yields the total project amount, which is $900,000.

15. B. This is an important point. The project manager is the one responsible for integration. 'A' is incorrect because the team should be doing the work. 'C' is incorrect since senior management is not involved in integration management. 'D' is incorrect since the sponsor pays for the project but is not directly involved in integration.

16. A. You should not add the deliverable. The reason is that this represents "gold plating," or adding functionality over and above the scope. It is not a good idea since this introduces risk and a host of other potential problems on the project. 'B' is incorrect because you do not know how it will affect risk, quality, or other factors. 'C' is incorrect, because you are ahead of schedule with an SPI of 1.5. 'D' is incorrect because it is the project manager's job- not the role of senior management- to deal with this kind of change request.

17. D. The scope management plan is the document that specifies how changes to the scope will be managed.

18. C. Any of the words "evaluate," "investigate," "understand," or "assess" should automatically put that answer at the top of your list to evaluate. 'D' is incorrect because the customer has hired you to evaluate and approve or reject change.

19. D. This question was very hard, but there was a way to reason out the answer. When initiating a project, you are essentially performing two processes: Develop Project Charter and Develop Preliminary Scope Statement. During those processes, it makes sense that you might use project selection methods, your methodology, and expert judgment, but choice 'D', earned value analysis, would probably not be useful until (much) later when the work was being performed.

20. A. Another bias of PMI is that you should confront situations directly whenever possible. If you see a choice that represents things like confront, problem-solve, or deal with the situation directly, that is a good hint that you may be on the right answer. In this case, all of the other choices do not deal with the actual problem. Although the first choice may not be pleasant in real life, you should deal with the situation head on and solve the problem.

21. A. The project manager is officially named and assigned in the project charter, which is one of the outputs of Develop Project Charter, an initiating process.

22. B. The project charter is a document, and that document should include the business need behind the project. This is a general description of why the project was undertaken.

23. B. The scope statement is an output of scope planning. The next step is the Scope Definition process, and the output of that is the work breakdown structure (WBS).

24. A. Hopefully your instincts kicked in here and said "this won't happen until later." If so, then you were right. Accepted deliverables are an output of Scope Verification. During Scope Planning, you are still trying to determine what the scope will include.

25. D. The project manager's job during integration is to solve problems and make decisions. It is not the team's job to do this! Their job should be to execute the work packages. The project manager should be fixing the problems that come up and keeping the team focused on the work.

26. A. Work package descriptions are contained in the WBS dictionary.

27. B. Scope Verification is where the customer and sponsor verify that the deliverables match what was in the scope.

28. D. The statement of work (SOW) describes the pieces of the project that are to be performed by an outside vendor. It often starts off general and is revised as the project progresses. 'A' is incorrect because the SOW does not meet the strict qualifications of a contract. 'B' is incorrect because it is too broad - the SOW is only about the pieces that will be outsourced. 'C' is incorrect because an SOW is not needed for all projects- only those that will be procuring parts.

29. C. This meets the definition of a cost-reimbursable contract. 'A' is a made-up term. 'B' is incorrect since the price is not fixed. 'D' is also incorrect. If you selected this one, you should review the difference between the time and materials and cost-reimbursable contracts.

30. B. In the code of conduct, PMPs are instructed to avoid conflicts of interest. Your strategy in selecting answer 'B' should have been to look for the one choice that both solved the problem and avoided the conflict of interest. 'A' is incorrect because leaving the project represents avoidance and does not deal with the issue directly. 'C' is

incorrect because the conflict of interest remains. 'D' is incorrect because by saying nothing, you are not dealing with the situation directly.

31. C. Documenting the relationship between the product and the business takes place before Scope Definition. A justification of the business need is included in the project charter. 'A', 'B', and 'D' all fall under the definition of Scope Definition.

32. B. The procurement management plan includes performance reporting specifications. If one of the choices had been "the communications management plan", that might have been a better selection, but given the choices provided, 'B' was the best one.

33. A. Cost-revisable is not a valid choice (and from the sound of the name, it does sound particularly safe to either party)! Choices 'B', 'C', and 'D' are all valid contract types.

34. B. In this case, make-or-buy analysis is the most appropriate. It is where you decide whether your organization should create the product or whether you should go through procurement. Choice 'A' would be an activity that was performed later in the process. Choice 'C' is a tool for measuring whether a project is worth pursuing and is performed in initiation. Choice 'D' is also an activity performed after the decision has been made to go through procurement.

35. C. With activities, the important order is define – order – estimate. The activity list is an input in Activity Duration Estimating.

36. B. Many people may formally accept the product, but in this list, the customer is the only one that fits the definition.

37. A. The activity list is a decomposition of the WBS. It takes the work packages and breaks them down into activities that can be sequenced, estimated, and assigned.

38. B. Float is how long an activity may be delayed without delaying the project. Choice 'B'

is the only one that fits this definition.

39. C. Procurement management is used when you go outside of the project for components of the project.

40. B. These best choices here may be narrowed down to 'B' and 'D'. Think of this senior management request as a change request. Why would you simply ignore it without investigating further? 'B' is the better of the two answers since it solves the problem. If 'D' appears to be a better choice, consider that it actually represents conflict avoidance, which is almost never a good choice.

41. D. Communication is the most important activity because the project manager spends an estimated 90% of his time communicating.

42. A. The formula for standard deviation on a PERT estimate is (P-O) / 6. This equates to 5/6. If you guessed 'B', you were probably thinking of the formula for a PERT estimate.

43. D. At first glance, many people think that the wording of answer 'D' is impossible, but it is the correct choice. This problem should be solved in the usual 3 steps with one additional step at the end. (Did you notice that this was the same project network diagram as shown earlier? This is not uncommon on the PMP). First, draw out the network diagram based on the table. Your representation should resemble the one below:

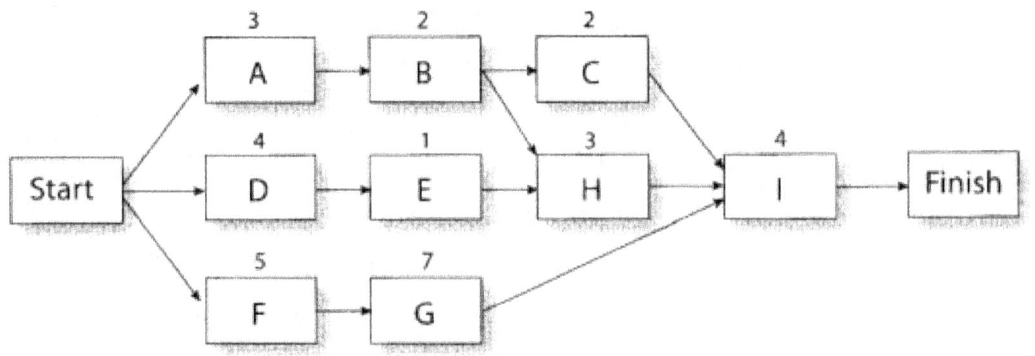

The next step is to list out all of the possible paths through the network. In this example, they are:

Start-A-B-C-I-Finish

Start-A-B-H-I-Finish

Start-D-E-H-I-Finish

Start-F-G-I-Finish

The third step is to add up the values associated with each path. Using the paths above, they are:

Start-A-B-C-I-Finish = 11

Start-A-B-H-I-Finish = 12

Start-D-E-H-I-Finish = 12

Start-F-G-I-Finish = 16

The final additional step is to increase the value of H from 3 to 7 and evaluate the impact. That would change the list to:

Start-A-B-C-I-Finish = 11

Start-A-B-H-I-Finish = 16

Start-D-E-H-I-Finish = 16

Start-F-G-I-Finish = 16

The answer, therefore, is that the critical path would change (there are now 3 critical paths), but the end date remains the same.

Moving from 1 critical path to 3 increases the schedule risk considerably, invalidating choice 'C'.

44. D. The project's deliverables would not be produced before the budget had been created.

45. A. Milestone charts are useful for communicating high-level status on a project. This represents the best of the 4 choices. 'B' would be far too much detail for an executive status meeting. 'C' would not be a bad thing to bring, but it would not be the best document for showing status. 'D' would not provide the status that the CIO seeks.

46. B. Resource requirements should be developed against the schedule activities. That is why they are called "activity resources requirements."

47. B. Analogous estimates, also called top-down estimates, use previous project costs as a guideline for estimating.

48. D. You cannot calculate the three-point estimate without knowing values for

optimistic, pessimistic, and realistic. In this case, you are only provided with 2 of the 3 numbers, therefore the answer is unknown.

49. A. The standard deviation measures how diverse the population is. It does this by averaging all of the data points to find the mean, then calculating the average of how far each individual point is from that mean. For a very diverse population, you will have a high standard deviation. For a highly similar population, your standard deviation will be low.

50. B. This is the best answer. Modern quality management stresses planning and prevention over inspection. This is based on the theory that it costs less to prevent a problem than it does to fix one. 'A' is incorrect, since prevention is stressed over inspection. 'C' is incorrect since contingency planning is part of risk management - not quality management. 'D' is incorrect since Quality Planning seeks to satisfy the quality standards. It is not focused on exceeding customer expectations.

51. C. Phases within processes is a made-up term and is not stressed in quality management. Many people mistakenly select 'A' for this; however, quality management does stress that the customer's specifications should be taken into account (the implication being that if the customer's specifications are satisfied, the customer should be satisfied with the product). B' is incorrect because Deming's T Q M philosophy stresses that the entire team has a responsibility toward quality. 'D' is incorrect because prevention over inspection is a big thrust of quality management, stressing that it costs less to prevent a problem than it does to fix one.

52. A. Perform Quality Assurance is the process that uses these five tools.

53. C. Organizational process assets, enterprise environmental factors, the project scope statement and the project management plan are all used as inputs into Quality Planning, but the quality baseline is an output.

54. D. Contract type selection is a tool used in procurement planning. The type(s) of contract(s) used are included in the procurement management plan.

55. C. The project manager is ultimately responsible for the quality of the product. If you guessed 'A', you were on track because Deming said that the entire team is responsible, but the word "ultimately" is the key here. The person ultimately responsible is the PM. 'B' is incorrect because the quality team is not a team identified by PMI. 'D' is incorrect because functional managers may be very involved in the quality management process, but they are not ultimately responsible.

56. A. This one is tricky because all of these tools are used in quality management! 'B', 'C', and 'D' are all used in Perform Quality Control, but 'A' is used in Quality Planning and Perform Quality Assurance. Benchmarking is used to establish quality standards based on the quality attributes of other projects, and it is not used in Perform Quality Control.

57. C. The purpose of a control chart is to statistically determine if a process is in control.

58. D. Stakeholders may want the project to succeed or fail! They may benefit or lose if the project succeeds. This is contrary to the way the word is used in many circles, and it is hard for many people to think of a stakeholder as potentially being hostile to the project.

59. B. A probability impact matrix is a tool of Qualitative Risk Analysis, and it is used to rank risks qualitatively, that is, based on their characteristics.

60. B. The opportunity cost is what you missed - not the difference between them. Because you invested in Project X, you missed out on the net present value of Project Y, which equals $75,000.

61. D. A risk register is created during Risk Identification, and it is updated in the

subsequent risk processes (Qualitative Risk Analysis, Quantitative Risk Analysis, Risk Response Planning, and Risk Monitoring and Control).

62. C. When an anticipated risk event occurs, the plan for addressing that is followed; however, some risks cannot be anticipated. In that case, a workaround is needed.

63. A. The first thing you should do is to plan for the new risks this situation presents. Remember that you should look for a proactive approach to almost everything. 'B' is incorrect because you cannot simply decide to withhold payment if you are in a contractual relationship. 'C' is incorrect because even though that may be something you would do, it is not the FIRST thing you should do. 'D' is also not the FIRST thing you should do, because this problem should be dealt with by the project manager. Running off to apprise senior management of the situation would not be the first thing a project manager does. It would be far better to do that after the project manager had assessed the situation and planned thoroughly for it.

64. B. In a functional organization, most of the power rests with the functional manager. 'A' is incorrect since that describes a projectized organization. 'C' is incorrect because that describes a matrix organization. 'D' is incorrect since there is no explicit model in which power rests with the project office.

65. D. Human Resource Planning is the process of understanding and identifying the reporting relationships on a project. An output of this process is the project's organizational chart.

66. D. The responsibility assignment matrix (RAM) does not include reporting relationships. Those are included in the project's organizational chart. 'A' is incorrect because the RAM does show who is responsible for what on the project. 'B' is incorrect because it shows roles on the project for the various team members. 'C' is incorrect because the RAM can be for either individuals or for groups (e.g. engineering or

information technology).

67. C. McGregor's Theory X manager distrusts people and believes that they must be watched every moment. 'A' is incorrect because that is more descriptive of the opposite, Theory Y manager. 'B' is incorrect because that is more descriptive of the practice of MBO (management by objectives). 'D' is incorrect because Theory X is not directly related to quality management.

68. A. The withdrawer is someone who does not participate in the meeting and therefore is not a constructive role. 'B' is incorrect because someone who is trying to gather more, good information is contributing in a positive way. 'C' is incorrect because a person who clarifies communication is adding to the meeting as well. 'D' is tricky, but it is incorrect. In project management terminology, a gate keeper is someone who helps others participate and draws people out. Gate keepers would help withdrawers become active participants.

69. D. In this case, the conflict is of a technical nature, so the best way the project manager could solve the problem is by using his or her technical expertise. 'A' is incorrect because legitimate power might stop the fight, but it wouldn't solve the problem. 'B' is incorrect because it also might stop the fight, but would not solve the problem. 'C' would probably be the least effective approach to solving this particular problem, since referent power is relying on personality or someone else's authority.

70. C. The three-point estimate is calculated by adding the pessimistic estimate + 4 X realistic estimate + the optimistic estimate and dividing by 6. In this case, it is

$$\frac{60+ (4 \times 30) + 18}{6}$$

This reduces down to 198 / 6 = 33.

71. C. Forcing does do away with the conflict... but only temporarily. It is when the manager says "This is my project, and you will do things my way. Period, end of discussion." The root of the problem is not addressed by this approach, thus the solution is only temporary.

72. D. Did you guess 'A' or 'B' on this one? The communication plan does not include the performance reports, and it may be either formal or informal, highly detailed or general, depending on the project and the organization. 'C' is incorrect, because it does not include the project's major milestones. 'D' is the right answer in this case because it details how you are going to gather and store information on the project.

73. B. One of the main reasons conflict arises on a project is over communication, and one of the results of a project's communication lines being broken is that conflict increases.

74. C. To answer this one, you should have asked "which choice solves the problem?" The problem is that an employee is giving you incorrect information, and you cannot ethically pass that information on. The only choice that directly deals with the problem and fixes it is 'C'. The solution is not painless, but it is the best choice of the four.

75. C. If there were 10 people on your project, that would yield 45 communication paths. Add 5 more, and you now have 15 people, which yields 105 communication paths. The question asks how many more paths would you have, thus the answer is 105 - 45 = 60.

76. A. It is the role of senior management to resolve organizational conflicts and to prioritize projects, and either of those may be at the root of this problem. 'B' is incorrect since this is a matter internal to the organization, and the customer should be buffered from it. 'C' is incorrect since the stakeholders cannot always bring influence to bear inside the organization. 'D' is incorrect since the sponsor functions much like a customer internal to the organization. The sponsor does not prioritize projects and would not be the best person to go to in order to sort out an

organizational conflict.

77. B. E-mail is informal written communication. Formal written communication involves such things as changes to the project plan, contract changes, and official communication sent through channels such as certified mail. As e-mail evolves in its usage and protocol, test takers should be aware that although they may use e-mail in a formal manner, it is not considered to be formal communication.

78. C. These are all considered to be forms of conflict resolution, even though none of these is considered to be an effective way to resolve conflict.

79. B. Again, as yourself "what is the choice that solves the problem?" In this case, 'B' is the best choice that solves the problem. Before you do anything else, you would want to meet with the person directly and discuss the problem. 'A' may be appropriate at some point but would not be considered best in this case. 'C' is not a good choice, since it is your job to resolve the problem and not the customer's job. 'D' is incorrect since it is too passive a choice and does not really deal with the problem.

80. B. This question is not only difficult, but there is a lot of information here to distract you. In this case, you should go to the sponsor and let him know, since he has approved the budget. If the project stands to deviate significantly (over or under), then the person paying for it should know as soon as possible. 'A' is incorrect because you are supposed to conform to the scope – not increase it! 'C' is also incorrect. You are working for the sponsor here, and it would not be wise to bill them more than the project costs. 'D' is incorrect, since you do not want to gold plate the scope by adding more than was originally planned.

81. D. Communication is important at all points in the project, but it is critical during integration. When performing integration management, the project manager's job is primarily to communicate.

82. C. Focus on direct, polite confrontation over practically any other method of conflict resolution. In this case, you are the one who needs to resolve the conflict, so you should take the initiative. Discussing the employee's actions with her before the meeting should actually produce a resolution to the problem. 'A' appears direct at first, but it is not really a direct way to deal with the problem. 'B' is simply making the problem someone else's problem - in this case, your boss's. 'D' is incorrect for the same reason 'B' is. This is a problem that you, the project manager, should solve.

83. A. With an SPI this far below 1,you have a significant schedule delay, and you should report this to management. 'B' is incorrect because you are doing quite well on cost, and there is no overrun. 'C' is incorrect since it is your job to deal with scope change- not management's. 'D' is incorrect because you cannot simply reject changes on the project. They must be evaluated thoroughly and fairly and sent through the scope change control system.

84. C. The team should be involved in the estimating process, and once they have bought into those estimates, you should resist pressure to automatically slash them. 'A' is incorrect since additional estimating is not what is needed here. 'B' is incorrect since this is just delaying the inevitable and perhaps making matters much worse. 'D' is incorrect since the dates need to come from your estimates and schedule development and not from the client.

85. D. To answer this, you should consider two things: 1. You need to avoid all conflicts of interest. 2. The solicitation process is supposed to keep all potential sellers on a level playing field. With these facts in mind, 'D' should emerge as the only choice that makes sense.

86. D. Another name for analogous estimating is top-down estimating, because it looks at projects as a lump sum and not broken down into pieces (which is known as bottom-up estimating).

87. C. This is a no-win situation, but you must obey laws in the country where you are performing the work, and 'C' is the only option that fully complies with the law. Refer to the PMI code of conduct and you will see that you cannot bend or break laws just to get the project done on time.

88. D. Critical path questions are solved in 3 steps. First, draw out the project network diagram. Your diagram should look similar to the one depicted below (did you notice that this is activity on arrow?).

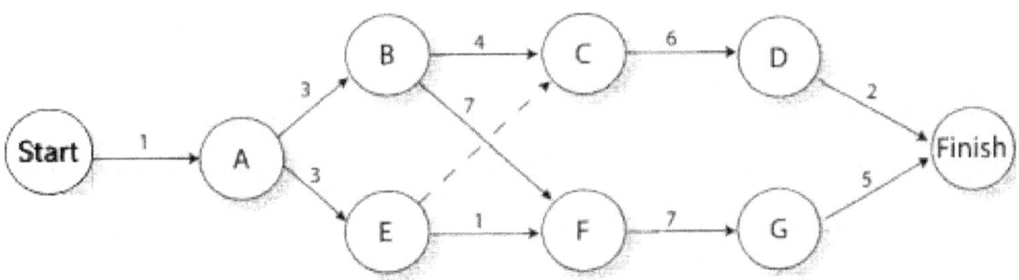

The next step is to list out all of the possible paths in the network. They are:

Start-A-B-C-D-Finish

Start-A-B-F-G-Finish

Start-A-E-C-D-Finish

Start-A-E-F-G-Finish

Finally, add up the length associated with each of those paths:

Start-A-B-C-D-Finish = 16

Start-A-B-F-G-Finish = 23

Start-A-E-C-D-Finish = 12

Start-A-E-F-G-Finish = 17

The critical path emerges as 23, represented by choice 'D'.

89. B. PMI's code of conduct states that you have a "responsibility to refrain from offering or accepting inappropriate payments, gifts, or other forms of compensation for personal gain." The best thing to do in this case is refuse the gift and let management know of the situation.

90. A. Activity Definition uses the technique of decomposition to produce the activity list.

91. A. In a fixed price contact, the seller is the one who bears the risk. If the cost runs high, the seller must deliver at the original cost. The buyer's costs are set (fixed), thus offering a measure of security. 'B' is incorrect since the seller has higher risk. 'C' is incorrect since the seller's risk is higher than the buyer's. 'D' is incorrect since cost reimbursable is another form of contract separate from this.

92. C. A project coordinator is someone who is weaker than a project manager but may have some limited decision-making power. 'A' is incorrect since a project manager does not have "limited authority" on a project. 'B' is a made-up term, and 'D' refers to the sponsor whose role on the project is to pay for the project, receive the product at the end, and give the project good visibility in the organization.

93. D. Risk Response Planning would be the point at which you determine an appropriate response to the risks that have been identified, qualified, and quantified. Only after the risks are fully understood and analyzed would you make a change to the scope.

94. C. Scope Planning is all about one thing: creating the scope management plan, and the tool of expert judgment is used to help create it.

95. B. Integration (in addition to most of the other processes) is not discrete. In other words, it isn't performed in a vacuum, but instead it is performed with all of the other processes in mind. This is even more pertinent for integration than any of the other knowledge areas. Also keep in mind that the word "integrated" has nearly the opposite meaning of the word "discrete." 'A' is incorrect because that is exactly what integration does. 'C' is also a purpose of integration, because the PMIS is the tool that the project manager uses to know what is going on with the project. 'D' is incorrect because that is also a definition of integration. The project manager is supposed to keep people focused on the work while he solves problems.

96. A. Maslow's hierarchy is based on the fact that your basic needs, like food and water, must be satisfied before higher needs, such as esteem, will become important. 'B' is incorrect because Herzberg's theory of hygiene factors is a different theory of motivation. 'C' is incorrect because this is a different motivational theory. 'D' is related to another theory of scientific management not covered in PMI's materials.

97. D. Definitions are very important, and this definition question is missed by many people. Understand that projects may or may not be strategic to the company. Although everyone wants their project to be exciting and strategic, more mundane projects also must be undertaken. 'A', 'B', and 'C' are all part of the core definition of a project.

98. D. In this case, your manager is auditing the process, and audits are used in Perform Quality Assurance. Audits are performed primarily to make sure that the process is being followed. 'A' is incorrect because quality control is inspecting specific examples and is not focused on the overall process. 'B' is incorrect because quality management is too broad a term to fit the definition of this process. 'C' is incorrect because quality planning is the process where the quality management plan is created.

99. A. Variance analysis looks at the difference between what was planned and what was

executed. Choice 'A' is the one that correctly identifies this.

100. B. In a matrix organization, power is shared between the project managers and functional managers. In a strong matrix, the project manager is more powerful, while in a weak matrix, the functional manager has more power. In no circumstances would 'D' be correct, as the project coordinator is, by definition, weaker than a project manager.

101. C. Team performance assessment is a tool of the Develop Project Team process. When using this tool, the project manager evaluates the team's performance with the goal of understanding strengths and weaknesses.

102. C. The project scope statement typically starts off general and becomes more specific as the project progresses. Progressive elaboration is a term that describes the way in which the details of the scope are discovered over time. 'A' is incorrect since it is more descriptive of the WBS than the project scope statement. 'B' is incorrect since the project scope statement is created by the project team and not by senior management. 'D' is incorrect because the project scope statement is a functional specification.

103. C. A devil's advocate is considered to be a negative team role. 'D', a gate keeper, is incorrect because in project management terminology a gate keeper is someone that draws non-participants and withdrawers into the process.

104. D. In Manage Project Team, the team is executing the work packages and creating the product of the project. Your job as project manager is to keep them focused on this. 'A' is incorrect because that may be your focus, but it is not the team's focus. 'B' is incorrect, because that is the project manager's job and not the team's focus. 'C' incorrect for the same reason. It is the job of the project manager and not the team.

105. C. The project management information system (PMIS) is the one described in the question. 'A' is not a good choice because you don't have to have information technology from beginning to end in order to successfully deliver many projects. 'B' is a made-up term. 'D' is not a good choice because the work authorization system (WAS) is used to make sure that the work is performed at the right time and in the right sequence.

106. C. On typical projects, most of the resources (both human resources and material resources) are expended during executing processes. Many people incorrectly choose 'B' because there are so many planning processes, but on the average, project planning takes less effort and resource than execution.

107. A. Inspection is a tool of Scope Verification, which is the process being described in the question. In inspection, the product of the project is compared with the documented scope.

108. C. "Quality Improvement" is something you may strive for, but it is not a process.

109. D. Schedule constraints would not contain leads and lags for activities. 'A', 'B', and 'C' would all make sense to include as schedule constraints.

110. B. Benchmarking is a tool of quality management for both the Quality Planning and Perform Quality Assurance processes. It takes the results of previous projects and uses them to help set standards for other projects. 'A' and 'C' are incorrect because they would be more closely aligned with Perform Quality Control. 'D' is largely unrelated to quality.

111. C. The activity list should include every schedule activity defined on the project. These schedule activities are then used to create the project network diagram.

112. A. The Close Project process should be performed at the end of each phase or at the

end of the project. It is the process where the project is formally accepted and the project records are created. It is important to understand that Close Project may happen several times throughout the project.

113. C. The release criteria for team members is defined in the staffing management plan.

114. D. The activity list is the output of the Activity Definition process. 'A', 'B', and 'C' are all inputs into Activity Definition.

115. C. There are many ways to mathematically solve this problem, but perhaps the simplest is to divide the schedule performance index against the length of the project. 20 weeks / 0.8 = 25 weeks. Therefore, we would expect the project to be 5 weeks late. 'A' should have been eliminated because with a schedule performance index less than 1, there is no way the project should be finished early.

116. C. The term slack is synonymous with float.

117. A. The project plan is a formal document. It is created during planning, and is used to guide the execution processes, monitoring and control, and closure. 'B' is incorrect because that would be closer to the definition for the product. 'C' is close to correct, although it is an incomplete list of what makes up the project plan and the question specifically asks for the BEST description. 'D' is incorrect because this is closer to the definition of the work breakdown structure than it is to the project plan.

118. D. As part of integrated change control, the project manager will need to know when change has occurred, manage the changes, and influence the factors that cause change, but the project manager should not take on the attitude of denying change whenever possible. Some change is inevitable, and all change requests should be evaluated and not automatically rejected.

119. C. Another no win situation. You cannot bend or break laws just to stay on schedule.

The end does not justify the means! You have to obey laws and observe customs in the country where you are performing the work. In this situation, you should look at options that do not involve working overtime.

120. A. There is a difference between the product scope and the project scope. The scope of the project may be much larger than the scope of the product! This question defines the product scope. 'B' would have a much broader definition than this. 'C' is the work that needs to be done to complete the project, but it does not deal with the attributes of a product. 'D' is a phrase often used in project management, but it is unrelated to this definition.

121. C. The supporting detail should be included with the estimates, and that supporting detail is included for situations just like this one. It will help you and the team member understand how the estimates were derived in the first place. 'A' is incorrect since the team member may well have a valid point. 'B' is incorrect because there is no reason to either stop work on the project or to send the team member scrambling for new estimates. 'D' is not a bad choice, but it isn't the FIRST thing you would do. As a starting point, go back and check the facts first. Then if it would be helpful to get another expert involved, you may elect to do that.

122. B. The term "baseline" causes grief for many test takers. Memorize that the baseline (whether it is the scope baseline, schedule baseline, cost baseline, or quality baseline) includes the original plan plus all approved changes. Once the budget change was approved, it should be added to the cost baseline.

123. B. Although it is unusual in the real world, a projectized organizational structure gives the project manager near total control of the project and the resources. 'A' is a structure where the functional manager is in charge of projects and resources. 'C' is a structure where the project manager runs the projects and the functional manager manages the people, and 'D' is not a real term for organizational structures.

124. A. This problem should be solved in the usual 3 steps with one small bit of reasoning applied at the end. First, draw out the network diagram based on the table. Your representation should resemble the one below:

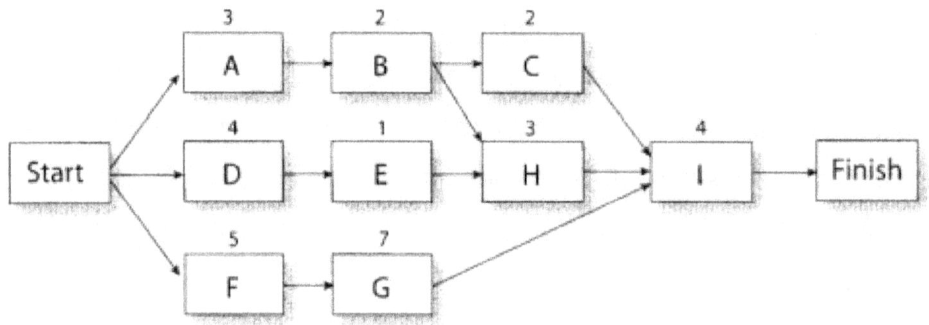

The next step is to list out all of the possible paths through the network. In this example, they are:

Start-A-B-C-I-Finish

Start-A-B-H-I-Finish

Start-D-E-H-I-Finish

Start-F-G-I-Finish

The last step is to add up the values associated with each path. Using the paths above, they are:

Start-A-B-C-I-Finish = 11 units

Start-A-B-H-I-Finish = 12 units

Start-D-E-H-I-Finish = 12 units

Start-F-G-I-Finish = 16 units

The question has asked for the path with the LEAST schedule risk, and that is represented by the shortest path (the one with the highest amount of float). The reason this path has the least risk is that tasks could slip the most here without affecting the critical path. In this case, it is Start-A-B-C-I-Finish, which corresponds to choice 'A'.

125. A. When a conflict of interest arises, it should be resolved in favor of the customer. In this case, your company's module has been determined not to be a good fit, so another solution is needed. 'B' is incorrect because conflicts should be resolved in favor of the customer. 'C' is incorrect because it has already been determined that it is not best for the project. 'D' is not correct because that is the project manager's job - not that of an outside party.

126. D. Leads and lags are adjusted as part of the Schedule Development process.

127. A. Just as important as understanding a formula is being able to interpret it. That is what this question is calling on you to do, and it can be quite hard. The schedule performance index (SPI) = EV/PV. Studying the formula, you can see that it compares how much value you actually earned (EV) and divides that by how much you planned to earn (PV). In this case, you have earned value at 1.5 times the rate you had planned.

128. C. The project organization chart, an output from the Human Resource Planning process, shows the reporting structure for the project.

129. C. Most of the planning processes have a logical order to them, and this question relies on an understanding of that. The work breakdown structure has to be created before cost estimates are performed, and the cost estimates have to be created

before the budget. This should make sense when you stop to consider it.

130. B. The list of risks, contained in the risk register, is an output of the Risk Identification process.

131. A. Activity Definition is the process where the work breakdown structure is further decomposed into individual activities.

132. B. A good rule is that if in doubt, select the "project manager." For this question, that would be correct. It is the project manager's job to resolve competing stakeholder requests and goals.

133. C. Your goal is to create a win-win situation in the negotiations. A win-lose agreement will usually not let you win in the long term.

134. C. The project manager's professional code of conduct instructs you to keep customer information confidential. You should ask permission before using or reusing any part of the project that is not owned by you or your organization.

135. D. The project budget, also known as the cost baseline, is an output of Cost Budgeting. 'A', 'B', and 'C' are outputs of the Cost Control process.

136. A. The way to approach this question is to remember how to calculate the schedule variance. It is EV-PV. If the schedule variance = 0, then EV must be equal to PV.

137. B. Schedule activities are a further decomposition of the work breakdown structure, and work packages exist at the lowest levels of the work breakdown structure.

138. B. This one is easier than it first appears. All "management plans," including the staffing management plan, become part of the project plan. 'A' is incorrect because it is an output of Human Resource Planning. 'C' is incorrect because it describes the

approach to staffing and not every detail. 'D' is incorrect because the organization chart for the performing organization would not be a typical component of the staffing management plan.

139. B. All of the identified risks should be added to the risk register. 'C' is incorrect since the risk management plan only contains the plan for how risk will be approached. It does not get down to the specifics of listing risks.

140. C. You have a responsibility to help mentor others in the field of project management. 'A' would be inappropriate, because your responsibility extends to the profession- not just to your organization. 'B' is incorrect because the best place to start is with your friend. 'D' is incorrect because a job change would represent withdrawal and would not solve the problem in any way.

141. B. Planned value is the value of the work you planned to do at a given point in time. The first step is always to calculate your budgeted at completion. It is 25 miles * 1,200,000 / mile = $30,000,000. Now you want to calculate the planned value for 3 weeks of the 5 week project. Simply multiple the budgeted at completion by 3 and divide by 5. This yields $18,000,000. The interpretation of this is that you planned to earn $18,000,000 worth of value back into your project after 3 weeks of work.

142. C. Read this one carefully. It is the definition of free float. 'A' and 'B' were synonyms, which should have thrown up a red flag for you. Those terms tell you how long a task may be delayed before it changes the finish date of the project.

143. D. Knowing the order in which these steps are performed is important and shows that you understand what is really taking place within the process. In this case, resources are estimated after you have defined the activities, but before you have created the schedule. This is the only answer that works, since the resources are based on activities, and you must have the resource requirements before you can

create the schedule.

144. A. Even when news is unpopular or unpleasant, you must deliver accurate statuses. 'B' and 'C' attempt to cover up or hide the news. Choice 'D' is unnecessary since you should report the current and accurate status to the customer.

145. B. Select Sellers is an executing process of procurement, and yes, it is important that you know this type of information for the exam. In this process, the buyer evaluates the responses from the potential sellers and chooses a seller to perform the work.

146. C. Know the roles of each person or group on the project. It is the sponsor who provides funding for the project.

147. A. Most importantly, the project should satisfy the scope, schedule, and budget. These are the principal factors of success. 'B' is important in many organizations, but you cannot control happiness as a project manager. The best you can do is to satisfy the scope of work. 'C' may be good to have, but it does not make the project successful. 'D' is also good in that you did what you planned to do when you planned to do it, but it is only an ingredient of project success.

148. C. Your primary objective is to satisfy the scope of work on the project within the agreed-upon cost and schedule. 'A' is the means to your end goal of a successful project, but it is only the way you go about it. 'B' is a great goal, but there are other parties here that need to be considered - not just the performing organization. 'D' is good, but you should focus on doing so by delivering what you promised on time and on budget.

149. D. Organizational process assets are anything you can reuse, such as a document, a previous project deliverable, or a methodology. Strong communication skills was not a good fit for this question.

150. A. It stands to reason that problems are fixed in the same process where deliverables are produced. In this case the process is Direct and Manage Project Execution.

151. B. An activity's late finish date is the latest an activity can finish without delaying the project. If it exceeds the late finish date, the critical path will change, ultimately resulting in the finish date slipping. Choice 'A' is close to the definition of free float.

152. D. Speeches such as this one are examples of formal verbal communication.

153. C. The contract change control system is defined in the procurement management plan and not in the contract itself.

154. A. Herzberg stated that hygiene factors must be present in order for motivational factors to work; however, hygiene factors do not motivate by themselves. They only enable the motivation factors to work.

155. B. The critical path represents the highest schedule risk on the project. If there is more than one critical path, schedule risk is increased. 'A' is not necessarily true. Two or more paths could be the critical path on the project all the way from beginning to end. 'C' and 'D' are ways to "resolve the conflict" when there is no conflict. There will always be at least 1 critical path, and having 2 or more critical paths by no means represents a conflict.

156. B. With a payback period, the shorter the time the better. The hardest thing about this problem is to reduce all of the times to a common denominator so you can see which one is the shortest! B is one year, and that is the shortest period of time of the choices.

157. B. A project network diagram (like activity on arrow) is an output of activity sequencing. The other 3 types of dependencies fall under the tool of dependency determination in Activity Sequencing

158. D. The processes are not set in stone so that once you have finished planning you can never return. Remember that projects are progressively elaborated, and that often times you will need to revisit processes again and again. There is no reason not to return to planning in the example given here.

159. D. The goal of duration compression is to accelerate the due date without shortening the scope of the project.

160. B. Conflicts of interest should be disclosed and avoided. 'A' is not a good choice, because you are not only bound to act legally, but also ethically! 'C' is not correct because you would not be keeping information confidential. 'D' would not be a rational choice. Resigning from both projects would cause more problems and solve nothing.

161. C. A standard usually should be followed, while a regulation has to be followed.

162. A. The evaluation criteria are created in the Plan Contracting process. In the Select Sellers process the evaluation criteria are used to select a qualified seller after responses have been received.

163. D. Richard is performing Risk Response Planning. After the risks have been identified, qualified, and quantified, they should be responded to. Risk Response Planning looks at how to make the opportunities more likely and better and the threats less likely and less severe. Remember that a risk is an uncertainty that could be good or bad.

164. A. In order to answer this question, you must first be able to identify in which process you solicit bids. It is the Request Seller Proposals process, and the output of Request Seller Proposals are the proposals from your potential sellers.

165. B. The work results of the project may be products, services, or results. 'A' is incorrect because work results could be a product, a service, or a result. 'C' is incorrect

because the work results don't have to meet quality to be considered work! 'D' is incorrect because the output of the work authorization system is not the work itself, but that a resource is authorized to do the work.

166. B. You have a responsibility to society, the environment, and the economy, and this comes higher than your responsibility to your boss. You are not to just follow orders, but you should think for yourself to ensure that the project does not do harm. Note that this is not merely a disagreement over the probable success of the project.

167. A. Some companies rest their reputation on the high quality of their product. Others do not. The quality policy defines how important quality is on this project from the performing organization's perspective.

168. C. Performance reviews review the current status against the plan.

169. D. By not facing the problem the team member is withdrawing. 'A', 'B', and 'C' are not terms regularly used in project management circles.

170. C. In a functional organization, the project manager has little formal power and may even be part-time! It is the functional manager who is more powerful in this structure.

171. C. A RACI chart is a type of responsibility assignment matrix created during Human Resource Planning. It is a chart that can show roles and responsibilities on the project. 'A' is incorrect because it shows resource usage levels across the project. 'B' is incorrect because it shows reporting relationships but not specific responsibilities. 'D' is incorrect because this general document tells how you are going to approach staffing the project.

172. D. Risk mitigation is when you try to make the risk less severe or less likely. By accelerating the construction, Kim is mitigating the likelihood of a hurricane

damaging her project.

173. C. The answer "fixed price" would have been the best here, but it was not in the list of choices! Of the ones listed, cost plus incentive fee would provide the seller with an incentive to keep their costs down. 'A' provides no incentive at all for the seller to keep costs down. 'B' would not provide the same incentive since the seller gets a fixed fee regardless of the project costs. 'D' is not a real contract type.

174. D. The more channels of communication on a project, the more difficult it is to control communications. 'A', "B, and 'C' are probably not true because these people are internal to the project, and the creation of the plan and analysis of the stakeholders, and communication with them, would not necessarily be more difficult.

175. A. Here is an example where you should ignore the trigger words "find out." You have a responsibility to comply fully with PMI in an investigation. 'B', 'C', and 'D' may be tempting, but you should cooperate with PMI.

176. B. Earned value is what you have actually done at a point in time. In this case the budgeted at completion for the project is $300,000, and you have completed 20% of that. All of the other facts in the problem are irrelevant. The answer is $300,000 * 20% = $60,000.

177. B. The project expeditor is the weakest role here. This person is typically a staff assistant to an executive who is managing the project. 'C' is the most powerful, and 'A' would be next. 'D' is not a term identified within PMI's processes.

178. D. This is a definition of configuration management. 'A' is incorrect because it is part of the Integrated Change Control process and not Develop Project Management Plan. 'B' is incorrect because that is more descriptive of inspection - a tool of scope verification. 'C' is incorrect because this is performed after initiation.

179. D. Perform Quality Control and Scope Verification are very tightly linked. Perform Quality Control is the process concerned with correctness, and Scope Verification is primarily concerned with completeness. In general, Perform Quality Control is performed just before Scope Verification, but they are often performed at the same time.

180. D. When a project ends or is cancelled, the project manager should perform the Close Project process. 'A' would come out of Close Project. 'B' is not defined as part of closure. 'C' might be appropriate in some cases, but you should lean strongly toward terms you have read in this book and not rely on experience.

181. C. The interpretation of this is important. A benefit-cost ratio indicates how much benefit you expect to receive for the cost expended. In this example you could to get $1.50 profit for every dollar of cost.

182. A. The activity list is the main input into Activity Resource Estimating. None of the others listed are inputs into that process, but consider that you could not perform Activity Resource Estimating without first having the list of activities against which you will estimate.

183. B. Life-cycle costing takes a broad look at the project, considering such things as operational costs, scrap value, etc. It doesn't just ask how much it costs to make a product, but it looks at the total cost of ownership.

184. D. The buyer is primarily assuming the risk here because they are in a time and materials contract. The seller gets paid for every hour they work.

185. B. A Control Account Plan (also known as either a CAP or a Cost Account Plan) is a control point placed on the WBS that is used by the project team to plan when the work packages cannot yet be defined.

186. C. Stakeholder analysis is done as part of Scope Definition so that you properly understand your stakeholders, their needs, and their expectations as they relate to the project's product.

187. A. Expectancy theory says that reward motivation will work if the team believes that the goal is achievable.

188. D. Organizational policies are not directly used in creating the project budget. 'A', 'B', and 'C' are all inputs to the cost budgeting process.

189. A. There is no way to get around that dependency, so it is a mandatory dependency.

190. D. A milestone chart only shows major events (milestones) on the project's timeline. 'A' and 'B' are incorrect because neither a Gantt chart nor a milestone chart are project plans. 'C' is incorrect because milestones are high level representations that do not show interdependencies between activities.

191. D. After the scope statement and the work breakdown structure, you should move into activity definition. 'A' cannot be done before the activity list since the activity list is used to create the network diagram. 'B' and 'C' cannot be done until the network diagram is completed. The correct order of these choices listed would be: 'D', 'A', 'C', 'B'.

192. B. Run charts show trends and variances over time and are used in quality management.

193. D. This is not the easiest question on the exam! You must first determine what you have done and then determine what is next. Look carefully at what you have created, and you will see that you have just finished the Plan Purchases and Acquisitions process and the Plan Contracting process. The next process to be performed should be Request Seller Responses.

194. C. Generally, any time you are looking at past performance, you are in a monitoring and controlling process.

195. C. The primary purpose of the project network diagram is to show the sequence of activities and their dependencies. 'A' is incorrect because the project plan is made up of all planning outputs. 'B' is incorrect because the percentage complete is not reflected on the network diagram. 'D' is incorrect because the project network shows the paths through the network, but it does not create them.

196. A. The quality policy is a document issued by the performing organization that describes their attitude regarding quality. As different companies place different values on quality, the quality policy will differ. For instance, a pharmaceutical company will almost certainly have a stricter quality policy than a maker of novelty toys.

197. D. Did you guess 'C' for this one? The risk would have been planned in Risk Response Planning, but if it was unforeseen and it occurred then you would have caught that in Risk Monitoring and Control. That is the process where workarounds are created.

198. B. Cost aggregation is adding the costs associated with work packages up along the structure of the WBS to get the cost of specific branches or the entire project.

199. A. The formula for the schedule performance index is earned value/planned value (EV/PV).To get EV, we need to know how much we have completed to date. The budgeted at complete is $300,000, and we are 20% complete. Therefore, EV = $300,000 * 20% = $60,000. Planned value is what we had planned to complete at this point. We are 1 month into a 4 month project, or 1/4 of the way through. 1/4 = .25 (25%), and our budgeted at completion of $300,000 * 25% = $75,000. Now that we have EV ($60,000) and PV ($75,000), we can calculate the schedule performance index. It is EV/PV, or $60,000 / $75,000 = 0.80. The interpretation of this number is

that the project is earning value 80% as fast as was planned, and any index that is less than 1 is a bad thing!

200. B. Collocation is a tool used in the team development process where the team is brought together in a single location.

Agile Project Management

Chances are you've already heard quite a bit about the world of agile. At times it might seem like everyone has either already signed up or is going down that route in the very near future. But in reality, there are plenty of people and organizations still pondering their first move.

PMI recently incorporated some agile questions into its PMP exam. Therefore, we have brought an extensive chapter providing an insight into this brave new world with enough to help you understand how this philosophy works.

In true agile style, it contains enough to start moving in the right direction and contains the least you need to know to get maximum return. Consider this chapter as a practical guide with key insights into Agile, Scrum and Kanban.

Introduction

THE STATE OF THE PROJECT NATION

The advance press is excellent but what exactly is the core problem with project management today? What is agile trying to fix? The bottom line is that projects regularly take longer than expected, cost more than budgeted for and quite often fail to deliver what was asked for.

That heady cocktail has resulted in a loss of business confidence. Traditional project management thinking has always acknowledged the importance of time, cost and scope and together they're known in the game as the Project Management Triangle.

But so many projects get lost in there that it's more like the Bermuda Triangle.

The project management triangle

Because of this rigid relationship, one side of the triangle cannot be changed without affecting the others. There are always consequences if the scope, time or cost parameters are adjusted. This happens frequently with projects when new requirements are added, timescales are reined in or budgets are slashed. Of course, it never happens the other way round! Over the years project teams have battled with trying to keep the three constraints balanced, invariably a case of Mission Impossible.

Straight from the kick-off of any project, the bean counters are out in force checking for any signs of overspending or of lost time. Project managers are under this double spotlight right up until the final delivery is made when the focus shifts immediately from how much and when to: is it any damn good? Late deliveries and extra cost are soon forgotten if what's delivered is what the customer wanted. This is a nightmare to navigate.

In the agile world, life is fundamentally different. The focus from the beginning is on delivering business value. Agile puts the conventional early obsession with cost and timescales to one side and concentrates on what the business wants or, more specifically, what it really needs. No more aiming for perfection, leading to simple ideas getting morphed into tremendously complex and elegant solutions. No more runaway budgets. No more customer dissatisfaction.

And remember, customers don't want better project management. They want better product delivery. All the agile tools and techniques exist solely to that end. Use whatever techniques get the best results but don't get hung up on the methods themselves. The ends are much more important than the means.

KEEP THE CUSTOMER SATISFIED

Projects never start with an intention to overegg the scope. One of the age-old challenges is seen as avoiding unnecessary gold plating and keeping the solution down to the minimum required to do the job adequately. However, in an attempt to be rigorous and thorough about pinning the scope right down there's a tendency for our customers to leave no stone unturned when asked: what do you want? This is unintentionally made worse when obsessing about getting the detail spot on, as if we're asking: is there anything else? It's akin to putting children in a toyshop and permitting them a wish list without any constraints.

Matters are made even worse when the customer thinks it's now or never and concludes that the only way to get a few nice useful bells and whistles is to demand them all up front and insist life is intolerable without the full package. This all leads to a crazy situation of non-essential requirements with an inflated budget and overlong timescales. In sharp contrast the first delivery with agile is aimed at being far more bare bones; just enough core features to get going and no more. The implicit understanding is that this will be added to over time in a measured way to build towards a fully featured solution.

With agile there's no mad dash to get everything at the January sales. Instead, let's start with a solid foundation and build from there.

SORRY, NO CHANGE THANK YOU

Under the old order once every 'i' is dotted and every 't' is crossed there's a huge emphasis on preventing change, or at least closely controlling it. Many popular frameworks for running projects, such as PRINCE2, focus heavily on pinning down the requirements and then introducing rigorous change control. Change is considered bad news – even frowned upon – and if anything does sneak through, the business gets charged through the nose for it.

Getting this formula to work is always a struggle because a desire for change is inevitable on any project. As a general rule of thumb this approach is pretty fruitful when it's implemented with rigour and the changes are few and far between. But even then, there are usually a number of battles along the way. Regrettably, the likely outcome is just on the right side of average, with a customer who is not unhappy with the outcome but far from delighted.

Additionally, not every project is on a road going directly from A to B where exact requirements can be buttoned down and where change is nothing more than a minor distraction. Most start life as a fledgling idea that needs validating in the real world and building on. Sometimes it's necessary to change track or even occasionally go back to the drawing board. That's the natural order for the evolution of most business ideas and it doesn't help when the approach to developing projects is swimming against the tide.

Agile is different. Agile embraces change and even encourages it. Change is not seen as the enemy, it's seen as an important part of the evolution of any good idea. Agile works toward delivering the bare necessities in the shortest possible time to market, so they can be tested early. Evolution is a natural part of the process and change is no big deal. This is exactly what customers need in most situations and it's no wonder this is like a breath of fresh air to them.

START SMALL, CHEAP AND QUICK

The intentions of the Agile Manifesto, principles and other elements of agile philosophy are great but how does this translate into action? How exactly is agile so different? Well, at the heart of the answer to this is that it starts off with a totally different approach to delivery and everything flows from there. Instead of beginning with a wish list as long as your arm and a restraining order on any change, agile begins by identifying the minimum needed to get going and builds from there. This is usually dubbed the minimum viable product (MVP) or the minimum feature set (MFS).

In practice, both describe the same thing. Namely the smallest possible delivery that addresses the business needs, creates the desired customer experience, and can hence be marketed and sold successfully. This reduces the time to market, is cheaper and gets the job done adequately. It can therefore be delivered more quickly than a fat, feature-rich solution. Less is more.

AN AGILE MIND-SET

It's usually dangerous to generalise but there are certain mental attributes that are well suited to an agile lifestyle. All of the agile frameworks are team-based and place a significant emphasis on teamwork, cooperation, collaboration and being adaptable. Getting on well with others and being fleet of foot sums it up for us. Hermits and dictators aren't likely to be keen advocates.

Certain organizational cultures struggle to embrace and adopt agile thinking. That's not a criticism, just a fact of life. Ultimately, it's down to individuals, and it's worth thinking about character traits and whether the whole ethos appeals. In that sense it's no different to other delivery frameworks such as PRINCE2; some are likely to float more boats than others. Anyone can try and adapt to an agile environment but some flourish. Certain personal traits help enormously:

- collaborative,
- committed,
- focused,
- open,
- respectful,
- courageous,
- honest.

These characteristics are sought after in any work colleague or anyone for that matter. A top rating in all of them identifies an asset for any team or project environment and the likelihood of taking to agile like a duck to water. Don't be too worried if the feedback indicates a mark of could do better because agile creates a supportive set-up that encourages these attributes. Running the risk of generalising again, agilists are prone to being very passionate. Some are even considered to be a little too evangelical or dogmatic but that's more about being over-enthusiastic than anything sinister.

It isn't a religious cult. Dipping into any of the agile forums can be a chastening experience at times, especially for a newbie, so don't be led to believe it's their way or the highway! Agilists don't bite but they do bark mighty loud sometimes.

GETTING AGILE

It's impossible to legislate for all the possible combinations of agile start points for organizations and their individuals. Occasionally, there are situations where a strategic decision is made by the Top Bananas to break with the past and go gung-ho for the Promised Land; when the right people are empowered to make it happen armed with an open chequebook and a specialist coach on tap. Great news if this happens but it's a rare event.

A more typical scenario occurs when an organization is dogged by failed projects and one or more people are convinced there must be a better way. Then the launch is accompanied by very little corporate buy-in initially and a minimal or non-existent budget. This is more likely to happen and allows us to think about the minimum prerequisites for any aspiring agile organization. It's a chance to pin down the critical success factors that must be given special and continual attention to bring about success.

- CSFS

The Critical success factors (CSFs) are those things that must be in place to ensure success. Those things that guarantee a right result. CSFs for an agile project typically include:

- An appropriate project. Don't get stuck into the Number 1 priority mission critical project that's behind schedule before it even begins. Best to start with a small one, focusing on proving the agile process works and ironing out any kinks. There's plenty of time to ramp up once on a roll.

- Suitable people. Not only to participate on the project itself but also to oversee the agile transformation. As a minimum, assemble people with an agile mind-set and the desire to make it happen. It's going to be a big team push and it will need a team effort.

- Realistic expectations. Be realistic especially in the short term. Expect immediate results but allow time for the benefits to filter through. Getting it right up front requires an investment and occasionally it can require a step back to make two steps forward. Set the bar at a reasonable height and build from there.

- Adequate training. Agile frameworks are easy to get to grips with and reading material on the web is a decent start. But factor training and mentoring into the plan. Agile coaches and mentors are geared up to dipping in and out of organizations – start with at least a day per week if the budget permits and if not then even one day a fortnight is better than nothing.

In the spirit of the occasion treat getting an agile project up and delivering to be part of the launch MVP. Pay attention to the CSFs. Choose the project wisely and surround yourself with the right people. There's no guaranteeing success but it's easy to skew the odds massively in your favour.

AGILE OUTCOMES

If the CEO is on his way to a board meeting and happens to drop by to ask for a couple of promotional sound bites – what's the bottom line? What can the board, senior management team, shareholders and fellow employees reasonably expect as outcomes? Why bother?

Put simply, if executed with a reasonably deft touch, agile will deliver immediately with:

- Early delivery of the MVP or MFS. Quick to market and early validation of the core concept. An end to playing the waiting game.

- Fit-for-purpose deliveries. Deliveries will 'do what it says on the tin', time after time.

No more crossing fingers and hoping for the best.

- Smaller initial investment. Beginning with a reasonable budget and investing further on proof of success. An end to high-risk endeavours and betting the farm for no good reason.

- All round flexibility. The ability to adjust and adapt to changing circumstances. No more meltdowns and recriminations at the whiff of a change request.

- Improved team performance. A virtuous circle of happy, engaged team members leading to improved performance. There's nothing wrong with smiling faces!

Most importantly, this isn't about creating low expectations. From Day 1 expect to see evidence that the goods are being delivered. Be reasonable of course but expect all of the above immediately!

SPOILT FOR CHOICE

Looking at the big picture and making sweeping observations is all very interesting but when it comes to running programmes and projects it's time to get down to specifics and chose an agile framework to work within. There are plenty of excellent options but we want to stay very focused by sticking with our three favourites.

- Lean

Considered to be one of the grandparents of the modern agile movement. Well worth checking out, especially by swotting up on the 7 Lean Principles. Excellent, thought provoking material but we're not sitting on the fence and it's not our first choice for running projects.

7 Lean Principles

1. Optimise the whole

2. Eliminate waste

3. Build quality in

4. Learn constantly

5. Deliver fast

6. Engage

7. Keep getting better

- **Scrum**

This is the current darling of the agile world and is in the process of taking the business world by storm, rightfully so because this framework is a game changer. It's sparking changes in the way businesses think and deliver projects. It's the agent for a quiet revolution and our favourite framework by far – the real deal. A great choice for projects of all shapes and sizes.

- **Kanban**

Despite our almost sycophantic support for Scrum, Kanban is right up there in our books and has plenty to offer in certain situations. It's an excellent alternative to Scrum and very easy to implement. Occasionally misrepresented and oversimplified but there's more to Kanban than first appears. Unbeatable at getting visibility of delivery in any environment.

VARIANTS AND OTHER OPTIONS

Of course, there are many other agile options, especially when the focus is on IT software development projects. Don't be surprised if you come across subtle variations on the agile themes, frameworks and organizations. At the framework level, interesting variants and combos are beginning to appear with Scrumban, Safe and others. All thought-provoking stuff but we suggest sticking with the quality brands initially.

TOO GOOD TO BE TRUE

The positive press about agile, and in particular Scrum, is a double-edged sword. The advantage is that it isn't a hard sell. The downside is that the expectations are sky high. Management teams are tuning into the sound bites – quicker, cheaper, better – and forgetting that there's no such thing as a free lunch. Managing these expectations is a challenge but no big deal if handled with a deft touch. There's nothing wrong with building on internal enthusiasm and tapping into the excitement. Just don't over-egg it.

Expect to deal with doubters and detractors. Genuine uncertainty is natural with any significant new venture, as is a degree of outright opposition and undermining criticism. An element of PR is required but on the whole it's best to let agile speak for itself.

Agile fundamentals

Some people think agile is new. Others think it's something old but repackaged. Then there's the it's just common-sense camp and those who say it doesn't work. You may even meet enthusiastic supporters who talk about it like it's some miracle cure for all business-related problems. There are many different opinions and perspectives but the headline news is it's an incredible sea change in the approach to running projects with:

- People interacting face to face, solving their unique problems creatively;
- Business involvement throughout the process from concept to cash;
- A reduced time to market to capture or retain competitive advantage;
- Early initial deliveries and quick wins to get valuable feedback fast;
- Fast incremental build of small enhancements to keep product fresh;
- Flexibility and responsiveness to change built in to the very ethos;
- An ability to change focus to remain ahead of the game;
- The user's and customer's needs at the heart of decision making;

- Shipping the product is everything!

Agile demands we find a simpler approach to getting product shipped, putting people and the interactions between them at the centre of everything. The core tools and techniques are simple to use. Flexibility and change are an integral part of the package. Yes, the movement is steeped in common sense but that's no bad thing because the more complex something is the harder it is to master. If used wisely, agile will deliver better results than any prefabricated processes.

Agile ways of working put the focus more on visibility, transparency and interactions between people and less on dogmatic process. The ambition is to empower people, setting them free to concentrate on delivery. Agile frameworks provide a structure to operate within, not a step-by-step guide. It's frequently said that agile is easy to do and hard to do well. This is mainly because of the visibility, organization and discipline it takes to make it work effectively.

There can be a feeling that agile ways of working are from the Wild West with little governance, no documentation, weak roles and ceremonies. But in fact, that isn't the case at all. Everything is there but there's a significantly defter, more frequent touch, and no process overload.

The most important reoccurring activity in an agile organization is the removal of any blockers or constraints that are preventing the team from getting on with the job. These impediments to delivery can range from minor irritations to complex organization handcuffs. They're productivity killers. Being able to focus exclusively on getting the job done without distractions is a breath of fresh air.

Getting ready: preparing to be agile

Many traditional project management methodologies are based on dotting all the i's and crossing all the t's before setting off.

Project requirements are drafted and redrafted, reviewed, revised, revamped and examined from every possible angle before any real work is done – whereas there's a view in some circles that agile is all about making it up as you go along, starting off with a vague idea and then winging it.

Agile ensures the end goal is defined up front and an enabling infrastructure is put in place for getting there but worries little about the fine detail of the journey before setting off.

DEFINING THE VISION

Without a vision, the people will perish. Moses said that about 4,000 years ago and from the looks of it most project managers still don't know what he was getting at. If we don't know where we're going, we won't get there and this is especially true with projects.

Having a clear vision is essential. Unfortunately, many visions go the way of bad mission statements and seem to say much but once put under the microscope reveal very little:

- We want to offer unparalleled quality.
- We aim to put our customers first and deliver value.
- We will be the best at what we do and loved by everyone everywhere.

- Defining a project vision
    - What is the name of the project or product you are making? Who is it for?
    - When will it be done by? What will it do? What will it not do?
    - What benefit does your business get for doing it? What benefit does your customer get by using it?
    - Write the vision in a way that your Mum or Dad can understand what you're planning to do.

A vision should be a tool that you can use. You can use it to communicate intent. You can use it to explain what you are doing and what you are not doing. You can use it to prioritize against. A vision for a project should be less of a strategic mission statement for the business and more of a tactical, practical device to help you stay focused. A vision must be open to change and get updated when it goes stale, which it surely will.

DRIVEN BY BUSINESS VALUE

Agile is totally focused on delivering business value. From the start of any project and all along the way, the business team will know exactly what they're getting for their money.

Every delivery, every feature, every nuance must be described in business-speak. Gone are the days of a person or persons unknown defining a list of requirements in technobabble or a foreign language the business doesn't fully understand before lighting the blue touch paper and retiring to a safe place.

Long gone are the days of the business putting its blind faith in people they hardly know. With agile the business describes what it wants and then works within the project team to ensure the vision is delivered exactly as requested.

If you don't know what it is, you're building (the vision), what benefit it will bring (value proposition) or who it's for (end user proposition), then you can have the best experts in the world and yet never deliver anything worthwhile.

BUILDING THE PROJECT TEAM

The agile team is a diverse, cross-functional group of individuals that has the ability and authority to deliver the vision on behalf of the business. Put simply, between them they have everything they need to get the job done properly.

The Product Owner leads the way in terms of the business vision but it's very much a team effort.

The team consists of people who have an agile mind-set, who are not afraid of change and don't need to use process and bureaucracy as a crutch to get by. Confident decision makers with a self-starter, can-do attitude are the best for this. Collectively the team must buy into the vision and co-own all aspects of the project delivery. If that isn't the case, there'll be big trouble ahead.

CREATING A BACKLOG

Once there is a practical vision in place and the business value is established, the next step is for the project team to pin down in more detail what's required. At the heart of any project is this type of requirements list and with agile this is known as the product backlog. It replaces the traditional, detailed, requirements-specification type approach and is in the form of a shopping list of ideas that's meaningful to the business.

Items on the backlog are always user-centric even if they have a technical slant. The litmus test is that they make sense to pretty much anyone.

A sensible place to start is for the project team to dig into the vision statement as a group – to make sure everyone understands it, its scope and what it's helping us to conceptualize. Making sure all parties are on the same page from the start is much easier than trying to fix a broken project two-thirds of the way through the process.

The diversity of the team is important, as they need to think about the project from all different angles. If necessary, specialists can be drafted in to help out.

DEFINE PRIMARY FUNCTIONALITY

The aim is to produce a summary list of what's needed to deliver the project vision. There are several ways to do this and our favourite approach is to think about each step of the customer journey to produce a workflow.

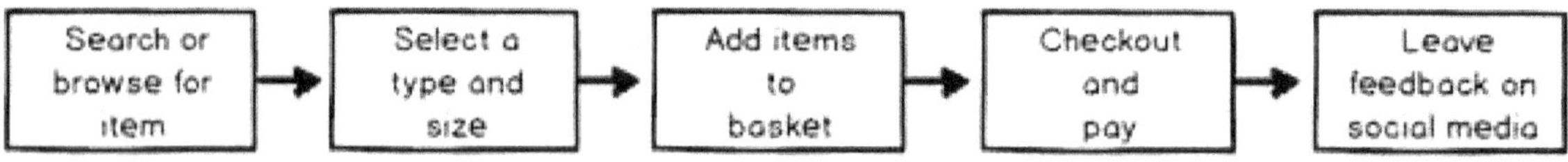

PRODUCE FEATURE GROUPS

Once the workflow is mapped out, gather together ideas about what's actually done within each step in the journey. Added together these items deliver the functionality of the step and are often referred to as feature groups.

Some of the items will be absolutely essential and some nice-to-haves. To begin with get all the thoughts down.

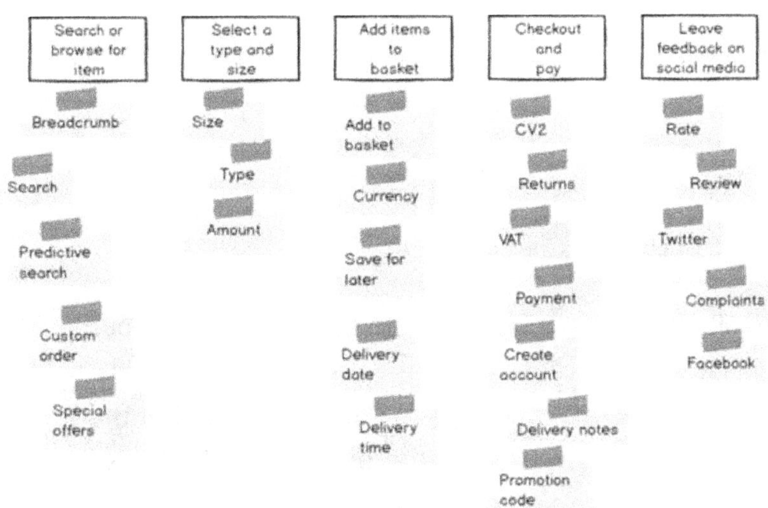

PRIORITIZING THE FEATURES

Using the value statement in the project vision and common business sense prioritize the ideas, in descending order with the most valuable item at the top of each list.

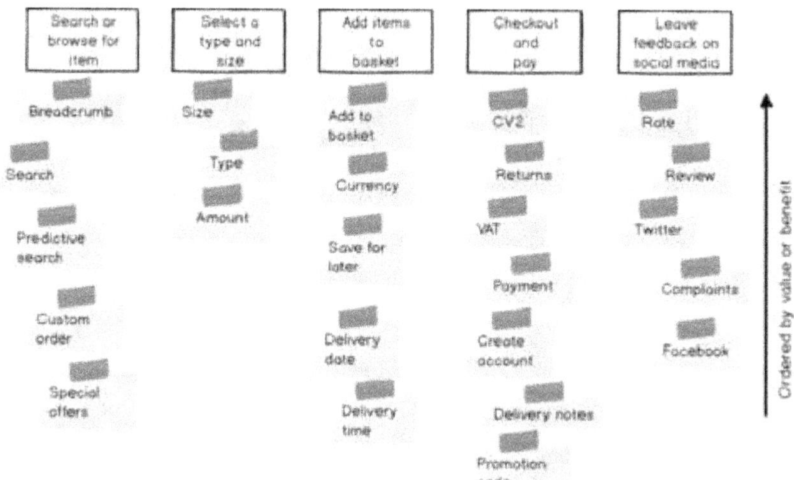

IDENTIFY THE FIRST DELIVERY

Once that's done think about whether each step in the customer journey is vital from Day 1 and what's the most valuable chunk of ideas within those steps. This selection process can be challenging and at the end of the day a matter of opinion, but the business, or whoever represents the needs of the business, is the best judge of all this. The end result is the minimum the project must achieve to deliver a useable outcome. This is usually referred to as the minimum viable product (MVP) or the minimum viable release (MVR).

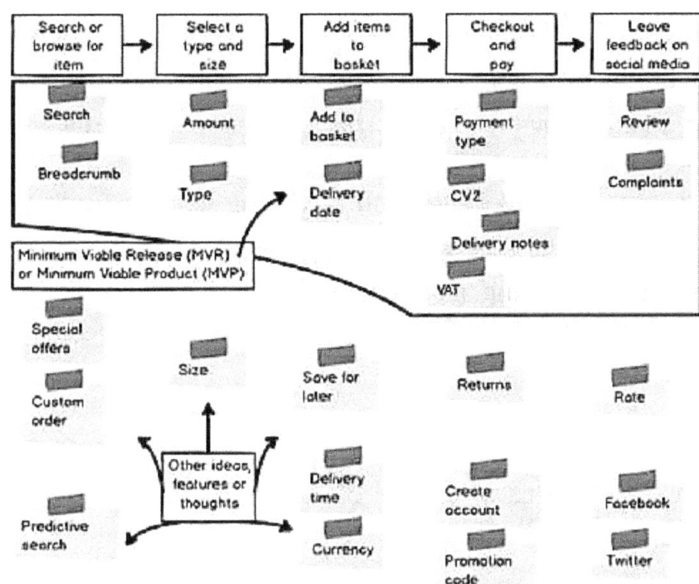

One of the biggest hooks for agile working is to get fast, meaningful feedback from the end customers and they need something tangible to provide an opinion about.

We need to be aware that the more there is in an MVP, the longer it will take to get feedback, whereas if we put too little in it, there will not be enough information to get feedback on. You have to strike the balance between return and risk – there's no golden rule. Try to find the point where you get good feedback on something useful, something that will help you make informed decisions. It can pay dividends to carry out market analysis beforehand.

Beware of loaded terms too. For some the word release means a publicly available product. For others it means something to see and test out on a closed audience. Remember, it's possible to release in little bits to a closed group and then, once this has built up, do a proper public release. Don't make assumptions that everyone means the same thing! No one approach is right, so pick the one that works best for you.

ADDING FEATURES

Once the MVP or MVR or whatever you want to call the first delivery is out there, the real fun begins. Additional functionality or even specific features can be delivered in bite-sized chunks or packaged up into bigger releases. This is called incremental delivery.

Businesses love incremental delivery. No more waiting for years and years for one huge delivery containing every imaginable bell and whistle. The agile delivery preference is for little and often. There is a balancing act here once again but the business decides what and when. The smallest possible delivery is one solitary feature that can be validated through practical use.

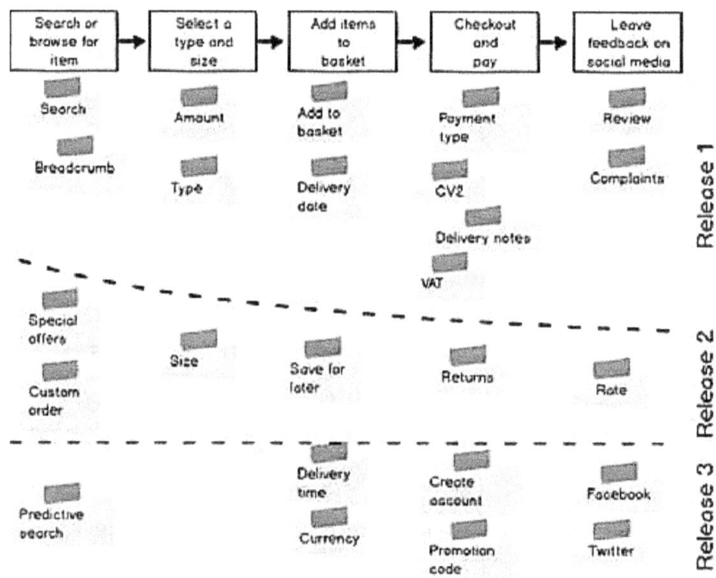

GETTING MORE INFORMATION

So far, the items in the first release, our MVP, are very lightweight and we'll need to get more detail on them. The reason they're high level at this point is because we're trying to formulate ideas and it's wasteful to elaborate requirements that may not go on to be used. That was more than enough to define the big picture and agree the MVP but not enough to get on with the work itself. The next step is to get more information on everything earmarked for the first release. The classic tool for capturing further details is the user story.

TELL ME A STORY

User stories are short, simple description of a feature told from the perspective of the person who desires the new capability, usually a user or customer of the system. They typically follow a simple format:

As a <type of user>, I want <some goal> so that <reason>.

Title	Assign a clear and concise title. This is a great way to summarize, index and search for stories.
As a <type of user>	We need to know who the end user will be. Who this feature is for?
I want <some goal>	What is the functionality the end user wants? Describe the 'what' not the 'how'.
So that <reason>	What is the reason for needing this feature (with some kind of business or customer benefit here)?

A user story is a very high-level definition, containing just enough information so that the team can produce a reasonable estimate of the effort to do the actual work. User stories are often written on index cards or sticky notes, stored and arranged on walls or wherever there's a space to facilitate planning and further discussion. They shift the focus from writing reams about features to discussing them.

A user story isn't a requirement. It is defined as a reminder to have a conversation and these discussions are often more important than whatever's written. It is these dialogues that spark the most important thinking points about the requirement. Remember, anything written today that won't get started on for a while can go stale. However, if we just gather enough detail to have a meaningful conversation, the reminder stays fresh and time isn't wasted penning copious detail. Win–win.

PIN DOWN ACCEPTANCE CRITERIA

How do we know when we are done? It's a key question and the first one that should be asked when starting a conversation prompted by a user story. In order to work efficiently, it's important to know when to stop. What are the boundaries of the work? How much do we do for it to be accepted? How will we avoid over-egging or gold plating the requirements? For a user story to be truly done it needs acceptance criteria – the prerequisites that have to be met for a story to be assessed as complete.

Acceptance criteria can take many forms, from simple conditions of satisfaction all the way through to rigorous and very exact checks. For the purposes of getting agile, simple binary statements written in plain language are the best place to start. Let's take the delivery date user story a little further by writing some simple acceptance tests for it:

- The delivery date will always be the next working day.
- The delivery date will be on a Monday if the order is placed on a Saturday.
- The delivery date cut-off time for orders will be 3pm.
- There will be email confirmation of the delivery date.
- All product lines have the same delivery date rules.
- We can never specify a time of delivery, only the day.
- The user can leave a note for the delivery driver.
- The anticipated delivery day will be shown on the screen when ordering.

Acceptance criteria written collaboratively by the team is the most likely way to cover all the angles. Led by the Product Owner or business representative, the team can talk through the stories, remove ambiguity and pin down the end game. These sessions are the best way to bring about team alignment and they don't need to be long, or laborious. They can be done just in time to start work; the aim is to start with the end in mind!

As a by-product, acceptance criteria provide a useful measurement to report progress against. Frequently, business confidence is undermined though being vague: 'I think we're nearly done' or 'I feel we're on track'. So these checks are an aid to being more precise by providing specific and measurable milestones: we're halfway through the acceptance criteria. This type of gauge will be easily achievable if the checks are properly formatted and alarm bells should be ringing if not.

SPLITTING STORIES

Sometimes, too much of a good conversation generates an over-abundance of material. Don't worry, this is a good thing – don't stop the dialogue, capture it all. Some ideas may be not appropriate for the story you're working through or may be too advanced for it. Not to worry, as once captured they can be filtered. Some of it can be added to other more appropriate stories.

Others may call for a new story to be written and this is known as splitting a story. A common example is where the Product Owner sees some of the acceptance criteria as unnecessary for the time being and wants to create a new story for the extended features – to be reminded to talk about them in the future. Once the new story is written, it can just be prioritized into the backlog along with everything else.

Keeping user stories to a manageable size is important. The more complex a story is, the more risk of something going horribly wrong. Huge reams of acceptance criteria are an indicator that a story has gone that way and must be split. There are times when this needs to happen multiple times and it's a legitimate way of breaking work down into reasonable-sized work packages.

YES, SIZE MATTERS

Now we have a vision, a backlog, an MVP and some well-written user stories complete with acceptance criteria. Great stuff. The next step to ask then is: how much work is all that? Traditionally project managers or specialist estimators carry out this task and then throw their predictions over the fence. But on an agile project the team, the people actually doing the work, produce the estimates. Apart from the huge benefit of more reliable projections, there's the advantage of getting team buy-in.

Of course, the size of each user story is required to predict when the MVP will initially be delivered and when the subsequent other features. But this is also a way of being forewarned of potential problems with individual pieces of work – head scratching, big intakes of breath and shaking heads are sure-fire signs of trouble ahead.

There are several estimating techniques well suited to agile projects and the following are worth considering:

- T-shirt sizing. Assign S, M, L, XL and XXL tags to everything and gauge roughly how much work each size involves. Easy to use and a good starter-for-ten but can lack precision.

- Story pointing. Use a Fibonacci scale for all the pieces of work in hand – for example from 1 to 100 points – where 1 is easy-peasy and 100 a raised eyebrow moment. Takes some getting used to but there are many devoted fans.

- Affinity prioritization. Sequence all of the stories in order of relative size – shuffling stories into smallest to largest order – and assign 1 to the smallest and the relative size to the next biggest and so on. Useful for getting to grips with the MVP.

LESS IS MORE

Outside of the agile world there's normally a cat and mouse game played at the start of a project. The business team know instinctively they have to ask for everything under the sun because there's only one shot at getting most of what they need. They also know that when things start to go pear shaped – as they regularly do in some form or other – the deliverables are going to be pared back. So, it's better to ask for the kitchen sink to improve the negotiating position.

Project teams know this goes on of course and are happy to have wiggle room built in for when the times get tough. The problem is that when the squeeze hits, many of the bells and whistles have already been delivered and it's too late to recapture that poorly invested effort. When the budget dries up or time runs out, the remaining work includes many non-negotiable essentials: for example, within a house renovation project, having a very high-spec kitchen with all the latest lighting and gizmos when the bathroom is still a bare shell.

It's common sense to start by delivering a barebones solution and build from there but there needs to be an understanding that the plug isn't pulled after initial delivery. There needs to be faith that there will be incremental deliveries from there on, building and honing the final product. Although much depends on trust, the whole ethos of an agile project is based on this premise, so nothing can go wrong unless the whole set-up is a total sham – which even agile can't insure against.

The MVP is the absolute minimum required just to get going. Every feature and nuance is non-negotiable without any nice-to-haves. The litmus test for anything on the first to-do list is that the whole MVP would be unusable without its inclusion. In practice, it doesn't matter too much if a couple of minor bits 'n pieces sneak in as long as the traditional gold plating doesn't happen. The objective is to end up with a lean, mean set of requirements that can be delivered quickly.

Once business teams and customers go through the loop, they immediately get how much better this works for them. The reduced time-to-market is a big winner and the key to competitive advantage. Plus, in practice it's usually very hard to predict the optimum final outcome and much easier to add to a working product. It is interesting to look back on any nice-to-haves after the first delivery is in place as normally other more important features come into contention. There's nothing stopping the team from having a long list of features waiting for the production line.

RISK AND EXPECTATION MANAGEMENT

Done properly, there's no risk of typical problems occurring on an agile project because risk management and mitigation is built into the framework – greater all-round visibility with a diverse team of specialists on hand helps enormously. The main risk with agile is going off-piste in some way or another:

- Don't deviate from tried and tested practices. Many people try to change too much, too frequently. Stick with the guidelines and make adjustments one at a time. If your changes don't work, drop them!

- Communicate, communicate and communicate. Bad communication is the root of all evil. Leaving information out is as misleading as giving bad information. Use the backlogs as the focus for regularly having the right conversations at the right time.
- Avoid large work items. The larger requirements are, the harder they are to understand. Break down any big items into smaller, more manageable chunks.
- Keep talking. The best way to manage agile risk is by continually having meaningful conversations with those people around you. Let them all talk!

MANAGING THE BACKLOG

It's hard to over-emphasize how important the backlog is. It is the cornerstone of your project. However, a good backlog can go bad very quickly if it's left unattended. The backlog has to be living, breathing and attention-seeking. Used well they're brilliant at helping to demystify what is coming up, helping communication with others, reducing risk and managing expectations. If they are neglected, they become a time-consuming distraction that sends people off course fast. Keep the backlog up to date!

At the start of a project, a backlog is usually full of functional requirements and features written as user stories. As the project moves on it will become filled with other items and a user story is just one form of product backlog item. Backlogs need to make all work visible and that includes faults, -non-functional requirements, improvements, enhancements, new feature requests – everything. Get them all written down and blend them in with everything else, ordering them by business value just as always.

The backlog belongs to the Product Owner, who remains accountable for it at all times. As such they should be constantly refining it by using it as the main focal point for discussions with all interested parties. Keeping the backlog up to date is hard work, but the benefit pays off through the visibility it provides and the conversations it initiates. Transparency builds trust on projects, and by far the best way to achieve this is to make your continually refined backlog visible to all.

The best way to make sure that your backlog is up to date is to get the team and Product

Owner to look at it every day. This can be part of a daily routine or part of any exercise where the team talk about what they're doing. Sometimes, the Product Owner spends a lot of time refining the backlog, sometimes just minutes. The important thing is that it's being used and referred to regularly.

CREATING THE RIGHT ENVIRONMENT

Success on an agile project is more about the individuals and the interactions between them working than anything else. It's hard to get people working together efficiently and their environment is crucial in promoting effective communications. The most successful agile teams are product focused, sit together and have easy access to their Product Owner. Plus, they're in sight of both the team task board and most crucially the product backlog. A team that sits together has fewer obstacles in the way of communicating, interacting or even just building rapport by chatting about the weekend.

Let's be realistic though. It's easy to say we should all sit together, laugh, work, be funny and good looking – but in reality, people make long commutes, companies have offices in different cities and partnerships with off-site teams too. Whatever the circum-stances, visibility, transparency, communication and interactions collaboration is key and quite often we need help to achieve this. Video conferencing, electronic task boards on giant touch-screen TVs, conference calling are never brilliant but always better than nothing. If, and it's a big if, there's a well-maintained and refined practical backlog, these tools can work.

There's no excuse for the team being uncertain about their objectives, what's coming up next and the part they have to play in it whatever the physical circumstances. If a team isn't communicating, it means something is seriously wrong and the bitter truth is teams usually perform badly as a consequence of a bad environment. We agree there's undisputed value in good communications and where there's a will there's a way.

IN RESUME...

- Start with a specific and meaningful vision; no waffle or vague targets.

- Business value is everything, so develop a shared understanding of what it really means.
- Love thy backlog. Develop a top-quality backlog with rock-solid stories that can be understood by everyone.
- Pin down that MVP! Everything rests on getting it right and out there quickly.
- Always works out the acceptance criteria up front; start with the end in mind.

Kanban

KANBAN FUNDAMENTALS

Kanban encourages teams to begin with the current status quo and build from there by consulting the people directly involved in the process. Change is therefore consensual, thus increasing the likelihood of Kanban being adopted enthusiastically. There are three guiding principles that sum it all up:

1. Start with what you do now.
2. Agree to pursue incremental, evolutionary change.
3. Respect the current process, roles, responsibilities and titles.

Stripped down to basics, there are five key steps for implementing Kanban: starting with producing a visual representation of the end-to-end flow of work, then placing constraints on the amount of work in play at any one moment in time plus finally steps to measure and improve the efficiency of the flow.

1. Visualize workflow. Kick off with a visual representation of the flow of work going from to-do to done status. Many prefer to add only one other step in between: in progress. Others prefer to break the workflow down into a series of procedural stages such as: plan, design, draft, build, test, deploy, with to-do and done as bookends.

2. Limit work-in-progress. Trying to do too many things at the same time is a proven recipe for disaster; it applies equally at an individual level and to teams. Kanban limits the number of items allowed to be on the go at any one time – known as the work-in-progress (WiP) – to ensure optimum efficiency. Common sense is enough to get started and then experience will help fine tune to pin down the optimal WiP limit.

3. Manage the flow of work. The aim is to achieve a fast, smooth movement from to-do to done. If so, it means the process is operating at optimum efficiency thus creating maximum business value in the shortest time possible. An important add-on is for it to be repeatable and consistent.

4. Make the process explicit. An unambiguous statement of how work gets done is essential for any objective review. With a common understanding it's easier to discuss issues impartially and reach a consensus on improvements. There must be a natural checkpoint at the end of each step with clear rules for moving on to the next one.

5. Improve collaboratively. Once the spotlight is on the workflow, ideas start to develop about how it can be improved. The WiP limit plays a key role in sparking discussions by forcing the team to focus on blockers to work in play when the limit is reached. An initial cap of no more than two tasks per person soon highlights problems that impede the flow; then the team simply faces up to those issues and resolves them.

Kanban's great for:

- Getting off to a low-risk, zero-cost, agile, fast start.
- Pinning down existing workflows and spotting glaring errors.
- Controlling multiple pieces of unconnected work.
- Keeping the numbers of jobs in play down to an acceptable level.
- Getting the team into an agile way of thinking.

At the heart of the Kanban method is a deceptively clever tool: the Kanban board. Calling these boards, a visual to-do list is an over-simplification but a decent starting point. The board is a graphic representation of the work to be done and the end-to-end flow from start to finish. The simplest and some argue the purest Kanban board consists of just three columns: things to do, tasks in progress and finally work done. This simple format is universal and matches any project or corporate workflow.

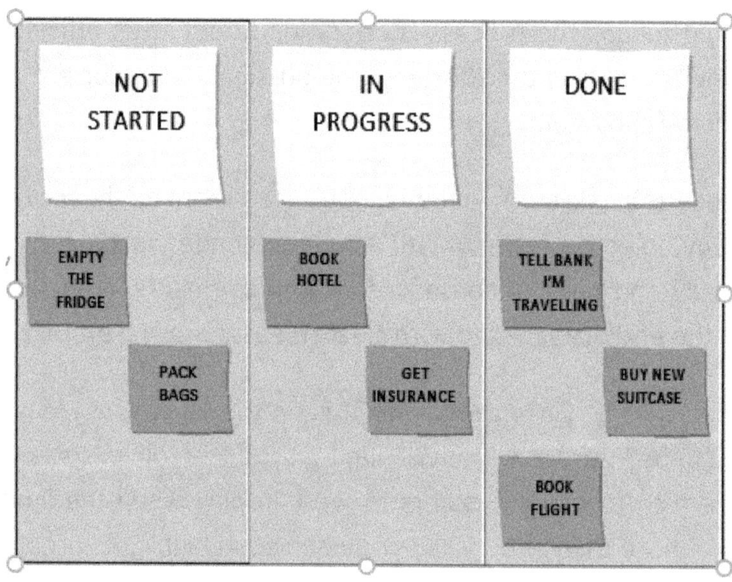

The Kanban board

After a while it's natural to be more thoughtful about the typical workflow of tasks and consider whether there are other steps en route to job done status. A popular variation on the theme is to separate out ideas that are being considered but are not yet definite runners. It's also worth breaking down the work-in-progress especially if more than one person along the way handles tasks. There must be a noticeable and measurable change in the status of work between each step in the process.

Keep it simple to start with and try out the four previously suggested stages: ideas, to do, doing and done. The demarcation between each status is clear and in turn generates the triggers for moving cards on:

1. Ideas – the maybe or maybe not stage when there's a question mark of any sort outstanding.

2. To do or in progress – when an idea is thought through and the only questions are who is going to do the work and when will it start.

3. In progress – once work is assigned to an individual or group and the task is actively proceeding.

4. Done – totally complete with nothing more to do except reap the rewards or accept the gratitude.

- **The definition of done**

One of the biggest challenges for any style of project management is pinning down the job done status work. It is essential to pin down the final exit criteria – also known as the definition of done – for each task up front to avoid disputes down the line.

Be precise about the criteria for accepting the delivery and settling the bill. There are often genuine misunderstandings about what's in and what's not and agile pays particular attention to this universal phenomenon. Engaging the customer or business is the only way to do this properly; yet another common-sense idea at the heart of agile philosophy.

Pinning down the definition of done is a collaborative process and there's usually an element of trial and error involved. Don't wing it; set adequate time aside to come up with the final exit criteria but don't agonize too much either. Running a few tasks through the process will iron out any small kinks.

- **More than just a board**

Once the starting format of the Kanban board is agreed, the first and almost pivotal decision to be made is whether to go for a physical board or an electronic one. Both have their pros and cons and there may be working practices that guide the final decision. A virtual board can't be beaten for accessibility and ease of sharing, as you're never more than a smart phone or iPad away. But in our opinion the most important thing is for the board to be highly visible, and nothing can beat a physical board for that. An old-fashioned corkboard, note cards, pens and pins are enough to begin with.

- **Low cost, high-tech alternatives**

Despite our absolute preference for an old-fashioned physical board, there are times when an electronic board either makes more sense or is even the only viable option. When individuals are regularly on the move or if the team is split over different locations, there are insurmountable physical issues to deal with and a tech option become more attractive.

But before giving up on having a physical board think carefully, especially when trailing agile for the first time. A tech alternative will work well enough from a functional perspective but is far less visible and engaging, so many soft benefits will be lost.

Electronic Scrum and Kanban boards are big business, so there's plenty to choose from. Trello is free, has excellent device coverage and is extremely easy to use. JIRA is more than a board and agile practitioners seem to either love it or hate it – there's an option to buy an add-on that prints user stories making it very easy to have a synced-up physical board too, giving the best of both worlds.

BUILDING A BACKLOG OF WORK

At the very heart of the Kanban board are the to-do items also known as the backlog in various agile frameworks. These individual items are all delivery focused and must deliver business value directly or indirectly. For example, setting up a Kanban board is a legitimate item but a meeting to discuss the options is merely part of the main job. The tasks are business delivery focused and not centred on activities. If an item on the backlog does not contribute to business goals, it should be removed.

A Kanban backlog is very similar to other agile work stacks, certainly in the way tasks are captured as user stories. However, there are subtle, yet important, differences with Kanban:

- All work is of a similar size. It's better to have smaller stories of roughly equal size. Splitting large pieces of work down into smaller, similar sized packages has been proven to improve workflow and results in more predictive end-to-end cycle times. Comparing like-for-like can also said the review process.

- The backlog is refined more regularly. The Kanban backlog tends to be exceptionally dynamic especially in a support type environment. Backlogs in other agile environments are far from static but they're just a touch less zippy. It's not unusual to review a Kanban backlog on a daily basis.

- Jobs are pulled not pushed. A Kanban team adopts a what's next policy rather than packaging up work into a connected delivery. The task assigned the highest priority is pulled into play when resource becomes available.

SHUFFLE THE DECK

Once all the ideas are thought through and the to-do list or backlog shapes up, the next step is to get work into a sensible sequence. Nobody in their right minds wades through their backlog in alphabetical order but it's surprising how many fall into a somewhat random approach, such as picking up whatever takes your fancy or responding to peer pressure or even a bit of both. It's surprising how many jobs are selected just because rather than for a scientific reason.

A core concept within all agile variants is the goal to be driven by business value, and simplistically that's the primary basis for assigning work priority. Work that delivers the most value gets launched before anything providing marginal benefits. Layered on top of this key driver is an assessment of the cost of delivery – if two things deliver the same value, it's common sense to deliver the easiest one first.

Don't get too hung up assigning business value as a comparative assessment is enough. In this context it's all about the relative importance of work, not a detailed analysis. Carry out a similar exercise for the cost of delivery, including all the relevant factors such as timescales, days of effort and hard cash involved. Multiply the two assessments together to get a total score that drives the priority sequence.

There are five simple steps for pinning down priorities:

- Rationalize the backlog into standard size work packages.
- Assign a business value to each item (between 1 and 10 where 10 is the highest business value).
- Assign a cost of delivery rating (between 1 and 5 where 1 is the costliest delivery and 5 is the cheapest).
- Calculate the business value × cost of delivery and sequence the results in descending order.
- Review and apply common sense!

CONTROLLING WORK-IN-PROGRESS (WIP)

In an ideal world each task plucked from the top of the to-do list progresses seamlessly to job done status before the next one is picked up. In reality life is never simple and most people end up with more than one thing on the go at any given time. This isn't multi-tasking, just switching effort whenever there's a temporary hold-up and this has been common practice from the year dot.

With Kanban there's an agreed limit to the total number of tasks in progress at any point in time – this is known as a WiP limit. It can be applied to individuals, the team collectively or both. A maximum of three things in progress per person is a popular individual ceiling. Or twice the number of team members as a maximum group threshold.

There's a tendency in certain circles to enjoy the early stages of any activity and for interest to wane towards the end, with a reluctance to tie up any loose ends. Kanban insures against this phenomenon with WiP limits. It helps avoid ending up with a sea of items in progress with many of them stuck at 99% complete. As part of the agile philosophy, the raison d'être for work is achieving business value and that doesn't happen until the job is finished.

Imposing a WiP limit is critical for implementing Kanban successfully and as such it is non-negotiable. Without this in place the board can veer towards a glorified to-do list. With a WiP limit there's a continual flow of work getting through to genuine completion and paying dividends. Individuals are forced to confront blockers and deal with them rather than put jobs on the back burner with a defeatist shrug of the shoulders.

Continual process improvement is an important part of the Kanban method and for the most part it happens naturally. The WiP limit forces teams to be introspective when necessary and look at what's clogging up the production process. This creates a culture of being on the lookout for tweaks and adjustments that make the machine run more smoothly, literally in the case of the motor industry where it all started.

MANAGING PROJECTS WITH KANBAN

At the very heart of agile is the concept of continuous delivery. No more long waits for anything of use to be delivered, as there's a continual flow of small, yet perfectly formed, packages. Kanban captures the very essence of this agile concept because every piece of work is a delivery in its own right. This becomes the ultimate litmus test for the perfect Kanban user story. Does it make sense in isolation? Will it put a smile on someone's face when it arrives on the doorstep?

Of course, projects can be delivered using Kanban too. A project can be broken down into smaller packages or individual user stories to be delivered incrementally. Far from being unsuitable, Kanban encourages the business community to think smaller and ask for deliveries that provide a return in their own right yet are part of a bigger picture.

Adding features to an established product is a great example. Kanban is a great option to get a foot in the door. It's easy to understand and simple to launch. Even though it's a stress-free entry point for non-agile organizations to test the water, Kanban isn't a compromise solution either. It can revolutionize teamwork and demonstrate what agile can offer. Introducing Kanban is a solution in itself or it can be a step towards Scrum or whatever agile framework tickles your corporate fancy.

IN RESUME...

- The Kanban board is at the heart of the operation.
- Get started by visualizing the as-is workflow.
- Limit the work-in-progress for maximum efficiency.
- Don't let the simplicity fool you; it's a deceptively powerful tool.
- Kanban can be a final destination... or a stepping stone for grander ambitions.

Scrum

Scrum has become popular because it works. It had its roots in product development and innovation in mid-1980s Japan before being built on and refined in the US during the 1990s. It's one of the safest way to get started with for those new to the agile discipline because it has clearly defined roles and responsibilities, ceremonies and artefacts, yet allows enough flexibility in their implementation to let its customers feel supported but not suffocated.

Scrum is a framework within which people can address complex adaptive problems, while productively and creatively delivering products of the highest possible value.

THE FRAMEWORK

Like all good ideas, Scrum is open to misunderstanding and misuse. To help make sure each implementation stays close to the original intent the creators, Ken Schwaber and Jeff Sutherland, created the much loved Scrum Guide. The guide is kept up to date and is freely available at www.scrumguides.org and it's seen as the single version of the truth.

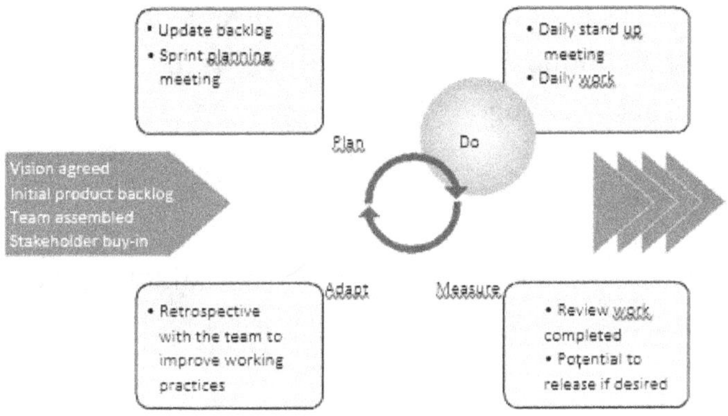

The framework of Scrum

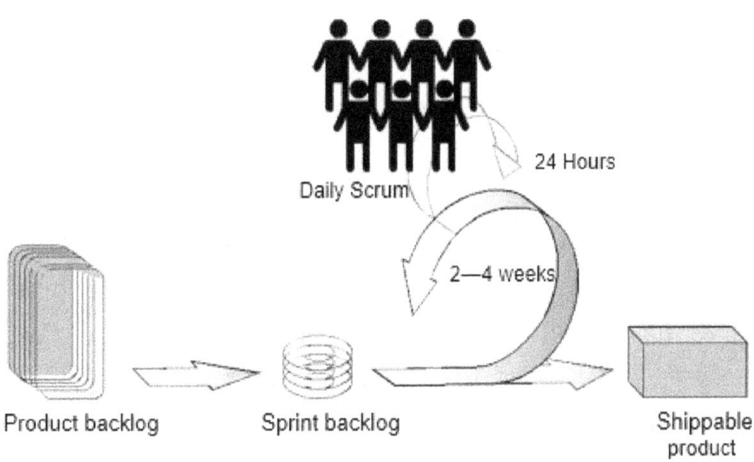

The sprint lifecycle

Scrum has always been about combining Lean and agile principles to help teams to deliver products. It isn't a project management tool, it's a framework for delivery and there's a subtle but important difference between the two. When Scrum is being done well it's a very natural process and it can be easy to forget it's even being used. Scrum is only there to support delivery – and we should always bear that in mind.

The end goal is to deliver good product using Scrum, not to try and get it working perfectly with delivery almost as an afterthought. It's there to serve and needs no glory, adulation or praise. It just helps us to get stuff done. However, we do need to let it do what it's good at, and to achieve this it isn't an option to pick and choose bits to use; it comes as a complete and perfectly formed package covering:

- Theory – not much to get to grips with, just fundamental, guiding principles;
- Team Roles – only a Product Owner, the Scrum Master and the development team, yet all the essential bases are covered;
- Events – the sprint itself, sprint planning at the start, the sprint review and the retrospective at the other end, with the infamous daily stand ups sandwiched in the middle:
- Artefacts – the master product backlog, the sprint backlog and plenty of others beside.

SELF-ORGANIZING TEAMS

At the heart of Scrum is the concept of a self-organizing team – a self-starting and self-governing team of experts that does whatever is needed to get the job done. The three roles within a Scrum team – Product Owner, Scrum Master and development team – are designed to make sure that everyone is able to work together seamlessly without stepping on each other's toes, yet not defined to such an extent as to make them inflexible and unable to adapt to change.

- **The Product Owner**

First and foremost on the team sheet is the Product Owner (PO) who is the single, final decision maker for what the project is all about and what the end product delivers. They represent both the business sponsoring the product being built and the customers it's being built for. Strategically, the buck stops here.

When a business has an idea for a new product or service it will have to invest money and equally importantly time into developing it. The management team or strategists who run the business may well be happy to stump up the cash but they don't usually have the time to get as closely involved in guiding the product development as they would like. Leaving any project team to their own devices can be a recipe for disaster and it could also be seen as un-agile. So the smart move is to nominate a business representative and delegate day-to-day ownership of a product to them. This isn't a new idea in the project world but agile has made this role of product owner into an art form.

The product owner represents the business in all senses, managing the scope, scale and direction of the new project being built. It doesn't stop there either as they also represent the needs of the end customer too. Product owners obsess about delivering value by representing the needs of the business and the end users, maximising the benefit delivered to both.

The business and the end customers are very different beasts. The business will have strong reasons for building something new and will be driven by a return on its investment of some description, perhaps straight financial profit or some other more noble reason.

- **The scrum master**

The Scrum Master is the chief organizer on a project and they're often fondly described as a servant-leader, someone who shares power, puts the needs of others first and helps people develop and perform as highly as possible. This is in sharp contrast to a project manager who typically exerts top-down control. The Scrum Master is an enabler, helping a self-managing team to complete the task of delivering working, valuable product. One thing's for sure: organizing a Scrum team isn't a walk in the park.

A Scrum Master facilitates all the ceremonies, making sure they happen on time, that the right people turn up, that the information to make the correct decisions is available and that the aims and outcomes are achieved. They need to be on tap to help the product owner manage the stakeholders if necessary and more than likely help keep the product owner on track. If that's not already enough, they're the guardians of the process making sure that things are done properly – like writing up proper user stories – and minimise distractions by managing any outside interaction with the team.

- **The development team**

In an agile environment the development team includes everyone else that's needed to get the job done. Their sole objective is to take the prioritised items off the backlog and turn them into working products. They're self-organizing, which is a nice way of saying that they're trusted to work together as sensible grown-up professionals who can focus on getting work done properly without needing micro-management.

The development team is the project engine house. They're multi-talented, multi-disciplined and totally fixated on delivery. Crucial to the team ethos is the concept of being cross-skilled, which means that although it's relatively rare for someone to be able to do every single job required, the team members use their individual skills to help each other out in order to get the job done.

Ultimately, the development team must be self-motivated and dedicated to building the product as best they can. They're in total control of their own destinies because they manage themselves. The product owner owns the vision and the Scrum Master provides support wherever needed, but apart from that the team make their own minds up and only answer to themselves.

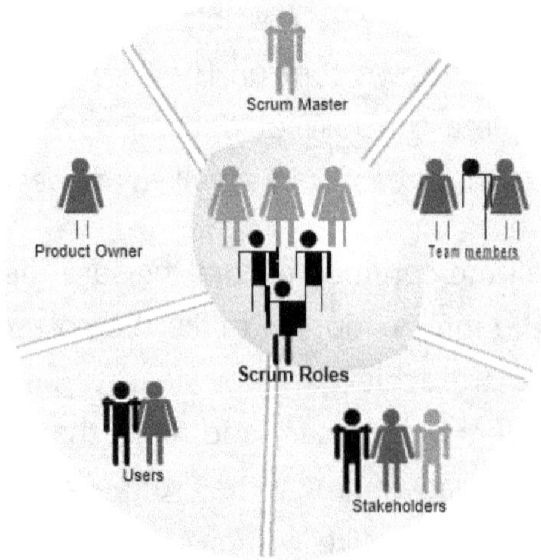

Scrum roles

KEY SCRUM EVENTS

Events in Scrum are simple, straightforward and always time-boxed. They're designed to give structure to the inspect and adapt ethos of the framework without constraining the participants to meaningless formality. Scrum needs all five events to be done in order to work properly; missing any of them out means you are not doing Scrum and leads to ineffective working practices, lack of visibility and confusion. Commonly, if teams are getting frustrated with a Scrum event and they have stopped finding it useful, it's a consequence of something else that's not being done properly.

The five Scrum events are:

- The sprint – provides a wrapper round the other events;
- Sprint planning – happens at the very beginning;
- Daily scrum – occurs on a daily basis without exception;
- Review – takes place at the end to showcase the outcomes;
- Retrospective – wraps everything up nicely.

1. The sprint

A sprint is simply a time-box of between one and four weeks that provides a space for work to be done. Essentially, it's a container for all the other Scrum events. Typically teams choose two weeks as a good sprint length but there's no perfect duration and all have strengths and weaknesses. If in doubt, start with either a one or two week sprint and see how that pans out.

Sprints can be seen as an opportunity to run mini-projects, and provide the regular heartbeat of product development so don't be tempted to vary the duration of every other sprint. Get into a groove. A sprint starts with planning to decide what will be built in it and ends with a review of the product built. A retrospective for the team affords them space to assess how they have worked together and consider any applicable improvements.

2. **Sprint planning**

Planning happens right at the start of each new sprint. It's an interactive session for the team to look at the user stories that the product owner has already refined, tuned and prioritised to see how many they think can be delivered within the time-box. There will be a frank and open discussion about complexities, risk, size, effort needed, business value and the details of what the business is asking for. The objective is for the team to fully comprehend and agree to what they're taking on.

A key characteristic of a Scrum planning session is that it's team led, albeit facilitated by the Scrum Master. In the old days a project manager would just tell the team what they have to get done and by when, whereas an agile sprint planning session lets the team decide what to take off the top of the prioritised backlog, allowing them to work at a realistic, sustainable rate. That's not to say they should set an easy target and the Scrum Master works with the team to ensure the end goal is challenging yet achievable. Ultimately the team have the final say when they have reached capacity.

The planning session is about confirming the understanding of each product backlog item before it's included in the sprint. It's essential that the business thinking be understood before work starts and any serious lack of clarity needs ironing out up front; that's what the product owner is there for. Resist the temptation to go into a session with a bunch of vague stories and promises, promises, promises.

3. **Daily Scrum**

The daily Scrum (or daily stand-up as it's often called elsewhere) is a time-boxed, micro-meeting that lasts no more than 15 minutes and occurs every day – preferably around a visual representation of current work. The more in your face this is, the better. The primary intent is to make sure the team members give each other a status check every 24 hours to ensure they're aligned to the goals set out in the planning session and on track to deliver the goods.

Sprints are mini-projects and like their big brothers they rarely go smoothly. Expect hiccups and use the daily Scrum to get blockers out in the open. This is where the Scrum Master gets down to business and picks up on any impediments that are obstructing progress with the expectation of facilitating a solution. This isn't a case of lobbing problems over the fence but of sharing them with the expectation of getting a very big helping hand.

The daily Scrum follows a very simple format, asking each team member three questions:

- What did you do yesterday? A progress update.
- What will you do today? Communicating intent.
- What impediments do you have? And this is the killer question.

Just getting everyone to the daily Scrum on time and keeping them focused on the script is a job in itself. Any meeting can go wrong, even more so when it's on a very tight schedule and with a very sharp focus. A task for the Scrum Master is to make sure that the team isn't distracted from attending this meeting in the first place, that it occurs and that it stays focused and relevant to keeping the team on track. It might sound easy but it sure isn't and everybody plays their part in keeping things on track. The Scrum Master isn't a nursemaid and shouldn't contemplate sending deviants to the naughty corner. Well at least not literally.

The daily Scrum is a barometer indicating the health of the sprint itself, the effectiveness of the team and a pointer to a host of other things. An experienced observer, such as a trainer or coach, will only need to attend one or maybe two daily Scrums to offer an expert diagnosis on how things sit generally. Admittedly even in a mature environment these sessions rarely go totally smoothly but a dysfunctional stand-up is an extremely serious warning signal, especially when it's not a one-off blip.

Vigilance is needed to keep the daily Scrum on track. Ultimately this is all about the people involved and there are certain personality traits that are capable of causing a serious derailment:

- Noisy observer: who is not part of the team but will try and butt in and cause disruption. Get the Scrum Master to explain that observers are there to observe only! Their comments can be made offline.

- Late arriver: who arrives for the meeting late and then asks to hear what everyone has just said again. Don't backtrack unless absolutely necessary as it only encourages this bad habit.

- Side-tracker: disrupts the meeting, often unintentionally, but always with negative consequences. The corrective advice offered is to stick to the script and update the team efficiently; we can hear all about your bad date another time.

- Habitual hater: doesn't want to be there and is uncooperative as a result. Well tough luck, so shape up or ship out because everyone in the team has to play by the rules.

- Silent types: even if someone is a bit shy they've got to chip in. The non-negotiable rule is that everybody contributes and silence isn't golden.

- Futurists: trying to star gaze into the future instead of focusing on the here and now is a distraction. Leave the vision to the product owner and concentrate on the next delivery.

- Problem solver: anyone who wants to spend the daily stand-up solving others' problems. Encourage problem sharing but save the resolution until after the meeting so it doesn't distract others.

Keep on the lookout for offenders, especially anyone who is guilty on several counts. And know thyself in case the main culprit is close to home.

4. The review

A product review, more commonly known simply as a review, is the opportunity for the team to show the fruits of their labours to the product owner. There are many opinions out there about what this should entail but in essence it's just a way to get vital feedback on what they're doing. The feedback serves a number of purposes:

- Facilitating pre-release product scrutiny: letting the team know whether what they're building is fundamentally right or wrong.

- Reviewing strategic objectives: checking the sprint goals are going to be met, with an opportunity to address any concerns.

- Massaging stakeholder expectations: this is at all levels and at an early stage; what they see is what they'll get! No last-minute surprises.

There are those who are adamant that at the end of the sprint, user stories should have been turned into working, tested, potentially usable product and this is both correct and admirable. But even if that isn't always the case, the product owner should review everything anyway. Honesty is the best policy and it's better to be open about a failed delivery and any problems encountered. At the very least it's an opportunity to review work-in-progress.

5. Retrospectives

At the end of each sprint, once the review is over and the product owner and stakeholders have gone on their merry way, the team gets together to reflect on how things panned out. This is called a retrospective and it takes place without fail, even if everything went exactly according to plan.

There's just as much to be learned from good execution as there is from a debacle. Don't ever be tempted to assume it's going to be a waste of time and rush headfirst into the next sprint.

Without wishing to sound like a broken record, this sounds easy to do, yet is challenging to do well. The core intent behind the retrospective is to bring the team together to talk about themselves, how they work, interact and deliver. This is by far the best possible illustration of the agile concept of inspect and adapt, looking at how the team worked together as a team and trying to find ways to make improvements. It's the cornerstone of being a self-managing, self-or-ganising team and must be taken seriously. It is as -non-negotiable as any of the other Scrum events but the most likely to be considered unnecessary. Don't commit that rookie error.

The outcome of the retrospective is a set of tangible and specific actions that helps the team improve their ability to deliver quality product. There's no need to generate reams and reams of ideas; a few well-chosen recommendations are far more likely to be implemented than a shedload of half-baked, hugely speculative ideas. Put a ceiling of five in place to maximise the chances of success.

SCRUM ARTEFACTS

Scrum artefacts are the items needed by the team to succeed. The non-negotiable big three are: the product backlog, the sprint backlog and the sprint burn down. Others are worth checking out down the line but stick to the essentials until they're properly mastered. In fact, many practitioners don't ever feel the need to travel further afield.

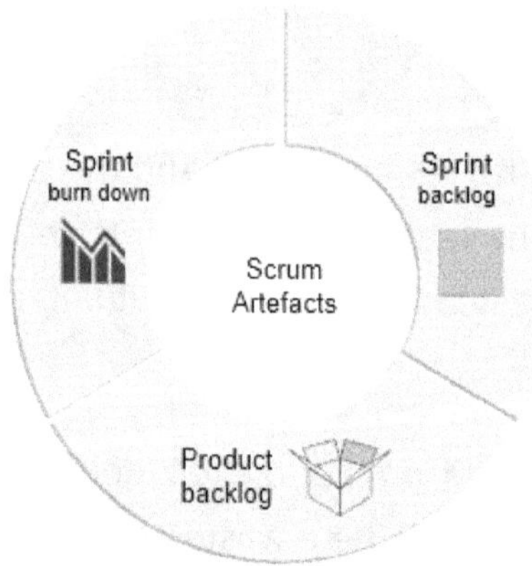

Scrum Artefacts

- **Product backlog**

The product backlog is a shopping list of ideas needed to deliver the end product: a list of desirements if you will. This backlog ensures the product owner's agenda is transparent, visible and easily accessible. It includes everything that might need to be done, not only the specific tangible stuff but all the non-functional things too.

The backlog is owned and often jealously guarded by the product owner who remains accountable for it at all times. of course they will, quite rightly, be canvassed, influenced and occasionally persuaded to reconsider the priority of the entries on an on-going basis. In fact the product backlog must be a living document, changing and developing all the time, with items getting added or removed as the product develops and more is learned about it.

- **Sprint backlog**

Within every sprint the team needs a clear plan of what's going to be worked on. The product owner prioritises the product backlog but the workers need to decide how many of the highest priority items can go into the sprint. The sprint backlog is a prioritised list of those items and is what the team agrees can be done in that time-box.

Unlike traditional projects, how much the team take into a sprint is a decision for the team to make, not the Scrum Master or the product owner. Taking the right amount of work in to the sprint is key to sustainability. Best practice, based on common sense, is to work on the sprint backlog in priority order.

- **Sprint burn down**

One of Scrum's key strengths is the way metrics are embraced and used to great effect. Gone are the days of speculating about progress or of getting stuck at nearly complete status. The sprint burn down chart is a simple visual representation of progress and the team uses it to track the product development effort remaining in a sprint.

IN RESUME...

- Scrum is popular because it works. Don't fall into the trap of thinking that bits can be missed out and it will still yield results.
- Have a clear vision and backlog; ensure there's only one product owner making decisions about the content and prioritisation.
- Deliver work with business value in every sprint. Working, usable product is the only

sign of success.

- Develop working practices, requirements, acceptance criteria, processes – everything as a team – with representatives of all the different disciplines.

- Don't put the product development team under too much pressure. Teams that work at a sustainable rate deliver better than those who are continually stressed.

- The best way to start is to, well, just start!

REMEMBER THE BASICS

After this extensive chapter, I will summarize for you the basics of Agile philosophy. Agile is above everything else a mind-set:

- Focus on tangible outputs. Good practitioners obsess over delivering business value and benefit; it's everything. Having the courage to say when something isn't right, passionately trying to make the lives of users better and always keeping the business vision in mind is at the heart of it all.

- Make what you do visible. The best way to guarantee success is to help others see what's going on so they can chip in with their skills and experience. Hidden work doesn't get done. Problems that are concealed don't get addressed. The team can't help with things they know nothing about.

- Share everything. Without sharing there can be no inspect and adapt, there can be no continuous improvement. Sharing is also about listening, encouraging and developing, so don't knock back ideas out of hand. Sharing is the basis for learning.

- Be cooperative and collaborative. The power of agile is in the strength of a self-managing team working together with a common goal in focus. Being open and honest, being supportive and operating as a collective will guarantee success. You're only as good as the people around you.

www.ingramcontent.com/pod-product-compliance
Lightning Source LLC
Chambersburg PA
CBHW080822220526
45467CB00008B/2169

* 9 7 9 8 7 1 8 1 8 6 3 4 5 *